Holography Marketplace
second edition

The original version of this book contained holograms from
the vendors and many of them are no longer in business.
Therefore this version of the book contains everything that
was in the original version but it has no holograms.

Edited by Franz Ross and Elizabeth Yerkes

PREFACE

The Holography Marketplace is an annual publication which provides an overview of the holography industry, detailed information on holography, and a database of businesses in the industry.

In this second edition of the HMP we expand our coverage of Artistic holography, and encompass the fields of Embossed holography, Holographic Optical Elements, and Holographic Non-Destructive Testing (Interferometry) .

In compiling our database we accommodated all questionnaires returned by our deadline. US and overseas addresses included in this edition are nearly twice the number of the first edition. Ninety percent of all North American holography businesses listed in the first edition who did respond to the questionnaire were telephoned to verify their listings. We therefore feel confident that this is the first and only database of holography in the world.

If your business is new or did not receive our questionnaire, please mail us information about your company on your letterhead to the address below.

In this second edition, all ($) signs represent US dollars, and (£) represent British pounds sterling.

We realize we are not infallible in our knowledge of the industry. If you feel we have omitted anything or made any factual errors, please write us. We do read the mail and will make every effort to improve and expand in subsequent editions.

Franz Ross   **Elizabeth Yerkes**

ROSS BOOKS
P.O. BOX 4340
BERKELEY, CALIFORNIA 94704
USA

Telephone: **FAX:**
(1)(415) 841 2474   **(1)(415) 841 2695**

CIP INFORMATION

Holography Marketplace. Second edition.

Bibliography:
Includes index.
1. Holography.
2. Holography industry Directories.

ISBN 978-0-89496-095-6

For information about purchasing mailing labels of the database, please write to the Holography Editor at the address below.

Ross Books
P.O. Box 4340
Berkeley, CA 94704 USA

TABLE OF CONTENTS

Section II: Detailed Coverage Of Holography

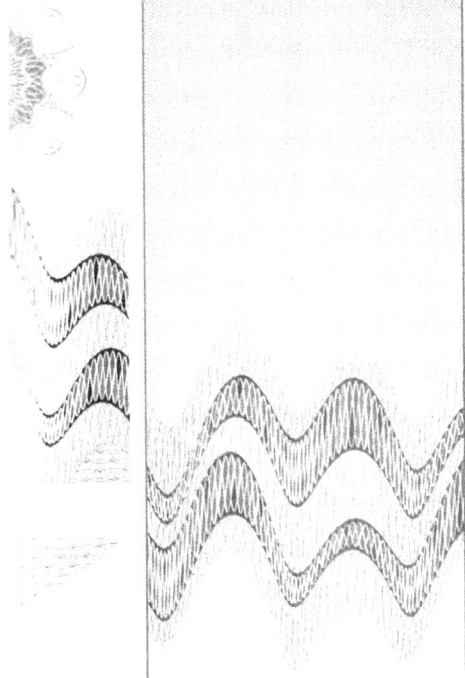

SECTION I
INTRODUCTION TO HOLOGRAPHY
GENERAL OVERVIEW

- Making a Hologram
- Viewing a Hologram
- Transmission Holograms
- Reflection Holograms
- Emulsions/Recording Materials
- Types of Holograms
- Color in Holography
- Business/Distribution of Holography

If you are new to the holography industry, this introduction is for you. In this first section we present an overview of holography as non-techincally as possible, and define some of the fundamental terminology. By giving you an overview of holography, we hope you will be better able to make your purchasing decision and carry on an informed conversation with suppliers.

If you have not done so already, it is wise to see samples of a variety of holograms. Although this book describes different holograms and their best qualities, there is no substitute for seeing holograms first hand. If there is no gallery or place nearby where you can see holograms, there are a number of mail order businesses listed in the Names and Addresses section that would be happy to send you a catalogue so you may order samples by mail. Holograms are unlike anything else and there are many types with distinct and striking variations.

For those who desire more coverage, Section Two covers several of these topics in greater detail.

What Is A Hologram?

We want to discuss holograms from a user's point of view. How are they made? What are the different types? What are the good and bad points of each type?

A hologram is a three-dimensional picture that is made on a photosensitive plate or film using a laser as the light source. The advantage holography has over photography is that it can have full parallax, while photography has none. By full parallax we mean that the viewer, while looking directly at the hologram, can move to the left or right and see around the side of the subject matter. With some holograms you can also see above or below the subject by stretching up or ducking down. Some holograms have full parallax and some have only horizontal (left to right) parallax. The amount of parallax depends on how the hologram was made. With a photograph, of course, you see only the flat image regardless of how you move the print. Over the years there have been a number of items marketed that are optical illusions and try to fool the brain into thinking it is seeing a three-dimensional image, but holograms actually are three-dimensional images.

Perhaps the best way to familiarize you with some of the basic terms used in holography is to describe making a simple hologram.

Making A Simple Hologram

We are in a holography studio and we see a table. On that table is a laser, some mirrors and a photosensitive plate in a plate holder. They are arranged as in Figure 1.

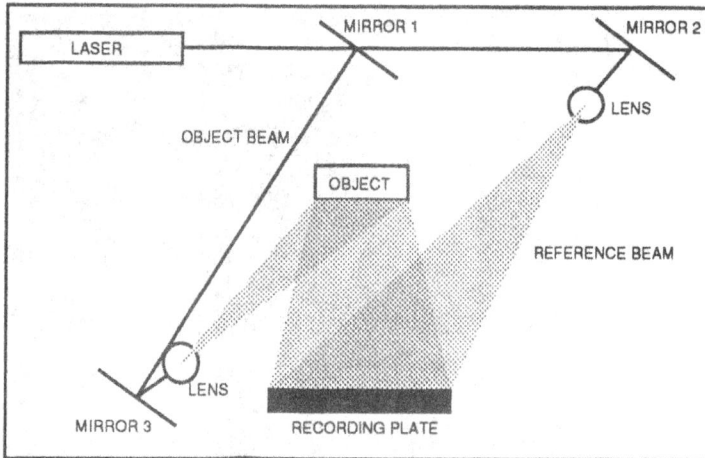

Figure 1. Transmission Hologram

This is a simple set-up. Let's say we are ready to make our hologram exposure. We turn on the laser. The laser beam strikes mirror 1 which is a partially mirrorized mirror. Because it is only partially mirrorized, part of the beam goes through mirror 1 and part of it is reflected. Mirror 1 is referred to in the trade as a beamsplitter for this reason.

The beam that goes through mirror 1 is called the reference beam. After passing through mirror 1, the reference beam hits mirror 2. Mirror 2 reflects the reference beam toward the photosensitive plate. A lens is placed in the path of the reference beam after striking mirror 2 so the beam spreads out and covers all of the plate. We call this the reference beam because it never strikes the object being holographed on its way to the photosensitive recording plate.

The other beam is called the object beam. It reflects off of mirror 1 and strikes mirror 3 which aims the beam toward the object. A lens is placed in the path after mirror 3 to spread the beam out so it illuminates the entire object. The light then reflects off the object (hence the name object beam) and strikes the recording plate.

Providing the length that the object beam travels on its way to the recording plate is equal to the length that the reference beam travels on its way to the recording plate, you will get a hologram when the plate is exposed for the proper amount of time and developed.

Viewing the Hologram

Viewing A Finished Hologram
In Laser Light
Our exposure is done. Next we take the exposed photosensitive plate out of the plate holder and develop it. We now have a finished hologram. Inspect-

ing the hologram plate, we see that the plate is transparent except for some patterns made by the object and reference beams.

To view the finished hologram, we :
• Put the hologram plate back in the plate holder .
• Take the object, and mirrors 1 and 3 off the table.

With the laser light turned on, the only thing now illuminating the plate is the reference beam, as in Figure 2. Since the plate is reasonably transparent, the laser light shines through the plate to us on the other side. Looking into the plate, you see the object just as though it were really there and at the same depth and position it was when originally shot. Moving from side to side while looking through the plate, you can even see around the image. When we look through the hologram in the upper right side, for example, we see a view from the upper right side of the Object. When we look through the lower left side of the hologram, we see the object from the lower left side, etc.

A detailed explanation of why this happens would be exhaustive. A simple explanation might go like this: the object and reference beam strike the photosensitive recording material at the same time and they create patterns in the photosensitive recording material referred to as interference patterns. After development, we aim only the reference beam at the plate and at exactly the same angle that it originally exposed the plate. As the reference beam goes through the plate the interference pattern causes part of the reference beam to change direction. This is called diffraction. The interference pattern, because it was created by light from the original object, diffracts the reference beam passing through it back in the direction of the original object from which the pattern was

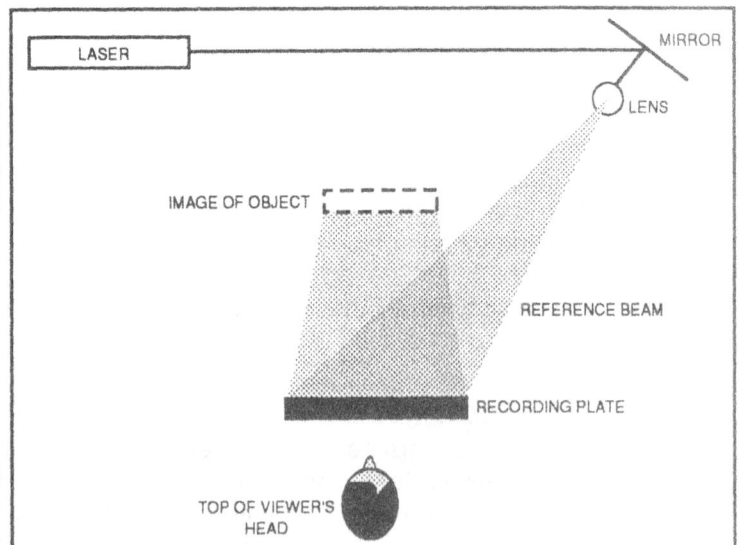

Figure 2. Viewing a transmission hologram (bird's eye view of set-up)

A.H. PRISMATIC

presents the

HOLOGRAM CENTRE
A Captivating In-Store Display for All Our Merchandise

The Best Products . . . The Best Designs . . . The Best Packaging . . . The Best Prices

A.H. PRISMATIC

Enquiries in the U.S.A. to:
A. H. PRISMATIC INC.
285 West Broadway
New York, New York 10013. U.S.A.
Tel: (212) 219 0440
Telex: 6973539 PRISM NY

All enquiries (excluding U.S.A.) to:
A. H. PRISMATIC LTD.
New England House, New England Street
Brighton, Sussex BN1 4GH. England
Tel: (0273) 686966
Telex: 877668 PRISM G

Pictured above: Floor Standing Unit displaying Film Holograms, Laser Jewellery, Laser Discs, Laser Spex, Hologram Jigsaws, Stickers and Boxes.

made. This deflected beam has exactly the same form as the beam that was originally reflected from the object. Consequently, the light recreates a three-dimensional image of the original object.

To record a clear interference pattern of our object and reference beam on the plate we use a laser which is a single beam of light at one wavelength. We cannot use regular light as our source of light because regular light contains many wavelengths of light which change all the time. If we make the exposure using regular light, the interference patterns would be completely blurred and useless because of the changing wavelengths.

In viewing a developed hologram plate, it is important that the reference beam shines on the hologram plate at exactly the same angle that it originally exposed the plate in order to clearly see the image of the subject in its original position.

Viewing A Finished Hologram
In White Light
Although it is necessary to use a laser to make the hologram, it is not always necessary to use a laser to see a hologram. In fact, most holograms can be seen in sunlight. Let's say that you have made one of these sunlight viewable holograms, which are also called white light viewable holograms. Once made, the hologram recreates the original object's image when the wavelength that was used to make the hologram passes through it to your eyes at the angle the reference beam struck the plate, as in Figure 2.

Sunlight consists of many wavelengths and includes the wavelength at which the hologram was originally exposed. If you shine sunlight through the hologram, the original wavelength that exposed your hologram passes through the hologram plate and recreates the image.

It is not necessary to use sunlight to illuminate your white light hologram. There is a range of light sources with which to view a hologram and some light sources are worse then others. Just remember that with white light holograms, you want a light source which contains your desired wavelength and can cast sharp shadows, like a spotlight or an average clear light bulb with a single filament (sometimes referred to as a point source light). Some very bad sources, such as fluorescent lights, which have an extremely diffuse light source, may render some white light holograms unviewable.

It is generally true that holograms which project an image far out in front of the plate or have great depth, almost always require a dominant source of light from a spotlight or a point source such as sunlight or a single filament lightbulb. If you have a hologram with an image

that projects far out in front of your plate, and you illuminate it with spotlights coming from different sources and different angles, the hologram forms projected images at all the different angles dictated by the spotlights. The mixture of projected images "blurs out" the image you are trying to see. Consequently, holograms made to be viewed in a wide range of lighting, including light from different sources, use subjects that have very shallow depth because if there is little depth to the object, all the images appear to be focusing in the same place. Thus, when you enter a shop that sells holograms, you usually find that the shop has subdued, overhead lighting with spotlights focused on the holograms. This serves the dual purpose of creating a lighting environment that is pleasant to be in as well as providing a single spotlight for holograms. People who display holograms in their homes find an inexpensive way to illuminate them is with a clear light bulb having a single filament. These bulbs are available at any shop with a large selection of light bulbs.

Viewing A Multi-Channel Hologram
This is an interesting double exposure (or multiple exposure) effect. It is possible in holography to make a first exposure just as we described in the earlier diagram and then take a second object to be holographed, put it in the position of the first object, change the angle of the reference beam and make a second exposure on the same plate. You then develop the hologram.

Holding the finished hologram up to a light source (which acts as the reference beam), you see the first object when the angle between the light source and the hologram plate is the same as the angle between the first exposure's reference beam and the hologram plate. Turning the finished hologram, you see the second object when the angle between the hologram plate and the light source is the same as the angle between the second exposure's reference beam and the hologram plate.

In other words, when tilting the hologram back and forth you see different objects appear. This is a multichannel hologram because there is more than one channel or angle between the hologram plate and the reference beam that plays back images. This is very common in embossed holograms. Many embossed holograms show more than one scene when tilted back and forth. Multi-channel holograms can be used to show before and after scenes or whatever a designer can dream up.

Reflection Holograms
The transmission hologram in Figure 1 is viewed by letting light transmit through the plate to our eyes. It is

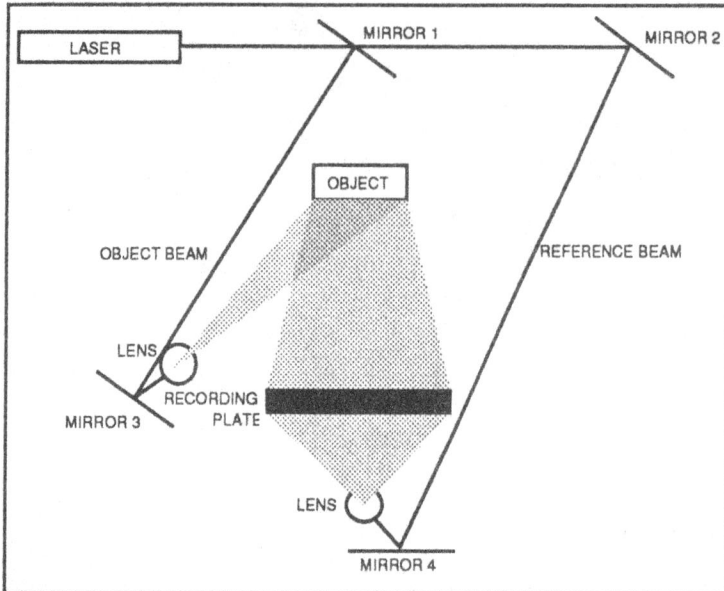

Figure 3. Reflection Hologram

called a transmission hologram because the light passes through the plate to us. It is also possible to make a hologram where the light reflects off the surface and back to our eyes for us to see the image. This is called a reflection hologram.

How are reflection holograms made? Look at Figure 1. If we transfer the reference beam around with mirrors so it illuminates the recording plate from the back instead of from the same side as the object beam, we create a reflection hologram, shown in Figure 3. It is that simple. All reflection holograms are white light viewable and, as their name implies, all reflection holograms are illuminated from the same side as the viewer. In other words, when viewing a reflection hologram light comes from your side of the plate, strikes the plate and reflects back to your eyes.

Summary

So far we have seen that there are three types of holograms:
• Laser viewable transmission holograms (viewable under laser light only)
• White light viewable transmission holograms (Viewable under any point-source light, also called sunlight viewable holograms)
• Reflection holograms (all reflection holograms are sunlight viewable)

Every hologram made is one of the above three types and adheres to the basic principles described above. The set-ups, however, can be wildly elaborate with mirrors and optics all over the place for special lighting and effects. This is a very open-ended and creative art form.

EMULSIONS AND RECORDING MATERIALS

Now we know that holograms are either transmission or reflection depending on what exposure setup we use, but what about the photosensitive materials that actually record the hologram patterns? We are tempted to call them emulsions but the word emulsion is reserved for particles suspended in a gelatin and it really only applies to the silver halide emulsion. Therefore we use the much broader term recording material which simply means something that captures or records our interference patterns. How many recording materials are there and what are they? On what are they coated?

Answering the last question first, the photosensitive recording materials are applied to either glass or film. Both are used in the industry. Film is obviously less expensive and easier to massmanufacture but glass is more popular for original masters.

We discuss recording materials in depth in Section Two. Here we give an overview of the different types of recording materials available and describe the mass manufacturing of each. There are four popular recording materials.

1) Silver halide emulsion
2) Dichromated gelatin
3) Photopolymer
4) Photoresist

Silver Halide Emulsion

Photography has used silver halide emulsions for decades. As you would expect, some major manufacturers of photographic film have adapted their formulae to provide film and plates for holographers. The appeal of the silver halide emulsions to the holographer is that they require less light for exposure than most of the other recording materials. This allows people to buy less powerful, hence less expensive lasers and still make good holograms. It is also convenient to buy precoated silver halide film and plates that have a reasonable shelf life which is not an attribute of some of the other recording materials. The drawback to the silver halide emulsion is that because it is granular, it scatters more light than the other recording materials.

Commercial Manufacturing Of Silver Halide Holograms

Before mass manufacturing a hologram you must make a model and have the master hologram made. The mastering cost is one of the things that is holding the industry back. Model costs range from $1,000 to $4,000, depending on the detail, and masters cost be-

INTELLIGENT

LIGHT FANTASTIC.®

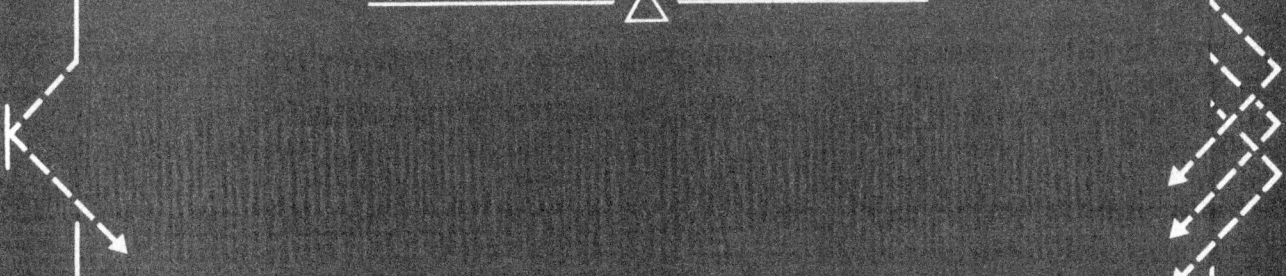

LIGHT FANTASTIC PLC
4E/F GELDERS HALL ROAD
SHEPSHED,
LEICESTERSHIRE.
LE12 9NH
ENGLAND
TELEPHONE (0)509 600220 △ TELEX 858893 G △ FAX (0)509 508795

HOLOGRAPHY

tween $2,000 and $10,000. Extremely short runs are generally not done with silver halide because the cost for model and master makes the unit price too high. Although most of the mass manufacturing of silver halide holograms on film and plates is not fully automated, there exist fully automated machines for silver halide film reproduction which are capable of runs in the tens of thousands. The mass manufacturing quality, which at first was poor, has improved considerably.

Dichromated Gelatin
The dichromate holograms (also known as DCG, or dichromated gelatin) are well liked by holographers because they give a very sharp, almost grainless image and they reflect light extremely well. The drawbacks to the recording material is that the dichromate chemicals do not have a long shelf life, the exposures are not as fast as silver halide and DCG is very sensitive to the environment. (The image disappears if it comes in contact with moisture).

The dichromate hologram is usually sold in a sealed glass sandwich. The DCG hologram's images are usually close to the hologram surface and display the image in almost any available light including fluorescent lights. The most common DCG hologram sizes range from small, inch-and-a-half circles, up to standard sizes of 4 x 5 inches. The DCG recording material holds metallic colors very well. Images of the inside of a watch look almost identical to the original watch.

Commercial Manufacturing Of
Dichromate Holograms
DCG holograms can be made in very short runs. Even orders of 1 to 100 are done commercially. Short runs are done by shooting the final DCG hologram directly from the object and thus skipping any mastering. This shortcut is used commercially because it allows businesses to test-market their product in short runs without the expensive mastering costs involved in a mass manufacturing run. Serious commercial runs, though, go through the whole mastering expense.

Photopolymer
Photopolymer is a relatively recent recording material. Polaroid is the largest company with a photopolymer that is marketed widely. Jeff Blythe in England sells a photopolymer he developed and DuPont says it will introduce a competing photopolymer in 1990. Photopolymer has some of the same attributes that dichromated gelatin has. It has a very sharp, almost grainless clarity and reflects light extremely well. The photopolymer recording material is much thicker than silver halide or DCG film. Despite its thickness, a method has been developed to apply photopolymer to products in large quantities.

Commercial Manufacturing Of
Photopolymer Holograms
Due to the model and mastering expense, it is not cost-effective to do an extremely short run of photopolymer holograms but the price is competitive with dichromate and silver halide in long runs. If you are a typical commercial customer, you will find that Polaroid is marketing its photopolymer system under the name of "Mirage Holography". Mirage holograms are furnished to the customer with either a black or clear backing. The clear backing allows you to place the Mirage hologram right over a printed area allowing type or graphics to show through. A black backing gives an image high brightness and contrast. On average it costs about $0.10 per square inch (2.54 cm) for a minimum volume production run. The price may drop as low as $0.05 per square inch for a large volume application. The master may cost from $8,000 to $15,000.

Polaroid offers a library of stock holographic images and prices vary depending on size and quantity. It takes about 75 to 90 days to produce a Mirage hologram, from artwork to shipping, and the sizes range from .66 x .75 inches to 10 x 14.5 inches.

Photoresist For Embossed Or Hot Stamped Holograms
This recording material is used exclusively for preparing embossed holograms. Photoresist is a photosensitive recording material that records the hologram like the other three recording materials. In a conventional silver halide hologram the interference patterns that create the holographic image are contained within the light-sensitive emulsion. If we use a photoresist rather than a silver halide emulsion, the pattern is in relief.

Since the surface is in relief, we can make a replica of the surface pattern in hard metal, and stamp out holograms in hot plastic in much the same way as we make audio discs. Embossed holograms can be delivered to you from the manufacturers as adhesive stickers or as hot stamping foil for stamping on books or magazines. Stickers can also be applied to books and magazines by machines.

Embossed holograms are used widely. They are on your bank cards, magazine covers, food containers, and many other products. Their primary uses are for identification (security) and for point-of-purchase display attention. Their advantage is that they can be mass-manufactured and machine-applied to products by the millions. We may see competition in the future from DCG and photopolymer holograms but the ability to apply large quantities of holograms at a comparatively low unit cost favors the embossed hologram at present by a wide margin.

Commercial Manufacturing Of Embossed Holograms

Once again, the model and mastering fees make this method appropriate for manufacturing in large quantities. Embossed holograms are literally stamped into plastic on a modified commercial printing press and consequently millions of holograms can be produced very quickly. The unit cost of embossed holograms is far lower than other types of recording materials if made in large quantities.

The model and mastering fee for the typical 5 x 7 inch hologram is about $5,500. Stock images can be purchased from major manufacturers of embossed holograms which eliminates the model and mastering fee. To give you an idea of production prices, one major manufacturer of embossed holograms says that if the mastering cost is paid and you plan a 2 x 2 inch embossed hologram to be delivered as stickers in a run quantity of 5,000 to 200,000 your price runs somewhere between $0.04 to $0.02 per hologram depending on the quantity you want.

Lasers

We will not go into depth discussing lasers in this edition but the difference between CW and pulsed lasers should be spoken about now. There are two major kinds of lasers; the Continuous Wave (CW) laser and the pulsed laser. The CW laser emits a continuous wave of laser light whereas the pulsed laser emits laser light in bursts.

Continuous Wave Laser

The power of a CW laser is measured in watts (w). The CW laser is by far the most common laser used in holography. In holography labs, most of these lasers fall in the 5 to 50-mw (milliwatt) range. One of the great problems with CW lasers is that they cannot make the extremely short exposures necessary to capture a live subject. Consequently, there must be absolutely no motion at all during the exposure with a CW laser. An exposure with a CW laser can take a fraction of a second to many seconds. Because there cannot be any motion at all during the exposure, we need eliminate any vibration coming from the ground. To do this we make or buy a vibration isolation table on which to put our laser, optics, and objects. Since it is absolutely critical that we have no motion at all, the subjects that we holograph with CW lasers have to be "dead" objects. Feathers might move in the breeze and living things move too much.

Remember that what we are recording on the plate is the reference beam and the object beam converging (or interfering) with each other at the plate. If the object moves even a microscopic amount (on the order of wavelengths) from one moment to the next, we will record two different interference patterns and the hologram fuzzes out or doesn't even show. The effect is like some photographic daguerreotypes of the 1800's- only much less forgiving.

Pulsed Laser

Pulsed lasers, on the other hand, emit extremely quick bursts of very powerful laser light. Consequently, the exposure time is much shorter than a CW laser. Exposures can be in nanoseconds. One nanosecond is one billionth of a second. You don't need a vibration isolation table for the pulsed laser. What can you shoot? Anything you want. You can shoot an entire room of people belly dancing in costumes of paper with feathers in their hair, and birds flying around the room. Why such freedom? Because your subject can't move significantly in a nanosecond.

What are the drawbacks of pulsed lasers? Why doesn't everyone buy one? The answer is money. They cost about $60,000. and require a lot of extra overhead and care. Lasers don't last forever and when a pulsed laser burns out it is expensive to fix. Holographers are anxiously awaiting, with cash in hand, a low-cost, easily maintained pulsed laser.

The reason this discussion on lasers is important is that who you go to for making your hologram depends on what you want holographed. If you have a corporate logo or some other object that does not move, a CW laser can do just as good a job as a pulsed laser. There is no need to pay extra to have your hologram made by pulsed laser if you do not need it. On the other hand, if you have a live subject, you must use a pulsed laser.

The H-1 And Master Hologram

It is important that we cover the topic of the H-1 and master in this introduction because it is a fundamental procedure in almost every job. H-1 stands for hologram one because it is the first hologram you make on the path to the desired final hologram. Sometimes the H-1 is the master hologram from which you make multiple copies. Frequently, though, there is more than one hologram that needs to be made before you get the desired master hologram from which you will make copies. If there is more than one hologram that needs to be made, the next hologram in the sequence is called the H-2, and then H-3, and so forth.

Why would anyone want to make an H-2? One of the big problems that holographers used to have was placing their object to be holographed exactly where they wanted it. Suppose we want the object in our final hologram to appear half in front and half behind the recording plate. How would you do it? This problem was solved by the following procedure (see Figure 4).

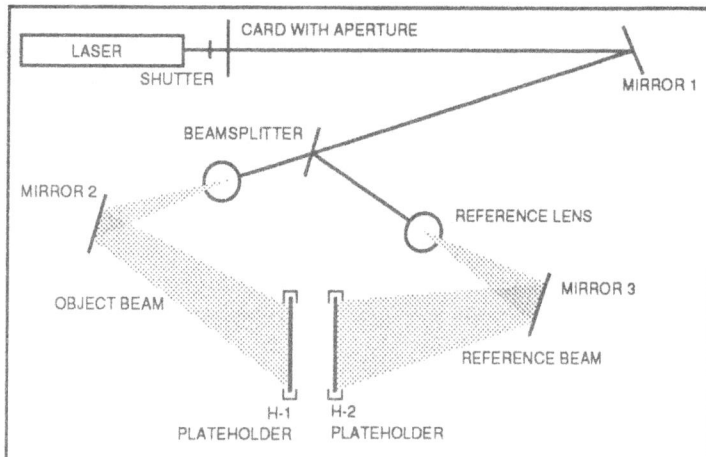

Figure 4. Reflection H-2 being made from H-1.

• Make a laser viewable transmission hologram described earlier (see Figure 1). We call this H-1 because it is our first hologram.
• Since the H-1 hologram creates an image of the object, why not use the image made by our H-1 as our subject and make a hologram of the image (which we will call H-2). In other words, make a hologram of our hologram. This H-2 hologram can be a transmission or reflection hologram depending on your need.
• It sounds strange, because you are making a hologram of an image and not an object. But it works. Now, since you can make a hologram of H-1 's image, take time to move the image around to wherever you want it positioned. Adjust the H-2 recording plate so that the image of the object is half in front and half behind the plate and then make your H-2. The problem of getting our object to be half in front of our plate and half behind our plate is solved.

In summary, here are three reasons the H-2 is made:

• The H-2 allows you to reposition the image of your subject. When you reposition your image from H-1, you may make your subject focus out in front of the recording plate, behind the plate or anywhere within the limits of your equipment (You are usually limited by the laser's coherence length and the quality of the optics). The creative potential here is enormous because you are able to move solid objects around like they are ghosts. You can have two objects occupying the same space, etc. The process of moving the image around to make the H-2 is called image planing.

• It gives the holographer a chance to brighten up the image. Since you may move your image anywhere, you can focus the image right at the recording plate. This concentrates the light directly on the recording material and brightens up the image considerably. This is commonly done in silver halide reflection holograms.

• A third advantage to making the H-2 is that if you don't like the position of your subject astride the recording plate, you don't have to find the original subject and set it up again. This can be important if you were shooting the belly dancers we discussed earlier.

Going through the pains of making H-1, H-2, etc. to get a good master is necessary for creating most holograms. It is technically possible to get to some desired holograms by Skipping this process but the results are generally very inferior. A master is almost always used for commercial jobs of value.

Some Holographic Procedures
Following is a short description of characteristics of some of the most common types of holograms.

Silver Halide Reflection Holograms
Silver halide reflection holograms are very common in most shops that sell holograms. Most silver halide holograms do not display well (or at all) under average fluorescent lights. Unfortunately, many shops use fluorescent lights. Most people who produce silver halide holograms know that their holograms will be displayed differently - under a spotlight or some other favorable lighting condition. Therefore, they produce holograms with considerable depth and projection. Consequently, when you look at what is generally available for sale, the average silver halide hologram has much more projection and depth than the average DCG or embossed hologram which is made to sell under any lighting, and consequently has little projection or depth. There are notable exceptions to this and the trend may change.

Common examples of silver halide reflection holograms are 4 x 5 inch holograms that project images two to three inches in front of the plate. They come in standard sizes up to 12 x 16 inches and the image can project or recede up to a foot in front or behind the plate under optimal lighting conditions. The creation of silver halide reflection holograms is limited mostly by the size of optics and the power of the laser needed to reduce the exposure time for larger shots. Another good thing about silver halide reflection holograms is how they are displayed. Since these are reflection holograms, they are viewed with the light source on the same side as the viewer. Therefore, they can be hung on the wall and treated like a painting. A nice clear light source or spotlight is usually arranged from overhead or sometimes the hologram is sold with its own light source and display stand. DCG and embossed holograms are also viewed as reflection holograms and can be treated the same way.

Silver Halide Transmission Holograms

Transmission holograms are popular among artists but are difficult to display because they require lighting from behind. Silver halide transmission holograms come as laser viewable transmission holograms and as white light viewable transmission holograms. Let us look at both.

Laser Viewable Transmission Holograms

Although sometimes displayed and sold for its artistic merit, the laser viewable transmission hologram is almost exclusively used for making the H-1

Laser viewable transmission holograms demonstrate amazing depth and projection when the correct equipment is used. It should be noted that the depth of the hologram is not so much a function of the power of the laser as it is the coherence length of laser light. You can read more about the coherence length from several of the books listed in the bibliography.

Theoretically, the maximum image projection in front of the hologram can be as great as the depth. Unfortunately, it is difficult for our brains to make sense of projected images. Because of this and the fact that there are some optical distortions in the image planing process, projected distances in transmission holography are usually kept under four feet. It should be pointed out that projected hologram images generate one of the highest shock and thrill responses from viewers. A good percentage of first time viewers respond by waving their hand back and forth through projected images in disbelief.

As we mentioned, the laser viewable transmission hologram is most often used as an H-1. Transfer copies (making another hologram using the image on H-1 as the subject) are then made from the H-1. These transfer holograms can either be other laser viewable transmission holograms, white light viewable transmission holograms or reflection holograms (always white light viewable).

Laser transmission holograms have the widest parallax (the ability to see around an object side to side) and resolve the greatest depth of objects. There are laser transmission holograms, for example, of people and objects in a 4000 cubic foot room exposed with a pulsed laser.

Pseudoscopic and Orthoscopic Images: Both laser viewable and white light transmission holograms share an interesting property. In Figure 1 we diagrammed the making of a laser viewable hologram. After developing the hologram, we put it back in the plateholder and, with no object on the table, we were able to see the original object in its original position on the table.

Now comes the interesting feature. Take the transmission hologram out of its plate holder, flip it over, and put it back in the plateholder. Step back and look at the plate. You see the image forming out in front of the plateholder (between you and the plateholder). It focuses in air the same distance in front of the plateholder as it originally sat behind the plate holder. You also see that the image is a pseudoscopic image. What is pseudoscopic? An image as we normally see it in everyday life is an orthoscopic image. When our viewpoint is moved to the right, a pseudoscopic image shows not more of the right view of the image but the left view, and when the viewer raises his viewpoint the lower part of the subject comes into view instead of the upper part.

A pseudoscopic image yeilds an exciting effect, but it can be confusing to the viewer. Some artists, however, have produced exciting pseudoscopic work using geometric shapes like wide spirals, pyramids and cones.

White Light Transmission Holograms

A white light transmission hologram, as discussed earlier, is a transmission hologram that can be seen in sunlight. There are several types of white light transmission holograms among which are achromatic and rainbow transmission holograms.

Achromatic Transmission Holograms: Achromatic transmission holograms are black and white transmission holograms. The achromats are reminiscent of nineteenth century daguerreotypes. For those who perhaps see the rainbow colors as too brash and want a more serious look to their hologram subject the achromat is a good choice.

Rainbow Transmission Holograms: One of the problems with most achromatic transmission holograms is that they have very little depth. This problem was solved with the rainbow transmission hologram (sometimes called Benton hologram after its founder Dr. Stephen Benton). The rainbow hologram is made by taking a laser viewable transmission hologram master and making a transfer hologram of it through a mask that has a narrow horizontal slit in it placed against the master. The resulting rainbow transmission hologram has no vertical parallax (you can't see over the top of the image) but has a wide horizontal parallax.

Rainbow holograms get their name from the fact that as you shift your viewing angle up and down, instead of the image changing its perspective, it changes colors, shifting through the colors of the rainbow. Rainbow holograms produce an image in a single highly saturated color which changes through the spectrum as the height of the viewpoint changes. By careful

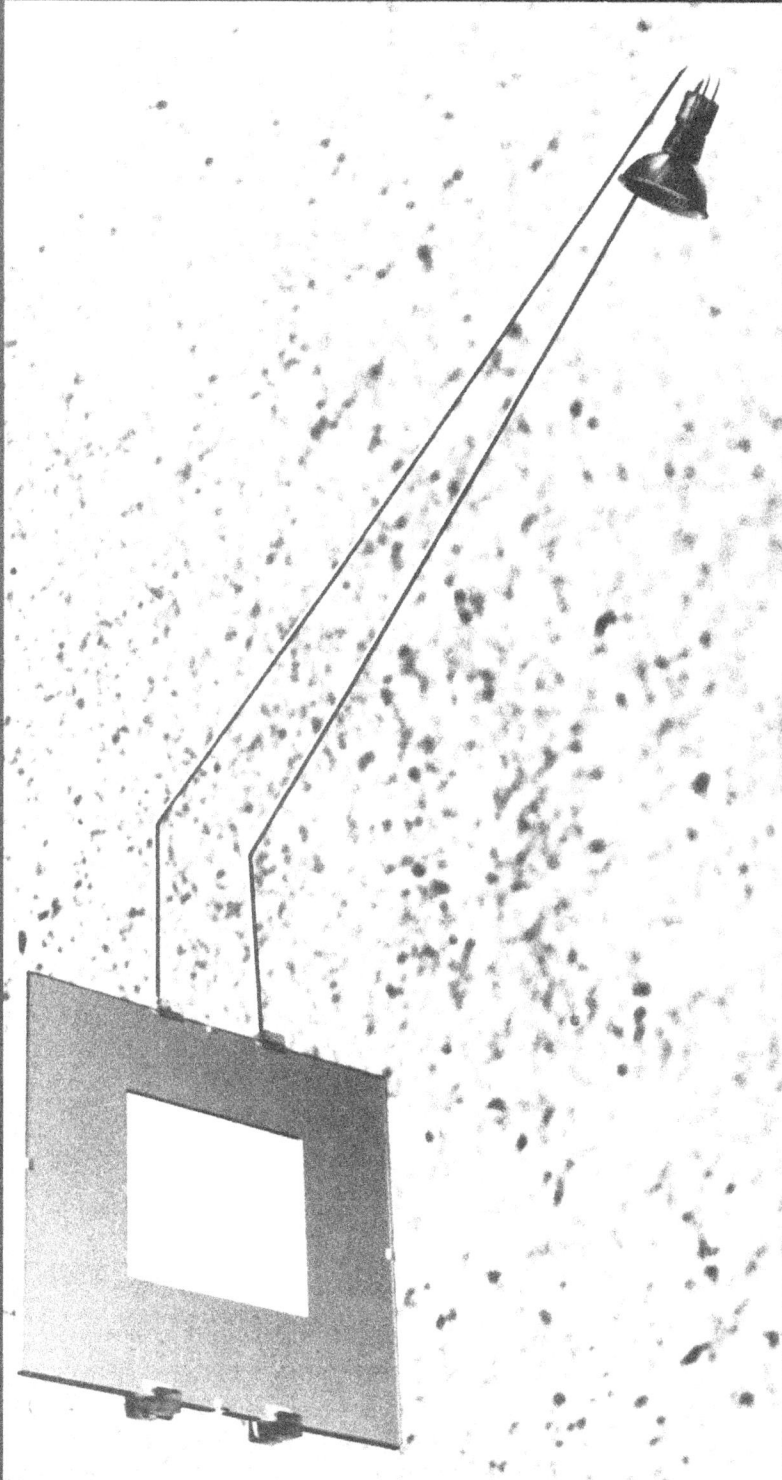

planning, the image may be made any desired color, or even combination of colors (a multi-color rainbow hologram).

Some rainbow transmission holograms are displayed in art galleries on glass plates and on film. However, they are much more popular in two other forms. The two most popular forms in which you see rainbow transmission holograms are as:

• embossed holograms. In an embossed hologram, the light goes through a rainbow transmission hologram that has been embossed in plastic, strikes a mirror backing and reflects back through the rainbow transmission hologram to your eyes.

• holographic stereograms. If you have seen "moving holograms" you most likely have seen one of these. This is one of the most exciting fields of holography. Included in this category are Cross holograms, computer-generated holograms and the recent Holodisk®.

Let's discuss each of these two types of holograms in further detail.

Embossed Holograms

We have a detailed technical discussion on making an embossed hologram in Section Two, but here is an overview. Embossed holograms are created in several steps. The most common way they are made is as follows:

1) The hologram master is made.
2) A rainbow transmission hologram is made on photoresist (photosensitive emulsion) from the master. 3) The photoresist is etch~d to relieve the hologram patterns.
4) The photoresist is plated with silver and a thick layer of nickel. The photoresist now behaves like a metal mold. The metal that forms on the photoresist is called the shim.
5) The shim is removed from the photoresist and now has holographic patterns on it.
6) This shim is used as a stamping die and stamps the holographic patterns into plastic.
7) The plastic with the holographic pattern stamped in it has a mirror-like backing so light comes through the plastic, strikes the mirror-like backing and, reflecting back out, displays the white light rainbow transmission hologram.

Let's look at these steps in a little more detail.

Steps 1-2: There is a possible shortcut here. Sometimes flat art can be directly exposed onto the photoresist material. This is recommended for simple projects only.

Photoresist is a very tricky medium to record on holographically. It is nowhere near silver halide in responsiveness and requires long exposures even with lasers of several watts of power. A small holography studio would have to be equipped with an expensive laser and heat-resistant optics. New alternatives to this medium of photoresist are being researched.

The typical turn-around time is four to six weeks to receive excellent photoresist plates that are fit for metallizing. The holographer should make several photoresist plates of good quality to cover any problems that might occur in the metallizing phase.

Steps 3-4: After the photoresist hologram is checked for clarity, brightness and overall quality it goes to the metallizing stage of production. A thin layer of silver is deposited on the photoresist. Silver by itself cannot withstand the stamping pressure, so additional coats of a nickel-based material are deposited to reinforce the back of the silver. When it achieves the desired thickness, the nickel-silver shim is pulled from the photoresist plate and this becomes a hard, stamping die.

Steps 5-7: The first shim must be perfect. This first shim can have several shims made from it and in turn several shims made from those. The heat and pressure of embossing thousands of holograms wears out the shims so extras should be made. Any deterioration becomes very obvious if the shims are not changed regularly.

Producing the final hologram is the job of the embosser. Embossed holograms are made by stamping the shim onto a heated polyester material which has a metallized backing. Although not used widely, colored metallized backing is an option. Most often the silver color is selected.

Applying embossed holograms: The press embosses the holograms on rolls of plastic stock which, in the most common commercial work, have an image area of six square inches for each impression. The finished roll consists of a roll of 6 x 6 inch embossed squares separated from one another by about one half inch. The holograms are often further divided on each 6 x 6 inch square as nine, 2-inch squares of different holograms, but you can use the entire 6 x 6 inch square for one hologram.

There are a number of options for displaying your final embossed product. Embossed holograms can be produced as stickers cut to your size specifications. Typical of the final holograms are individual peel-and-stick holograms. If the stickers are to be placed onto some surface be sure the back of the sticker is thick and strong enough not to conform to a textured surface

onto which it might be applied.

The best surface on which to apply a hologram is a smooth and rigid one so you can be sure that the hologram is flat and consequently able to reconstruct its image properly. There are applicator machines to apply the sticker to your product, but it is often done by hand.

Instead of stickers, you may emboss onto hot stamp foil and apply the hot. stamp foil directly to a surface like a book cover. If you choose this route, be absolutely certain of the strength and smoothness of the cover material. The embossing foil is very thin. If the hologram has ripples in it as a result of a bumpy surface, the image itself might appear rippled.

It is a good idea to have the surface approved by the holographer and the printer, together, before going into production. Also beware of coated paper stock and printed surfaces. These surfaces may present problems for adhesion of holograms.

Hot stamping the hologram onto a paper surface is a popular way to apply your hologram. In general, the hot foil application keeps precise registration better than methods of sticker application. Although hot stamping appears more expensive, when you consider the costs of a very large run using sticker application they both can come out about equal in cost. Smaller runs seem to favor sticker application.

Regardless of which method you choose, it is advisable to be there when the holograms are applied so you can check the quality.

Embossing pulsed holograms and holographic stereograms: Pulsed hologram images are used in embossing and have been very successful. Be sure the pulsed holographer knows that you want this to be an embossed hologram. It is a more complicated approach which may include a step to reduce the pulsed hologram to the six-inch square size of the embossing machine, or whatever limitations your embosser has. Do not use this approach if your turn-around schedule is too tight. Allow enough time to make sure all steps take place with breathing room if a problem occurs. The results of pulsed embossing are stunning and it is worth pursuing.

Flat integrals (a type of holographic stereogram) which we discuss next, may also be embossed. This also is a relatively new option for embossed holography and opens up new territory for creative advertising.

Holographic Stereograms

A stereogram is a diagram or picture representing objects with an impression of solidity or relief. Consequently, a holographic stereogram is a hologram of pictures or diagrams which gives the impression of solidity or relief.

Several techniques are used to make holographic stereograms. Names are given to the various methods and you will hear names like:
- Integram
- Cross hologram
- Lesliegram
- Multiplex hologram
- Benton stereogram
- Embossed stereogram
- Alcove hologram
- Holodisk®

As with any trade, one of the problems in holography is understanding the jargon. Sometimes several names are used to describe the same item and there are many special nicknames used to describe special techniques. The names you see above are all holographic stereograms and the first five are really the same type of hologram. To give you a clear idea about stereograms, we now describe how a Cross holographic stereogram is made.

How To Make A Holographic Stereogram

Generally, this is the way a 360-degree Cross holographic stereogram is made:

1) Make a small stage that rotates 360 degrees.
2) Put your subject on the stage.
3) Set up a regular movie camera on the stationary floor.
4) Film the subject as it turns the full 360 degrees. Slight motion is possible but the subject cannot make radical or jerky moves. This creates what is known as time smears in the hologram---places where the subject looks jagged. Slow, even movements when filming yield the best results. Make sure to shoot at least three frames for each degree of rotation.
5) Develop the movie film.
6) Make a hologram of the image in each frame of your movie. See Figure 5. .
7) The holograms will be on a roll of holographic film. Each hologram, as you can see, will be as tall as the width of the roll but of very narrow width.
8) Set up a mask in front of your roll of holographic film to get one narrow hologram (you will be shooting a white light rainbow transmission hologram).
9) Shoot the first exposure, advance the holographic film one frame, advance the movie film one frame, expose again and so forth for the entire movie film.

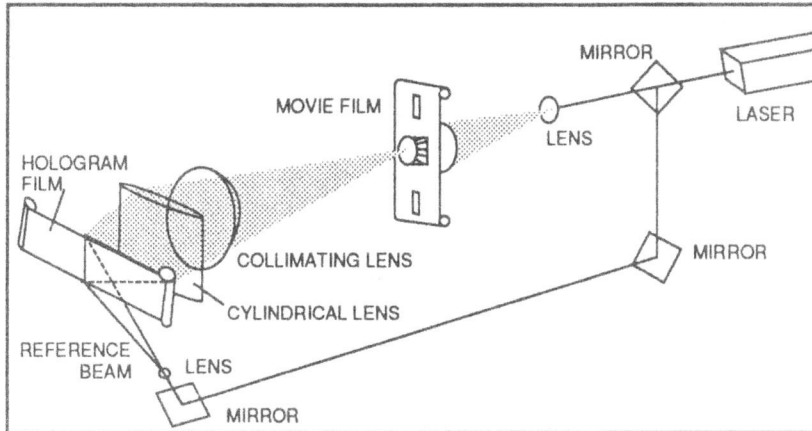

Figure 5. Optical set-up for a white light holographic stereogram.

After development, take the roll of holograms and wrap it around a strong cylinder of clear plastic that is mounted on a display which is able to rotate 360 degrees. Place a clear (unfrosted) light bulb with a single filament in the center just below the holographic film.

Viewing The Holographic Stereogram

Turn the light on and rotate the cylinder. Several things happen here:
• The movie frames are moving past the eyes.
• Each eye sees different images at the same time thus creating a stereo view.
• Since this is a rainbow white light transmission hologram, the image forms in the same position it was when originally shot (the image usually forms in the center of the cylinder).

What we have is a moving holographic stereogram and the viewer sees a three-dimensional "movie" of the image moving around in the center of the cylinder.

Other Holographic Stereograms

Less Than 360-degree Curved Cross Holographic Stereogram: The holographic stereograms like the example above can be made in a curved format less than 360 degrees. On the market today are 90-, 120-, 180-, and 360-degree curved holographic stereograms. The ones that are smaller than 360 degrees generally stay fixed and the viewer walks past them to see the motion. The smaller curved stereograms are inexpensive compared to the 360-degree stereograms.

Flat Holographic Stereogram
(Benton Stereogram): Instead of a curved stereogram as described above, you can make a flat holographic stereogram. The procedure for shooting is a little different.

The subject is not on a moving stage this time but on the ground at a distance from the camera. Now we build a straight railroad track on which to put the movie camera. Without aiming the camera directly at the subject, but facing the camera in the subject's direction, the camera moves along the track and takes photos at equal distances. We then develop the film and holograph it much the way we did in the Cross hologram just described.

The result is a flat sheet which, when held up to a light and tilted from side to side, displays an image in motion. With some changes in the holographic process, you may also make this a reflection hologram. The reflection holographic stereogram is viewed by a viewer standing under a light, holding the flat sheet and tilting it from side to side. The image can display above the surface of the hologram and move about as it is rocked from side to side.

Embossed Holographic Stereogram: Earlier it was mentioned that rainbow transmission holograms are used as the first step in the embossing process. Since a holographic stereogram is usually a transmission hologram, why not take a flat holographic stereogram and emboss it?

It turns out that this can be done and it is a popular application. Other techniques along this line include shooting live subjects with a pulsed laser. Very recent developments allow holographers to reduce the size of the image in pulsed shots. In theory you should be able to make a series of live shots, using a pulsed laser, and reduce them to fit an embossed holographic stereogram. Perhaps someday there might be curved embossed holographic stereograms as labels for canned products.

Alcove Holographic Stereograms: What we have done above is to take flat art, such as movie film, in which the subject moves from frame to frame and make holograms of it. As you know, computers can now make original flat art. Computer generated images are obviously easier to work with, particularly when making corrections. Why not generate all the art you need to make a holographic stereogram using the computer?

This is the concept behind computer generated holograms. One example is the alcove holographic stereogram. The group led by Steve Benton at the Massachusetts Institute of Technology's Holography Division of the Media Lab is working hard on this concept and they have already produced some remarkable results. The computer, in this case, creates the flat art. The flat art is then filmed from the video display terminal. The film is then used to make the holographic stereogram.

What also sets this hologram apart from the others is the way it is viewed. The image is seen within a clear, concave half-cylinder, an alcove, with close to 180 degrees of horizontal parallax. Along with the wide parallax, the image in the hologram can have depth going back through the hologram to infinity. Image distortions are corrected before output in the computer.

The alcove hologram is probably the closest holography has come to a totally synthetic, three-dimensional subject to date.

Holodisk®: This is a fascinating application of the reflection holographic stereogram. The process takes 360-degrees worth of photographic images and holographically integrates them onto a flat Holodisk®. It is lit from above and the subject can appear below, straddling, on, or above the disk surface within the limitations of the reflection hologram.

A slowly rotating turntable brings 360-degrees of views to a stationary viewer. The obvious application would be to put the Holodisk® on phonograph records. As you listen to your favorite song an image could dance before your eyes. Unfortunately the revolutions per minute are too high for the Holodisk® to be used on records, but research continues on this.

Holographic Stereogram Sizes

Sizes vary with each holographic stereogram. The standard curved holographic stereogram is about ten inches high and eighteen inches in diameter with a curved face in 120-, 180- and full 360-degree formats. The widest film roll you can purchase is a standard 42 inches wide. Attaching one end of a roll to another, these holograms can theoretically be made in indefinite lengths. Using motion picture footage, we could conceivably run hundreds of feet of continuous holographic film. Current technical restrictions, however, need to be overcome before this can be a reality.

Holographic Optical Elements (HOEs)

As we have seen, a transmission hologram behaves in some ways like a lens because it lets light pass through it and focuses that light to form an image. It is possible to make transmission holograms that act as very good lenses. In other words, instead of making an image, the hologram is designed simply to spread or focus light. In the same way, reflection holograms can be made into exceptionally good mirrors. This is what the field of HOE is all about. We discuss this further in Section Two.

Diffraction Gratings

A diffraction grating is a device which bends light. Diffraction gratings can be made by etching, by deposition, acoustically, or holographically.

Holographically, diffraction gratings are made by the interference of two (or more) beams of pure, undiffused laser light. Diffraction gratings are among the most simple holograms to construct and they may be designed to produce unusual bursts of very pure color or simply diffract a specific color. Diffraction gratings can also be made to reflect a very high percentage of the specific color you desire.

A number of items are used artistically to create variations in the diffraction patterns. Passing the laser beam through different kinds of lenses, pieces of plastic, or reflecting the beam off surfaces like crumpled aluminum foil or even ground glass can yield surprising optical effects.

COLOR IN HOLOGRAPHY

Color Variations

Reflection holograms are usually monochromatic under white light. However, it is possible to change their color. One way this can be done is by swelling the photosensitive recording material using a specific percentage of an inert chemical, and changing the reference angle a certain degree (to correspond with the color you want) and then making the exposure.

Let us say you want to have two subjects in the hologram. You want each one to be a different color and you want them seen by the viewer at the same time (ie. not multi-channel). For two colors, the first exposure is made, the subjects are exchanged, the recording material is pre-swelled to another color's percentage, the reference angle is changed and the second exposure is made.

The pre-swelling makes the whole recording surface fatter for the exposure. When the hologram is processed the swelling agent is washed out and the thickness of the hologram returns to normal. At normal thickness, the hologram selects not the red wavelengths by which it was made, but instead selects, for example, shorter wavelengths--ranging from an orange-red to violet, depending on how much it was pre-swelled.

For registration purposes the reference angles also must be changed with each pre-swelled color exposure. If you make a two-color hologram by preswelling the recording material and not changing the reference angle accordingly, the result will be a two

channel hologram. Looking straight at the plate you see a red apple, for example, and turning the plate a little the apple disappears and blue grapes appear.

There are many complications with the above procedure: swelling agent ratios, changing of reference angles, and multiple exposures. Added to this is the problem that the hologram cannot be copied without going through the whole thing again (it cannot be mass manufactured). Consequently, this method is used for a single hologram or for small-run, custom work.

Color Variations: Full Color Holography

This is not easy to do (or everybody would be doing it) but it is easy to understand. In the back of the human eye are light sensors called rods and cones. They are acutely sensitive to three colors: red, green, and blue. The human brain receives only this information and uses it to create our full color world.

The idea, then, is to make a full color hologram by using three colors of laser light (red, green, & blue) instead of one as your exposing light source. One of the snags to three laser color holography is that although three single wavelengths are very good at reproducing colors (as in TV), they do not in general record the colors of objects well; this is something color photography does much better.

It all sounds easy but it is very difficult. A full color hologram, with colors like you see in a color photograph of a person, are very difficult to produce. There have been a few serious attempts at full color holography. "DOJO" by Toshihira Kubota of Kyoto Institute of Technology and "Pencils" by Paul Hubel of the Dept. of Engineering Science at Oxford University, England are two holograms that come to mind. Although much better than any other attempts, these holograms still fall short of photographic type quality.

Recently some very good holographic stereograms have been made in color like "The Clown" by Sharon McCormack, and many people feel that this will be the

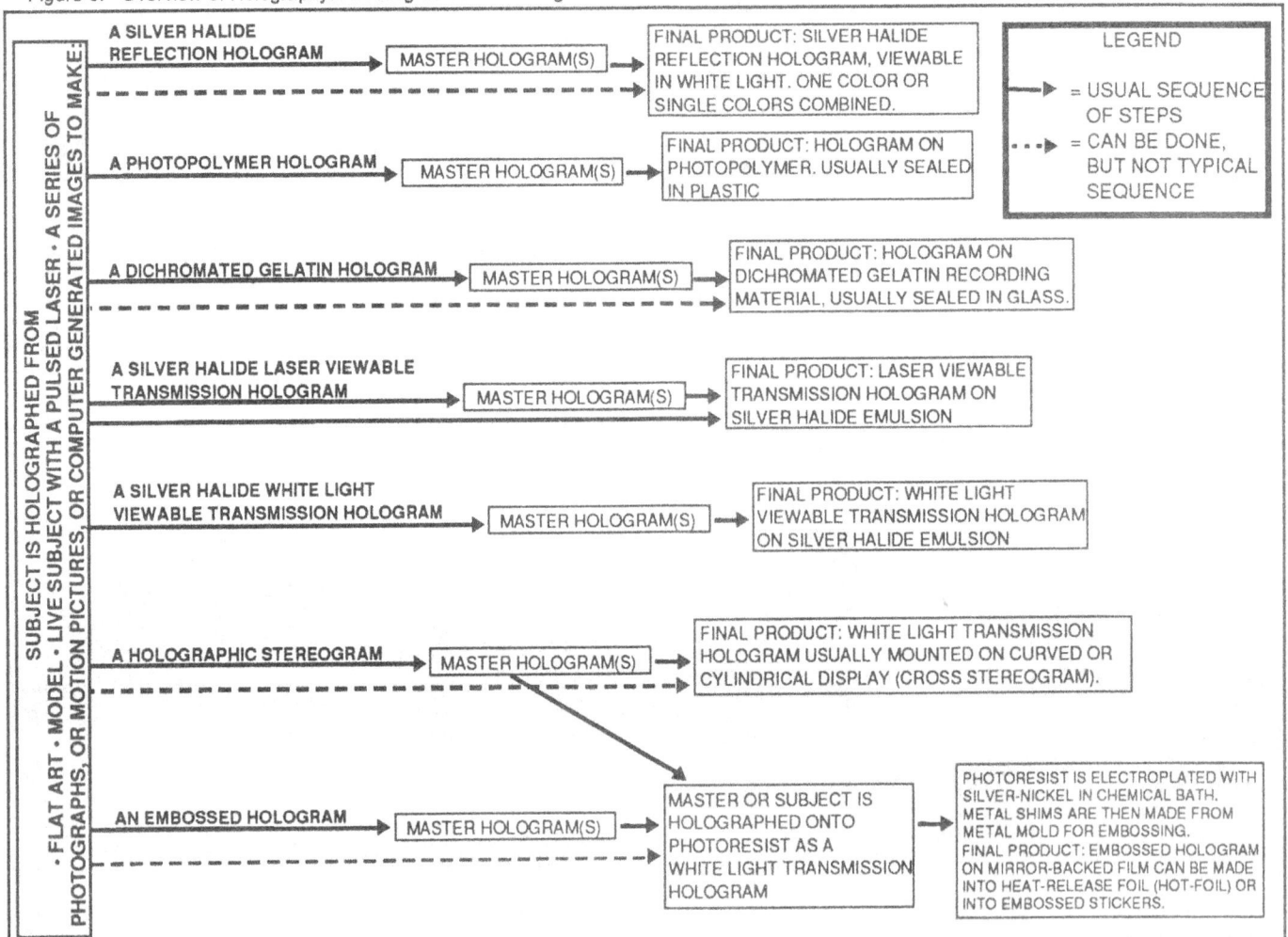

Figure 6. Overview of Holography. All holograms are white light viewable unless otherwise stated.

first full color holographic process to be perfected.

Bird's Eye View of Holography

Figure 6 is a chart showing the most common artistic holograms which we have briefly introduced here.

THE BUSINESS OF HOLOGRAPHY

What Is The Market For Holography?

The global holography market is composed of hundreds, perhaps thousands, of companies, universities, and other organizations involved in developing different aspects of this technology. As the listings in this publication make clear, it is also composed of hundreds of associated equipment and service suppliers, users and investors. This discussion focuses principally on the "core" industry: the group of companies and other organizations that are working with the technology of holography for commercial or internal applications.

That industry is made up of a wide range of players. There are a few publicly traded holographic companies. Applied Holographics PLC and International Bank Note are two of the most notable. There are a substantial number of major corporations with significant holographic efforts. Polaroid, DuPont, 3M, Pilkington, Dai Nippon, Hoechst Celanese, and ICI all fall into this category. Likewise, a number of medium sized corporations are actively involved. Newport Research Corp., liford, Agfa-Gevaert, Crown Roll Leaf, and Ealing Electro Optics are examples of this tier of the industry. Of course, much of the research and some of the commercialization work takes place within the university or nonprofit area. MIT and ERIM, in the U. S., and Loughborough University and others in Europe are a few examples of these.

In typical pyramid fashion, there are a large number of small companies that make up the bulk of the holography industry. Although the size and focus of these companies ranges widely, there is a tendency for them to group according to applications area. For example, there is a developing worldwide network of artistic hologram producers, distributors and retailers. Often, even relatively small companies tend to be globally oriented within their applications segment. It is not unusual for relatively small embossed hologram producers to have affiliated companies on several continents.

Section I: Introduction to Holography 23

How Can The Market Be Quantified?

Any attempt to estimate the size of the holography market poses a number of problems. First is a question of definition. It is clear that a company producing holograms and selling them as decorative or security items is part of the holography market. But is Nippon Electric Company (NEC) a part of the holography market? NEC is using a computer generated holographic optical elements (HOEs) as a key component of its compact disk player. Some businesses consider these aspects very much a part of the industry and attempt to develop meaningful information on them.

The second is the question of quantification. Frequently, sales figures are difficult to obtain. Most companies in the field are privately held and are understandably reluctant to release detailed financial data. Fortunately, there are a few public companies, notably, Applied Holographics PLC and International Bank Note (IBK), owner of American Bank Note Holographics, which are required to publish their financial reports. Likewise, in the European Community certain private companies are required to file annual reports with varying degrees of detail depending on their size, status and location.

A third problem is allocation. For example, the U. S. government and other countries have active military and domestic research programs which include small holographic activities. It is often difficult to identify meaningfully the portion of resources devoted to holography. A similar problem exists when dealing with large industrial end users. As noted below, Pontiac, a division of General Motors, is introducing a holographic safety device. While incorporating a holographic element, the bulk of the cost of the device is for nonholographic material.

How Large Is The Market?

Estimates range widely, but it seems probable the answer is around $200 million in 1988 and growing rapidly. As for the holography industry, there have been at least four industry market studies within the past few years. The table below shows a summary of the conclusions from these studies.

Figure 7.	Estimated and Projected Market Size. (All figures in Millions of US$).	
LITERATURE SOURCE	SIZE - YEAR REPORTED	PROJECTED SIZE - YEAR
Bus. Comm. Co.	$82 - 1985	$239 - 1989
Tech. Insights	$23 - 1985	$140-400 -1990
Int'l Res. Devel.	$153 - 1986	$253 - 1989
Fut.Tech.Surveys	$119 - 1988 (US Only)	$500 - 1993 (US Only)

A review of these numbers and independent analyses leads us to conclude that the market in 1988 was approximately $200 million, growing to about $230 million in 1989, in wholesale prices. Further examination of the data in Figure 7 and actual corporate performance data suggests that the rate of growth of the industry is significant: on the order of 25-35% per year.

The Market Sectors

The industry is quite fragmented. This is one reason why some professionals have difficulty recognizing its potential. The embossed holography segment is closely allied to the printing and packaging industry. The film and artistic dichromate gelatin sectors ally with the giftware and exhibition promotions industry. NDT and interferometric applications are tools within industries as diverse as aerospace and acoustic consumer products. HOEs are finding broad applications from consumer and military products to the telecommunications and computer industries. And, head-up-displays (HUDs) are installed in sophisticated fighter aircraft among other applications. The following paragraphs outline some of the key current and future applications areas.

Embossed Holograms

This market segment is one of the most universally recognized. In North America and Europe the embossed holograms of the Visa™ dove and the MasterCard™ globe/MC are in hundreds of millions of billfolds, while credit card companies in South America and India are now examining the option of issuing cards featuring holograms. Millions of people around the world also carry phone cards, debit cards, check cards, identity cards, and bus tickets bearing some embossed holographic image. The hologram is used in these applications as an anti-counterfeiting, security device as well as for decorative appeal.

This application area is significant. American Bank Note Holographics, Inc. for example, had sales for credit card applications alone totalling more than $13 million in 1988, and several other producers offer secure facilities so they can seek a slice of this market. In addition to anti-counterfeiting applications, embossed holography is used in the design, printing and labeling industries.

While National Geographic was one of the first, to date hundreds of annual reports, magazine covers, advertisements and product labels have displayed holograms. With the development of wide web production, embossed holograms are also beginning to be used by the packaging industry. Last Christmas, Bacardi Imports, Reynolds Metals and American Bank Note Holographics broke new ground by introducing

a national sales campaign based on the use of holographic holiday packaging.

Embossed holography is beginning to establish itself in the true security printing market as well. Australia has already introduced currency with embossed holograms as a key part of its security system. Several other countries, including the U. S., are currently introducing or studying the use of holograms on currency and official documents. Recently, Austria issued a run of 4 million nine schilling stamps bearing a hologram and is currently reprinting a second issue.

In aggregate, the worldwide sales of embossed holograms during 1988 may have exceeded $50 million. American Bank Note Holographics, whose parent company is International Bank Note, alone accounted for $21.9 million in sales. In recognition of this growing business area, several companies have begun to market their ability to transfer this technology to other organizations.

Artistic Holograms

The artistic hologram sector also includes silver halide, DCG and photopolymer glass and film products. Although smaller in annual volume than the embossed sector, there are numerous small companies and individual holographers active in this area of the industry. This segment probably contains the widest spread of producers around the world.

The products produced include the high value, large format holograms for trade show or point of purchase promotion, film and DCG holograms for the giftware sector, specialty advertising products and fine art work. Some of the products that have shown market strength include holographic watches, the holographic promotion of high-quality baseball cards produced by Upper Deck, Kellogg's successful use of holograms to build market share in the U. K. and sales of holographic tee shirts.

Despite somewhat flat sales for film holograms during 1988/89, Holographic Images launched a series of two-color silver halide film holograms, and Holos Gallery manufactured the first commercial holograms produced on Polaroid Mirage photopolymer film.

Interferometry - NDT
(Non Destructive Testing)

The interferometry/NDT applications of holography are more difficult to quantify because many of these applications are "in-house" projects by industry and government. The "invisible" nature of the market has begun to change recently, however, with various companies marketing ready-to-use systems. The increas nsing popularity of electronic speckle interferometry is also tending to give more of a commercial character to this market segment. Business Communications Co. estimated that this part of the industry accounted for $55 million in sales in calender year 1985. Technical Insights, on the other hand, put the figure at $3 million for the same year. A good estimate for sales is in the $20-30 million range for 1988.

Holographic Optical Elements (HOEs)

HOEs are one of the major growth areas of holography. With their ability to manipulate light, HOEs have applications both for the medium to large scale and for the microscopic scale. One of the best known applications is that of head-up displays (HUDs). These devices are used extensively in both military and commercial aircraft. The units themselves are produced by divisions of major corporations such as Pilkington, Hughes and Flight Dynamics. HUDs and related applications have begun to work their way into the automotive market. Pontiac, for example, recently announced that it would use a flat holographic element to replace the bulky "high-mounted center stop lamp" on its Grand Am SE model. It appears to be only a matter of time before a variety of mass production applications develop in the automotive industry.

Small format HOEs are currently being used in a wide variety of applications. Physical Optics Corp. and British Telecom are among those with active research in fiber optic multiplexing products using HOEs. NEC and Pencom International have both developed HOE based compact disk reading heads: thousands are currently being sold to consumers around the world. HOEs are used in laser printers, graphics scanners and bar code readers and are finding their way into various military products. In the future, small format HOEs may well be a fundamental building block of optical computers.

During late 1989, Du Pont de Nemours announced the introduction of its holographic photopolymer for HOE, HUD and artistic producers. In addition to the in-house activities of numerous major companies, several smaller companies, such as APA Optics, Holomax and Holotek, specialize in HOE production. This area of the marketplace is again rather difficult to quantify. Bearing this in mind, we estimate that HOEs may have accounted for more than $20 million in sales, exclusive of research applications, during 1988.

Research

In addition to the product oriented categories previously outlined, there are significant research efforts in a myriad of industrial, university and government laboratories. A 1981 Harvard Master's thesis concluded that the U.S. Government alone was sponsoring approximately $29 million worth of non-classified holographic research projects. Canada, Switzerland, Spain, China, West Germany, the UK and several other countries have major optical research centers within which holography is an important part. Several major international corporations have established research centers devoted in part to holographic applications.

In conclusion, the holography industry is in the early stages of its development but shows real signs of producing several areas of rapid growth over the next decade.

---Contributed by: The Editors of Holography News

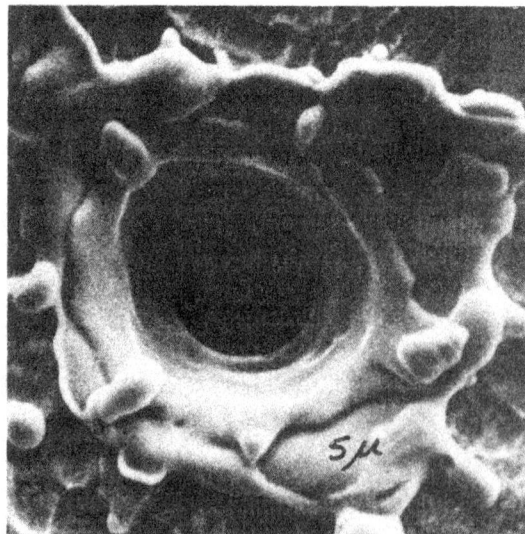

THE HOLOGRAPHY DISTRIBUTION PROCESS

Distribution

Holography is unquestionably a business for the nineties. Still in its infancy, it has an annual growth rate that is better than many long established enterprises. Although any business like holography that is on the cutting edge of a new technology offers risk for those venturing into it, there is much to be made for those who succeed. One advantage for the entrepreneur is that there is not a competitor waiting on every corner. Another advantage of this market is that while many mature businesses cringe at the thought of advancing technology destroying their markets, holography thrives on advancing technology.

We can divide the holography market into two broad categories, artistic holography and industrial holography. Artistic holograms can be defined as holograms created for the purpose of being seen and whose value derives, at least in part, from the image presented. Artistic holograms, therefore, can be anything from commercial art like bank card holograms to limited edition fine art holograms. Industrial holography encompasses items like supermarket scanners and holograms used for nondestructive testing of machine parts. Although still developing, there is a network of distributors, wholesalers and shops that sell artistic holograms. We outline how this market works from the top down.

The Copyright Holder And The Manufacturer

Copyright Holder

The distribution process starts with the copyright holder. As its name implies, copyright means that no one has the right to copy a unique work of art made by someone else without that person's permission. All works of art can be copyrighted and for any unique work of art there exists only one copyright holder. The copyright holder may be a group of people such as a Business. This means that every hologram, if properly copyrighted, has only one owner, the copyright holder. Holograms that are not copyrighted can be copied by whoever has the master as often as he wishes.

Sometimes the copyright holder is a business that pays a manufacturer to make a master hologram and multiple copies. On the other hand, sometimes the manufacturer is the copyright holder and can then make masters and copies itself. In order to keep our discussion as clear as possible, we describe each of the businesses in our distribution chain as separate entities. Therefore, the copyright holder is at the top of the chain and controls all distribution.

There are statutory limits stating that after a number of years, a piece of art transfers into the public domain and anyone may use the image. In the USA this is the life of the artist plus 50 years.

The Manufacturer

The manufacturer is only a manufacturer and gives the buyer the best price for creating multiple copies of the hologram in question. The manufacturer's price is important because the copyright holder estimates the retail price of the product based on the unit cost from the manufacturer. To get the unit cost, take the total bill from the manufacturer and divide it by the number of copies made. Everything depends on volume. The more units you manufacture, the lower the unit manufacturing price. The lower the unit manufacturing price, the lower the retail price you can offer the product for. It is a broad generality, but many copyright holders take their unit manufacturing cost and mark it up approximately six to eight times to get the retail unit price of their product.

The Chain Of Distribution

Let's assume the copyright holder decides to make a hologram. What does the distribution channel look like from this point on?

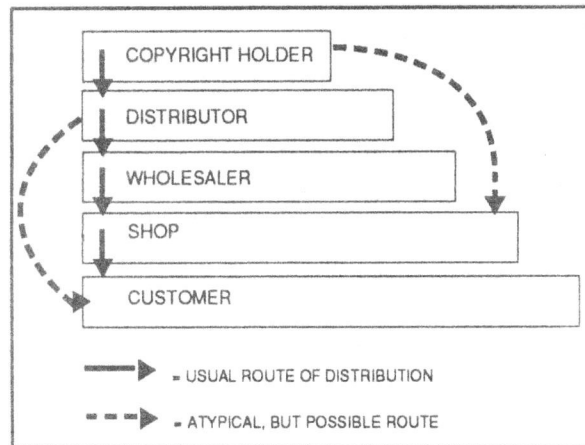

Figure 20. Chain of holography distribution

The levels shown above exist in the holography market and there are some rough numbers that show, approximately, the relationship between the different levels. It should be pointed out, however, that this is a young industry and frequently a business engages in several of the above functions. There exists one business, for example, that owns copyrights to some holograms that they manufacture; they import and act as a distributor for certain products; they sell products to other wholesalers; they wholesale to stores, and have a storefront themselves. They are everything but the customer! There are other businesses, however, that engage in only one level like a shop or distributor. We discuss each of the above businesses as separate entities in the distribution chain.

We feel it is our duty to give some idea of how discounts work in this industry. It is a difficult task since there is no established order of business. After some research, however, we have come up with what is a rough idea of how an item is discounted. Figure 21 breaks down the discounts for an item costing $50.00 to the final customer.

BUSINESS	(RETAIL PRICE = US $50.00)			
	DISCOUNT OFF RETAIL PRICE	GROSS AS % OF RETAIL PRICE	COST TO BUSINESS (UNIT ITEM)	GROSS PER UNIT ITEM
COPYRIGHT HOLDER	80%	30%	$10.00	$15.00
DISTRIBUTOR	70%	10%	$15.00	$5.00
WHOLESALER	60%	10%	$20.00	$5.00
RETAILER	50%	50%	$25.00	$25.00
CUSTOMER	00%	00%	$50.00	$00.00
		100%		$50.00

Figure 21. Typical costs, margins, and profits for holograms

A rough description of the above businesses follows.

The Distributor
The distributor buys the product in large volumes and frequently has a contract from the copyright holder that gives him some kind of exclusive territorial right for a product. The copyright-holder, on the other hand, sometimes demands that the distributor not sell competing products. The distributor generally handles all import problems such as customs and sometimes handles translation of written material that accompanies the product. Because there are two more discounts that have to be taken before the product reaches the customer, the distributor gets a very large discount from the copyright holder and has the obligation to order and sell large quantities of the merchandise.

The Wholesaler
The wholesaler is the link to the retailers. The essential function of a wholesaler is to get the product into shops. Wholesalers hire sales representatives or commission agents who visit shops and persuade them to buy the product. A good sales representative or agent should help the retailer set up the display, keep inventory current and generally help to ensure that the holograms look good and sell to the consumer. It is important that the wholesaler's representative visit the retailer regu-

larly. A wholesaler normally holds stock from a number of different suppliers, offering a full range of holographic items to retailers. This is a convenience to the retailer who does not have to deal with many different hologram suppliers. Frequently the wholesaler has two catalogues. One is used for shops and one is used for direct sales to customers. Wholesalers do not generally have exclusive rights to a product.

The Retailer
Retailers are the point of contact with the public, the place where the holograms are displayed, seen, and bought. Retailers of holograms include specialist hologram shops and galleries, gift shops, department stores, chain stores, stationers, jewelers and toy shops. Some are independently managed by the proprietor, some are national chains with a central buyer.

Only the buyer in the specialist hologram shop, usually the proprietor, is predisposed to buy holograms. In all other stores the person making the buying decision has to be persuaded that the saleability and profitability of holograms (or products featuring holograms) is preferable to the other goods that are on offer to him. In sales, the personal touch counts, so regular visits and product updates are important features of the wholesaler's function.

Discounts
The exact percentage discounts between businesses varies. In some businesses, like publishing, the final retail price of the product is established by the copyright holder and discounts follow a fairly established procedure. Holograms, however, are sold much like gifts in the gift trade. A retailer generally receives a catalogue that tells him only what it costs the retailer to buy an item. The price at which the item is sold is for the retailer to decide. This is called the retail price. The wholesaler frequently suggests that the retailer to mark the item up 100% (which means that the retailer buys at an average of 50% discount off the retail price).

Wholesale confusion: There is some necessary confusion in the industry about the word wholesaler. Retailers frequently find that the catalogues quote "wholesale prices". To the uninitiated, this implies that the prices being quoted are the prices that a distributor gives the wholesaler (wholesale prices) when actually the prices being quoted are prices at which the retailer buys the merchandise.

From the distributor or wholesaler's point of view, this is an unfortunate but necessary confusion. Distributors and wholesalers point out that the retail price of a product is established by the retailer. Since wholesalers have different catalogues for retailers and end customers, they need to designate whether the prices being quoted are for the public or not. What do you call the price being quoted to retailers? You can't call it the retail price! So you have to call it the wholesale price even though it might cause some confusion.

Prices that are quoted to a distributor or wholesaler from the copyright holder are usually given as a percentage discount from the wholesale price (the price quoted to retailers).

Repackaging And Customizing

Holograms can frequently be bought by any dealer in the distribution chain in an unfinished state for a reduced price. There is a small savings that can be made if you are willing to put in the extra work, but the big appeal is that it allows you to repackage or customize the holograms for sale.

Fine Art Holograms

The distribution chain for a fine art hologram, in theory at least, is similar to that of artistic holograms although more straightforward. The day may come when holograms are a fully accepted art medium and part of the established art market, regrettably this is not true today. This means that the traditional gallery distribution system of the art world is not open to most artist holographers and they do not have the benefit of the guiding hand of an agent, manager or dealer.

The fine art holographer's point of contact with the public is generally through exhibitions, either in specialist or public art galleries. There are several specialist galleries around the world which hold exhibitions and sell work, and many not-for-profit galleries. The specialist galleries are generally commercial operations which put a large mark-up on the artworks (or demand a large discount from the artist), whereas the not-for-profit galleries put a 15-25% mark-up on sales.

While the non-profit organizations "cost" less on any individual sale, they do not usually have continuing displays and representation for the artist. Specialists, on the other hand, do keep a permanent exhibition and often buy some of the artist's work themselves. Again, because their continued commercial existence depends on sales, the for-profit galleries tend to work hard marketing fine art holograms.

Because the market for fine art holograms is not yet fully developed, artists often sell directly to the public. Most of them know the few serious hologram collectors, and are pleased to deal with other buyers who contact them. In "dealing direct" there is often pressure on the artist to extend discounts to collectors because there is no middle man involved. Extending such discounts, however, undermines the growth of the distribution system and interferes with efforts to build a growing awareness of holographic art.

Custom Designed Holograms

The markets for security and advertising holograms are similar to each other. They differ from the giftware and art markets in that the end buyer is not a consumer buying one hologram, but a company or group of companies usually buying in large quantities. The exception is the company buying only a few holograms, usually large-format, for use in promotional events such as trade shows or point-of-purchase advertising.

As discussed previously, the end user can deal directly with the producer. Often, however, intermediary agencies bring the producer and client together. These intermediaries fill different functions from those in the giftware business. Instead of buying and selling as principals like a distributor or wholesaler does, they tend to be agents being paid either by the purchaser or the producer. They usually work on a commission in the range of 15-25%.

The intermediary ensures that the client gets a suitable hologram, well-designed, delivered on time and with the required quality. The intermediary works with client and producer to design the hologram and supervise all stages of its production, from model-making or artwork, through master and proof, to installation in its final environment. This may be hot-stamping onto a credit card, insertion in a brochure or installation on a trade-show exhibition. Everyone wants the hologram to look good - the specialist agency has to ensure that it is well integrated into its display environment and, if it is a promotional piece, that it is designed and installed to fill its potential. This involves liaison with producers, designers, printers, print finishers and above all, the client.

The hologram producer may handle all this in-house and many have large sales teams. As with any industry, producers find it valuable to deal directly with clients, especially large clients who bring repeat business. But often the purchaser needs the support and advice of an independent agency. The agency can recommend the best producer for a particular job. It can encourage a client to make the most of holograms by using large format film in a point-of-purchase application to tie in with a brochure using embossed holograms. The agency should also be aware of the many pitfalls that can delay the delivery of the order and should work with the suppliers on the purchaser's behalf to ensure that problems are minimized.

If the custom hologram is for a security or anticounterfeiting application then the purchaser will want to ascertain that all stages of production are in secure facilities. Credit card companies and banks especially cannot risk holograms finding their way into the wrong hands. Several embossed hologram producers are now secure, but some have higher levels of security than others.

Distribution - Sign Of A Maturing Industry

Ten years ago the few companies producing holograms dealt directly with their buyers. Today the industry has grown and diversified. It is a sign of the maturing of the industry that these distribution mechanisms are now in place. The distribution chain smooths the path for the customer who wishes to buy holograms, whether that customer is a large corporation or an individual consumer.

SECTION II DETAILED COVERAGE OF:

· Recording Materials
· Color of Holography
· Embossed Holography

· Holographic Optical Elements
· Computer Generated Holograms
· Non-Destructive Testing

RECORDING MATERIALS

In this section we outline the features of several of the most popular recording materials used in holography. The most widely used recording materials are:

1) Silver Halide
2) Dichromated Gelatin
3) Photoresist
4) Photopolymer

No material is perfect and here is a chart of the advantages and disadvantages of each type of material:

MATERIAL	ADVANTAGES	DISADVANTAGES
Silver Halide (Bleached)	High Speed	Only Fair Quality Complex Processing
Dichromated Gelatin	Highest Quality	Poor Shelf Life Complex Processing Poor Image Stability
Photoresist	Easily Reproduced (Embossed)	Low Speed, Limited Spectral Range Transmission only
Photopolymer	High Quality	Short Shelf Life Complex processing

Figure 8. Popular hologram recording materials

Silver Halide

Silver halide is probably the most widely used recording material. The primary reason for its popularity is its speed. Fast emulsions reduce stability problems and consequently the need and cost of high powered laser equipment.

There are five atoms which, because of their atomic similarity, are called the halides. They are chlorine, bromine, iodine, fluorine and astatine. Silver halide emulsions are made using either silver chloride, silver bromide, or silver iodide. The other two halides are not used because silver fluoride is insoluble in water and astatine is radioactive.

Some of the more common brands of silver halide emulsions used in holography are listed below:

PLATES	FILM	SPECTRAL SENSITIVITY	EMULSION THICKNESS PLATE	FILM	GRAIN SIZE
Agfa-Gevaert					
10E56	10E56	Blue-Green	7 μm	5 μm	.090 μm
10E75	10E75	Red	7 μm	5 μm	.090 μm
8E56	8E56	Blue-Green	7 μm	5 μm	.044 μm
8E75HD	8E75HD	Red	7 μm	5 μm	.044 μm
Eastman Kodak					
649F	649F	All	15 μm	6 μm	.060 μm
120-02	SO-173	Red	6 μm	6 μm	.050 μm
	SO-253	Red	9 μm	9 μm	.070 μm
	SO-424	Blue-Green	7 μm	3 μm	.065 μm
Ilford					
SP673	SP673	Blue/Red	6 μm	6 μm	.040 μm
SP675T	SP672T	Blue/Green	6 μm	6 μm	.040 μm
HOTECR	HOTECRRed		6 μm	6 μm	

Note: (1) HOTEC R is a new product Ilford will make available in 1990.
(2) Soviet and Bulgarian emulsions exist with a grain size of .015 μm but they are not readily available to the market.

Figure 9. Common silver halide emulsions.

Emulsion Creation And Exposure

A typical silver halide emulsion is made by adding a solution of silver nitrate to a solution of potassium bromide and gelatin. Silver bromide crystals form and the mixture is called an emulsion. The emulsion is heated for a certain amount of time which is called the ripening process. During this time the grain size increases and the speed of the emulsion is increased. Some doping agents may be added to the emulsion at this time to foster proper crystal growth. After the ripening process, the gelatin is allowed to cool, then it is shredded and the soluble potassium nitrate is washed out of the emulsion. The emulsion is then heated again, with more gelatin added and then cooled and applied to a base. The thickness and hardness of the emulsion is important in holography because emulsions too thick tend to deform during development. Emulsions that are too hard can either retard chemical reactions or create vacuoles in the emulsion left by migrating atoms. These vacuoles tend to scatter light.

Let's assume our emulsion is made and we now want to expose our emulsion to light. It sounds surprising but a perfect crystal structure of silver bromide does not react to light in any appreciable way. However, a crystal with defects does react with light. Fortunately, as they form, most silver bromide crystals will create defects which consist of some interstitial (out of order) silver ions displaced in the crystal structure.

The process of the photochemical reaction is not known in exact detail but it is believed that when light strikes a silver bromide crystal, enough energy is available to remove an electron from an occasional bromide ion. The electron produced is able to migrate through the crystal until it comes in contact with an interstitial silver ion. The silver ion takes the electron and becomes silver metal. Silver atoms formed by this mechanism apparently act as a nucleus for the formation of aggregates of ten to 500 silver atoms. This formation is known as the latent image because it is too small to be seen by the naked eye.

After exposure, the emulsion is developed. The developer goes to the site of any silver bromide crystal with a latent image and causes all the silver in that particular silver bromide crystal to be reduced to silver metal and deposited on the already existing latent image of silver metal. This causes a worm-like grain of silver metal to form which is limited in size by the amount of silver available in the silver bromide crystal. This growth is considerable, amplifying the size of the latent image silver metal by a factor on the order of 106. If the developer is left in contact with the emulsion long enough it eventually attacks all the silver in the emulsion. But the speed of development is slow enough that you can use a timer to take the emulsion from the developer just after the latent image, and not the unexposed silver bromide crystals have been developed. At this point the developer has

converted silver ions to silver metal if and only if they belong to a silver bromide crystal that was exposed to light. We now place the emulsion in a fixer solution which attacks all silver bromide crystals that were not exposed to light. The fixer makes these silver bromide ions soluble and removes them from the emulsion.

So the result is an emulsion with black spots where light has struck, and clear spots where no light struck.

Ideal Silver Halide Emulsion

An ideal silver halide emulsion depends somewhat on its use but there are three main factors to consider in any emulsion: thickness of emulsion, grain size of silver halide crystals, and sensitivity (or density of silver halide crystals) in the emulsion. We can generally state the following:

Thickness of emulsion: It is generally agreed that emulsions of more than 10llm are neither practical or theoretically necessary to produce most volume holograms. It is pointed out that thicknesses above this size cause problems in development.

Grain size of crystals: Grain size becomes an important issue in holography because we are recording fringe patterns that are wavelengths apart. Too large a grain size may create excessive scatter which may fog or destroy your hologram and too small a grain size makes the emulsion have no useable sensitivity. It is generally agreed that the most ideal grain size is in the range of $.01\mu m$ to $.035\mu m$.

Sensitivity of emulsion: The ideal exposure would probably be 100 - 300μJ/cm2 to give a useful density (0=2-3). If exposures are much longer than this, the main attraction of silver halide emulsion, its speed, comes into question and other emulsions become more attractive.

Dichromated Gelatin

The dichromated gelatin emulsion takes a completely different approach to image recording. In this process the substance you suspend in the gelatin reacts, upon exposure to light, with the gelatin itself. The exact chemical constituents of gelatin cannot be listed because gelatin is a biological product made by boiling animal parts such as hooves, bone, etc. and processing it chemically. Gelatin has many uses. A great deal is used in medicine, in food (jelly, etc.) and in photography. Gelatin, therefore, is mass produced for a variety of uses, has a molecular structure that cannot be fixed with any certainty and its properties vary from batch to batch. This makes scientific control difficult.

It was discovered that gelatin with a small amount of dichromate, such as $(NH_4)_2Cr_2O_7$, becomes progres-

LASERS by LiCONiX
BETTER by DESIGN

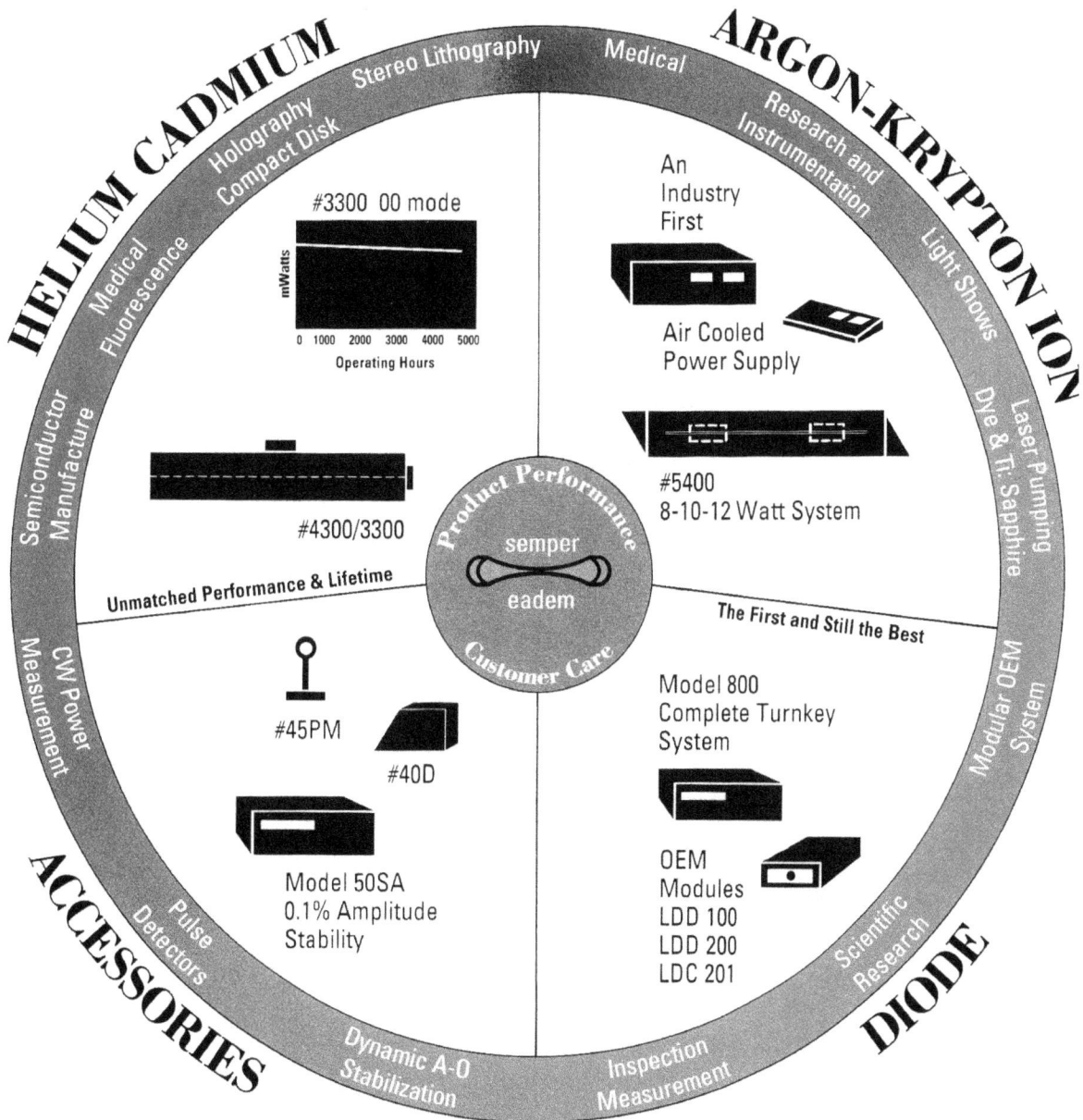

HELIUM CADMIUM · Stereo Lithography · Medical · **ARGON-KRYPTON ION**

Holography · Compact Disk · Research and Instrumentation · Light Shows

Medical · Fluorescence · Laser Pumping Dye & Ti: Sapphire

Semiconductor Manufacture · Modular OEM System

CW Power Measurement · Scientific Research

Pulse Detectors

ACCESSORIES · Dynamic A-O Stabilization · Inspection Measurement · **DIODE**

#3300 00 mode

mWatts

0 1000 2000 3000 4000 5000
Operating Hours

#4300/3300

Unmatched Performance & Lifetime

An Industry First

Air Cooled Power Supply

#5400
8-10-12 Watt System

The First and Still the Best

Product Performance
semper
eadem
Customer Care

#45PM

#40D

Model 50SA
0.1% Amplitude
Stability

Model 800
Complete Turnkey
System

OEM
Modules
LDD 100
LDD 200
LDC 201

LiCONiX

3281 Scott Blvd.
Santa Clara, CA 95054
Tel: (408) 496-0300
Fax: (408) 492-1303
Telex: 9103796475

Distributors: Optilas: France 33-1-6-77-4063, Germany 49-89-801035, Benelux 31-01720-31234
Laser Lines: UK 44-295-67755 Seki Technotron: Japan 03-820-1711

LASERS for HOLOGRAPHY
HELIUM CADMIUM and ARGON/KRYPTON ION

LiCONiX is the principal supplier of HeCd lasers for the holography market.

Users of dichromated gelatin emulsions, photoresist and silver halide film appreciate the power of the HeCd laser at 442nm in the deep blue. Up to 80 mwatts are available with very good gaussian mode and coherence length of 10cm. A UV line at 325nm is also available for fabrication of small pitch diffraction gratings.

The HeCd is an air cooled laser, operates at 117VAC, and draws less than 6 amps. Cooled by natural convection, there is no fan or water induced vibrations.

LiCONiX also supplies high brightness, long coherence length Argon ion and Krypton ion lasers with output in the blue (488nm), green (514.5nm), and the red (640nm). Various power levels are available. A coherence length extending etalon accessory gives outstanding depth of field possibilities.

LiCONiX can supply all your laser needs for holography!!!

Call us at 408/496-0300

sively harder upon exposure to light. After investigation, it was found that the Cr2 element of the molecule absorbs light to become the ion Cr3+. This ion then stimulates cross linking of molecular chains in the gelatin. The cross-linked parts of the gelatin then have a different refractive index than the unlinked gelatin. This difference produces the image that you see.

Although there is no such thing as a perfect recording material, dichromated gelatin has won the respect of almost all reviewers. Because it is almost grainless, it has very little scattering and very high resolution. Dichromate holograms are also among the brightest holograms available because of their large refractive index. The negative aspects of the dichromated gelatin are its short shelf life, complex processing chemistry, and poor image stability (disappears on contact with water) . But the pluses outweigh the minuses for many holographers and dichromate enjoys a large industrial market.

Photoresist

Photoresist is used extensively in the computer industry to make circuit boards. A similar procedure is used in holography to produce a stamping die that is then used to emboss the holographic fringes in plastic to produce embossed holograms.

The most widely used photoresist is Shipley AZ-1350. The sensitivity of this emulsion is greatest in the ultraviolet and drops off rapidly toward the blue end of the spectrum. Therefore holograms are best recorded with either an Ar+ laser at 458nm or with a HeCd laser at 442 nm. A typical exposure on factory supplied emulsion is 250 mJ/cm2 at 458 nm. A commonly used developer is Shipley AZ-303 A. It is usually diluted 4:1 or more. The exposure can also be controlled by pre or post- exposing the resist to an incoherent source such as a fluorescent lamp. Some people coat their own plates although the emulsion is available on plates from the manufacturer. If you do coat your own plates, it is recommended to have an antihalation backing of some kind on your plate. The emulsion is usually applied by spinning. The layer of emulsion is around 1 to 2μm thick. This is then baked at 75°C for 15 minutes to ensure removal of the solvent. Figure 10 shows exposure versus thickness of emulsion for those who want to coat their own plates.

Photopolymer

Although there may be other offers in the future, Polaroid is presently the only large company effectively marketing a photopolymer for holographic use. It is called DMP-128. Unlike other recording materials, DMP-128 has single molecules (monomers) suspended in the recording material that are capable of linking up (polymerizing) with each other when the proper

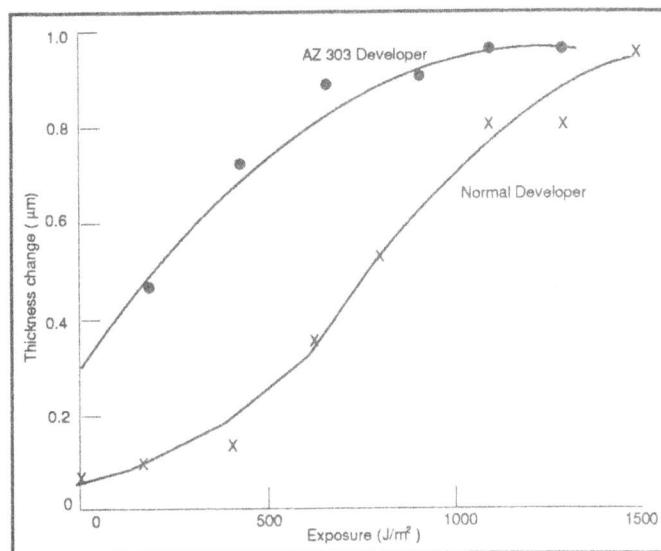

Figure 10. Depth versus exposure characteristics obtained with AZ-1350 photoresist (Hariharan, p. 107).

catalyst, also suspended in the emulsion is exposed to light. When a photon is absorbed by the catalyst numerous monomers polymerize. It is estimated that 100 to 1,000 monomers react per absorbed photon.

DMP-128 is provided as a coated film on glass or selected flexible film in thicknesses between 3 and 20 microns. Under proper conditions, the emulsion has a shelf life exceeding one year. DMP-128 is insensitive to light and must be activated before exposure by incubating the film for several minutes in an environment of 50% relative humidity to allow absorption of a controlled amount of water vapor. The emulsion is then exposed and the suggested exposure is 4-8mj/cm2 for transmission holograms and 15-30mj/cm2 for reflection holograms. It is recommended to follow the laser exposure with a uniform white light exposure to complete the photopolymerization throughout the film. A single developing and fixing bath follows exposure and additional processing baths can be executed to affect the emulsion.

Tests that have been conducted on reflection holograms made using Polaroid's DMP-128 photopolymer indicate the existence of alternating solid and porous planes throughout the film (the effects of liquid immersion on optical properties are consistent with a porous hologram structure). Therefore, holographic properties may be changed in a predictable manner by filling the pores with a material having a specific refractive index. These changes may be controlling color changes or inserting hardening chemicals to make the emulsion more resistant to change.

Diffraction efficiencies of over 90% have been obtained with an index modulation of 0.033. The diffraction can be reduced to less than 10% using an index matching fluid and, upon evaporation, the diffraction efficiency returns to its original value.

COLOR IN HOLOGRAPHY

Full Color Holography

Although research continues on producing a full color hologram, no one has yet perfected a method for doing this. It seems to be the shared opinion of most holographers that it is not a question of if but when it will happen. Currently, almost all holograms produced are pseudocolor holograms. Some very impressive pseudocolor holograms have been made, however, including some recent multiplexed color embossed holograms.

Pseudocolor Holograms

Holographic color comes in two types, depending on whether the hologram is a "thick" hologram such as a OCG hologram or silver halide reflection hologram, or a "thin" hologram such as an embossed hologram or silver halide rainbow hologram.

A hologram is considered to be a "thick" hologram if the thickness of the recording medium is considerably greater than the spacing of the interference fringes. Interference fringes are created wherever the wavefronts meet and cancel each other out. When an interference pattern is recorded, the fringes are not only on the surface of the recording medium but go all the way through the medium. These fringes are called Bragg planes. Each fringe that you see on the surface is actually the intersection of a Bragg plane with the surface. In a reflection hologram, formed by the reference beam and object beam striking a plate from opposite sides, the Bragg planes slice through the medium at very shallow angles as indicated in Figure 11 A;

and the planes are very close together. In a transmission hologram, formed by the reference beam and object beam striking a plate from the same side, the Bragg planes slice through the medium at steep angles as indicated in Figure 11 B, and the planes are farther apart.

If a light ray, passing through the medium, passes through several Bragg planes, the hologram is considered to be "thick". Such holograms are also called Bragg holograms, because multiple reflections from the Bragg planes dominate the interference effects upon reconstruction. "Thick" holograms have an intrinsic color depending on the spacing between the Bragg planes, which is determined by the laser color used in making the hologram and the exposure and processing conditions which may shrink or swell the recording medium. "Thin" holograms act like prisms, producing all of the colors of the rainbow and directing each color into a different direction. Neither "thick" nor "thin" holograms actually record the color of the objects or artwork from which they are made. Instead, they merely record the object as it appears in the laser light.

Whether a hologram is a phase hologram or an amplitude hologram has no relationship to the color of the hologram, or whether the hologram is thick or thin. A phase hologram is one in which a light ray passing through a fringe is delayed or advanced relative to a ray passing between fringes; while an amplitude hologram is one in which a light ray passing through a fringe is dimmed relative to a ray passing between fringes. Phase holograms are usually much brighter than amplitude holograms.

Almost always, 2D, 2D3D and 3D rainbow holograms are recorded as transmission holograms and are embossed. An embossed hologram has essentially no thickness--it exists only on the surface of a sheet of plastic. Consequently, embossed holograms always have the properties of "thin" holograms. The color in 20 holograms results from different parts of the hologram directing the diffracted spectrum into different directions. Figure 12 illustrates this.

In a "thin" hologram such as a 2D, 2D3D, or 3D rainbow hologram, different regions are recorded in such a way that they direct the spectrum into slightly different directions. Thus, only the red part of the spectrum might be visible to a viewer from one part of the hologram, while from an adjacent part only the blue part of the spectrum is visible. When the hologram is tipped forward and back, all the colors seem to shift up or down the spectrum because different parts of the spectrum come into view.

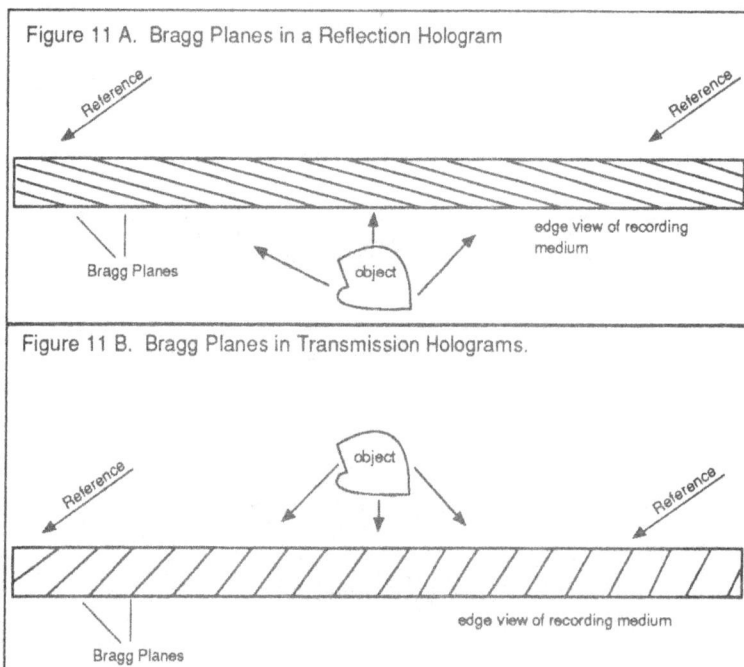

Figure 11 A. Bragg Planes in a Reflection Hologram

Reference Reference

Bragg Planes

edge view of recording medium

object

Figure 11 B. Bragg Planes in Transmission Holograms.

object

Reference Reference

edge view of recording medium

Bragg Planes

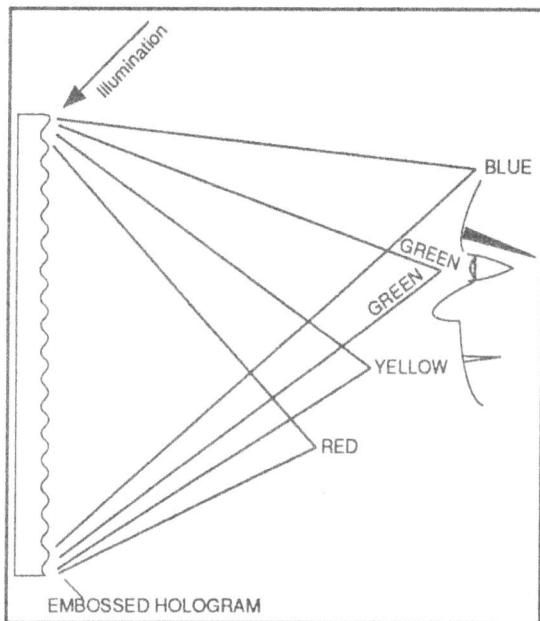

Figure 12. Color spectrum in one color component of a 2D, 2D3D or 3D. Rainbow hologram. In this position, the viewer sees the color green.

Because the colors shift when the hologram is tipped, it is not possible to make "thin" holograms in which the color remains the same at all angles. In practice, it is common to take advantage of the pure (unnatural) spectral colors available in holography to produce brilliant, dynamic effects impossible in other graphic art media.

Because the colors in holography depend on viewing angle and illumination angle, it is normal to specify colors at one viewing angle and illumination angle. Viewing and illumination angles are shown in Figure 13. Unfortunately, there is no agreement among holographers on standard viewing and illumination angles. The customer normally just specifies what colors are desired and the holographer makes sure that at some particular angle the colors are appropriate. Normally the specified colors can be red, yellow, green, blue, white, or black (black is simply the absence of color) . However, in some particular applications where the viewing geometry must be specified (for example, a hand-held electronic game, a video game or a security device), it is possible to shoot the hologram to reconstruct particular colors into precisely defined directions.

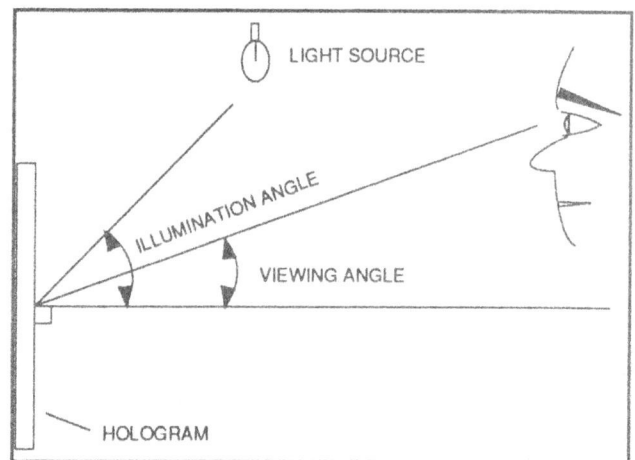

Figure 13. Viewing and Illumination Angles

References:

Silver Halide Emulsion
Phillips, N.J., (1985). The role of silver halide materials in the formation of holographic images, Proceedings of the SPIE, 532, 29-38.

Photopolymer
Hay, W.C. & Gurenther, B.D. (1988). Characterization of Polaroid's DMP-128 hoographic recording photopolymer, Proceedings of the SPIE, 883, 102-1 05.

Ingwall, R.T., Stuck, A. & Vetterling, W.T. (1986). Proceedings of the SPIE, 615, 81-7.

Ingwal, R.T. & Fielding, H.L. (1985)., Optical Engineering, 24, 808.

Ingwall, R.T. & Troll, M. (1988). The mechanism of hologram formation in DMP-128 photopolymer, Proceedings of the SPIE, 883, 94-101.

Emulsion And Exposure
Hariharan, P. (1987). Optical Holography, Cambridge University Press, 88.

Larimore, L., (1965). Introduction to Photographic Principles, Dover, 101-2.

Photoresist
Bartolini, A.A. (1974). Characteristics of relief phase holograms recorded in photo resists. Applied Optics, 13, 129-39.
Bartolini, RA (1972). Improved development for holograms recorded in photoresist. Applied Optics, 11, 1275-6.

Hariharan, P. (1987). Optical Holography, Cambridge University Press, 106.

Saxby, G., (1988). Practical Holography, Prentice Hall, 285.

Norman, S.L. & Singh, M.P. (1975). Spectral sensitivity and linearity of Shipley AZ-1350J photoresist. Applied Optics, 14,818-20.

Livanos, A.C., Katzir, A., Shellan, J.B. & Yariv, A. (1977). Linearity and enhanced sensitivity of the Shipley AZ-1350 B photoresist. Applied Optics, 16, 1633-5.
-

EMBOSSED HOLOGRAPHY

Due to the commercial popularity of embossed holography, we treat embossed holography separately and in depth here. In this section we cover the following items:

• General description of some successful uses of embossed holograms
• Outline the main steps used to create embossed holograms
• More detailed explanation of each step used to make embossed holograms
• Costs of making and printing an embossed hologram

Although embossed holography is used in HOE and other applications, we assume in this chapter that most people are interested in using embossed holography for its commercial artistic effect.

Successful Applications

Clearly the most well known application of holography, and one of the most lucrative, is embossed holography. As security devices, embossed holograms appear on currency, bank credit cards and general sales merchandise. As product advertising they appear on cereal boxes, books, and magazines. They are also used extensively in laser disks and numerous HOE devices. Here are some commercial examples of how successful embossed holograms can be.

Food Boxes

A good example along this line is Ralston-Purina. They developed three distinctive hologram images and embossed them on every box of their "Ghostbusters" cereal. This initiated a brand sales increase of 300%. They are repeating the experience with another campaign.

Magazine Tip-Ins

Uddelhom Steel Company used a holographic image in its first penetration of the US market. This was the focal point of an insert run in "Iron Age" magazine. A subsequent survey of their readers revealed the ad received a 221 % effectiveness rating.

Credit Cards

Probably the most obvious example. The hologram integrated into each Mastercard significantly reduced the passing of fraudulent credit cards which had previously resulted in counterfeiters charging in excess of $35,000,000 annually.

Magazine Covers

Applying an eagle hologram to the March 1984 cover of "National Geographic" magazine resulted in over 400,000 new subscribers to the publication and over 14 months of readership response. Later, an image of the Taung child skull was integrated into another cover, resulting in a 25 year record for advertising revenue in a single issue. Numerous magazines have put holograms on their covers since then.

Labels

Adhering holographic security labels to Prince Manufacturing tennis racquets forestalled a counterfeiting problem that kept sales at an equivalent level for six successive years. With the protection provided by the holographic image, Prince's sales demonstrated a threefold increase.

Book Covers

Zebra Books, one of the nation's largest mass paperback publishers, put a hologram on the cover of every book published in one of its romance series. The program was so successful that they are now adding a second line. This means literally millions of covers with embossed holograms on them.

Currency

Everyone is well aware of the fact that color copy machines are quickly getting to the stage of perfection that will make counterfeiting very easy. In the words of a security printer, this means that manufacturers of currency will be forced to take measures to enlarge the expense and time necessary for counterfeiting'. Embossed holograms are a primary candidate for filling this need. The Reserve Bank of Australia has already issued the first currency with an embossed hologram (a $10 banknote). Although the first printing was flawed and the image tended to wear off, subsequent research and a second printing has made the banknote durable and successful. The Austrian National Bank has followed by issuing a 5000 shilling note (approximately £250 or US$380) with an embossed hologram

Steps Used In Making An Embossed Hologram

An overview of the steps used to make an embossed hologram are as follows.

  1) Create artwork
  2) Make a laser transmission hologram
  3) Copy hologram onto positive photoresist emulsion
  4) Deposit conductive layer on photoresist
  5) Electroforming of metal masters (1 st, 2nd, etc.)
  6) Emboss plastic by roller or hot stamping
  7) Apply hologram to product at bindery or by hand

Now we describe each of the above steps in detail.

1) Prepare the artwork for an embossed hologram.[2]

Embossed holograms made for advertising or artistic effects have a common problem. Most of them are intended to be seen under any available lighting condition, which includes very diffuse lighting like fluorescent lights. This means that the hologram you make cannot have a great deal of projection or depth, because fluorescent light does not display depth well. This means that 30 objects, if they are used, must be very shallow. The most common objects used are either flat artwork or very thin, miniature models.

3D Holograms: A 30 model, of course, needs to be made for a 30 hologram. Close consultation with the holographer before and during the making of the model is important. The following points need to be kept in mind:

• Get exact specifications on what volume of space your model can occupy before starting. Remember that an embossed hologram has very little depth. Discuss the material being used to create the model with the holographer.

• The final color of the model does matter. Colors that turn out well depend on the laser used to expose the object.

• Once your model is made and painted, you can view it under the light of the laser that will make the master. This will give you a rough idea of what areas of your object the laser light reflects strongly from and from what areas it reflects weakly. You should remember, though, that your hologram may be viewed in unusual lighting conditions and the actual final color depends on the lighting conditions while viewing.

2D3D Holograms: A 2D3D hologram is a hologram made up entirely from flat artwork appearing on two or more levels or distances from the plate. If the flat artwork is entirely one level, the hologram is called a "2D" hologram. If it is layered on two or more levels, the hologram is called a "2D3D" hologram. The attraction of a 3D hologram is that you have a model you can see before the hologram is made and, although shallow, there is some 3D depth to the final embossed hologram. Although one might think that a 3D hologram is the "best money can buy", 2D and 2D3D holograms have several advantages.

Advantages of 2D3D holograms from the holographer's point of view:
• 2D3D holograms are all made using exactly the same setup, so the holography setup never has to be changed
• The "object" for a 2D3D hologram is simply flat artwork, so there are no object motion problems or object mounting challenges.

• All of the image information for a 2D hologram or the 2D part of a 2D3D hologram is contained on the surface of the hologram, so the 2D part of the image can be made much brighter than an ordinary 3D hologram without loss of image detail.

Advantages of 2D3D holograms from the point of view of the graphic artist and the ultimate user:
• Graphic artists are well acquainted with two-dimensional media, so their techniques and skills are easily adapted to designing 2D and 2D3D holograms
• It is possible to make mock-ups of 2D3D holograms as an aid to making a sale, without going through the hologram mastering process and expense.
• The graphic artist who designs the 2D or 3D hologram can define all aspects of the image: artwork, imagery, colors, and image placement. The holographer becomes a technician who carries out the graphic artist's instructions without having to make artistic decisions in the process of making the hologram.
• 2D3D holograms can be viewed in any kind of lighting without difficulty, making them suited for the many applications in which there is little opportunity to control the lighting, such as packaging, point of purchase displays, advertising, anti-counterfeiting, textiles, clothing, toys, greeting cards, and decorations.
• The greater brightness and sharpness of 2D3D holograms compared to other types of holograms make them excellent eye-catchers.

20 and 2030 Artwork: The artwork for 2D and 2D3D holograms is essentially the same sort of flat artwork that any ordinary printer would use. The only difference is that, since a hologram allows a person to look around a foreground image, the background should be complete, without "cutouts" for the foreground. Color separations are usually prepared from line art, and the color for each separation is specified. It is also possible to produce 2D3D holograms from color photographs, though it is more complicated.

Line art for 2D and 2D3D holograms should have an unbroken line or boundary for each region of any given color. Each layer in the image should have its own separate artwork. Color overlays should be provided with line art to designate the colors of the various parts of the image in each layer. Each layer should have its own color overlay.

2) Make a master hologram

The master hologram is generally a rainbow transmission hologram. If you intend to make your own master hologram and have another company make the next step, the photoresist, then close cooperation is necessary. Some companies require the reference beam to be at a specific angle and perhaps even a specific

wavelength. Most likely, the people that do the photoresist step will also want to see a final transfer hologram to use as a guide. Quality, of course, is absolutely essential. Any defects in the master will be magnified in the subsequent steps. If you are creating a multiplex embossed hologram, the Benton method is preferred since the images are truly multiplexed and not slits.

3) Make the photoresist

If you intend to make your own photoresist, there is a good article by Burns that you should read because it describes both photoresist and electroplating in detail[3]. There is also a worthwhile discussion by Saxby[4]. The theory of embossed holography has been covered in technical literature as early as 1973[4].

Photoresist: Most holograms to be embossed are recorded in Shipley AZ1350J photoresist because it is the only commercially available positive photoresist with sufficient resolving power for holography. This photoresist is designed for the microcomputer industry and consequently does not have the sensitivity that one would wish for making holograms. Hopefully Shipley or a competitor will supply an emulsion more suitable to the holography industry.

Some individuals have tried to make their own resist coating. It is not recommended as resist plates of excellent quality are readily obtainable up to 7 x 7 inch-es[5]. Emulsions are generally applied by either spin coating or dipping. Spin coating is the preferred method and you can find suppliers that will coat plates for you but you should be very cautious about the results. Your coating should be mirror smooth, free of defects and striations when viewed in yellow safe light.

Exposure: A frequently used exposure is at 457.9 nm using an argon laser[6]. This is the argon wavelength to which this emulsion is most sensitive but, unfortunately, it is also one of the weakest argon wavelengths. If you use an etalon the available power is reduced by about 50%. At our desired wavelength of 457.9nm, an 18 watt argon with an etalon can put out about 700mw and a 5mw argon with an etalon produces 120mw. Burns reports the Shipley emulsion is a factor of about ten times more sensitive to the 441.6nm wavelength, which is available from Heed lasers, but Heed lasers are not readily available with more than 40mw of power. In addition, the coherence length of Heed lasers is poor relative to argon lasers equipped with etalons. So an argon laser with an etalon is preferred and Heed lasers are preferred only for contact copies.

To give you an idea of the sensitivity, Burns reports that the Shipley photoresist requires a threshold energy to be effectively exposed. At 457.9nm, it generally requires a total exposure energy of 250,000 ergs/cm^2 to 2,500,000 ergs/cm^2 depending on the thickness of the photoresist. Burns found he could achieve this exposure by delivering 150 ergs/cm² per second for a 30 minute exposure or 420 ergs/cm² per second for ten minutes. Either way it is a long exposure and system stability could be a problem.

The exposure Burns finally used delivered 1,000 ergs/cm2 per second and the exposure time was 4 minutes 38 seconds. Development was for 40 minutes with tray agitation at a temperature of 53° F using Shipley 303A developer diluted 6:1 with distilled water.

4) Deposit electrically conductive layer onto photoresist

We now coat the photoresist with a layer of electrically conductive silver. Three of the most popular methods for coating and their respective benefits and disadvantages are described below. Regardless of the method, the thickness of coverage of the conductive layer only needs to be several hundred angstroms. You want a nice, even thickness with no pinholes or cracks. Watch the temperature to make sure it does not exceed the melting point of the photoresist.

METHOD	COST	SPEED	QUALITY PROBLEMS?
Silver Spray	Inexpensive	Fast	Possible quality problems
Vacuum Deposited Silver	Expensive	Moderate	Excellent quality
Electroless Nickel	Inexpensive	Moderate	Depends on operator

Figure 14. Popular methods for coating photoresist with electrically

Silver Spray: This method offers high production rates with low start up costs. It requires only a spraygun system, spray booth, and solutions. You can use a two nozzle spray gun to mix the two reagents with a commercially available two part silver solution or mix your own solution from existing formulas[7], [8].

Vacuum-Deposited Sliver: To proceed with this method, you need to buy vacuum metallization equipment. In the procedure, the photoresist is affixed at the top of a vacuum bell jar and a small quantity of silver is evaporated from a hot filament in the bottom of the jar[9]. The benefit of this method is that it requires little skill on the part of the operator, and therefore leaves less room for error.

Electroless Nickel: This is a three step immersion process that is inexpensive. It requires dip tanks (one of which must be heated) and solutions that may be mixed in house or purchased commercial[10],[11]. The steps are first to sensitize the photoresist with stannous chloride solution, then dip it in palladium chloride, and finally the electroless nickel deposit which takes place in a heated tank. There is a wash step between each of the three solutions. The electroless nickel process does take some operator experience since the type of agitation, temperature of solution, immersion time, etc. can all change the outcome.

5) Electroplating the conductive photoresist (making the stamping shim)

The metal master or shim, as it is referred to in the plating industry, is the first generation of masters for your embossed hologram. Nickel sulfamate is used by almost everyone for the electroforming bath. It is commercially available and relatively inexpensive[12]. The size of the plating tank is, of course, dependent on the size of your master. The other components of your plating system are:

• DC rectifier
• Nickel anodes
• Filtration pump
• Solution heater
• Solution agitation system

Burns used a 12 volt, 100 amp DC rectifier with less than 3% ripple. The anodes were Inco nickel rounds contained in a titanium basket. A submersible filtration pump, two quartz immersion heaters and a rockerarm agitator were also used.

A number of factors must be controlled to have an efficient plating bath[13]. Among the items are:

• Current density
• Bath temperature
• Bath PH
• Bath specific gravity
• Agitation quality
• Bath cleanliness

Organic materials like rubber tubing that go into the bath must first be leached for at least 24 hours in a heated 10% sulphuric acid solution. Metallic materials that might have surface contaminants should also get this treatment. Continuous filtration of the system keeps down the contaminants, agitates the solution and helps stabilize the temperature. A continuous flow of fresh solution should pass over the object being plated.

Sometimes tensile stress causes the nickel deposit to curl away from the photoresist and there are additives for the solution to control this[14]. These additives also tend to increase the grain size and make the master brittle[15].

Jigging: Before you immerse the photoresist in the electroforming bath, you need to make a conductive frame or jig to hold the photoresist. The idea is that the DC current from the rectifier goes through the jig to the conductive layer on the photoresist. So the jig should make a complete circuit around the perimeter of the plate and not damage the conductive layer. Jigs are usually made in-house but you can visit electroplating suppliers for ideas. In the electroplating industry, the jig together with the part to be electroformed is called the mandrel. It is immersed in the solution and is the cathodic element of the plating system. Some jigs are made to be used once and some are reusable.

One of the problems associated with the jig is that plating starts on the jig and works toward the center. The problem with this is that it can result in your plate having thick edges and depressions or voids in the center of the plate. To prevent this, your jig should be at least 25% larger than the desired finished embossed hologram.

Plating The First Generation Master: The plating process is like any normal plating process in that you have to constantly monitor temperature, current density, pH, and concentration of solution. The solution used in the tank is nickel sulfamate. The anode is a bar of pure nickel and the cathode is your mandrel. Be careful of long exposures because the thicker the plated deposit and the longer the plating cycle the greater the risk of having defects. A perfectly flat metal master without flaws is your goal[16].

Once your shim has been plated, you remove it from the jig and separate the metal shim from the photoresist. Sometimes you can pry the two apart with a razor blade and sometimes it takes a sturdier tool. Great care should be taken at this step because you are essentially working with a piece of metal foil which may be as thin as .05 mm (.002 inches) and it can be easily damaged. After separation, rinse the metal shim in a solvent to remove any resist that might be stuck to it. Inspect your shim to see if it is good or if you have to go back and make another photoresist. The smart thing to do is to make more than one photoresist to begin with, just in case you fail on your first try.

Duplicating Metal Shims: Now that you have your first generation master, you use your first master to make the next generation of masters. Here are some things to keep in mind:
• Each metal master can make approximately ten "next generation" masters.
• Each time you make a next generation master, your image is reversed left to right.
• Each time you make a next generation master, you lose about 1 cm (.25-.50 inches) from each side.
• The production run of each metal master varies widely with the operator and machine but by a conservative estimate you should get at least 2,000 holograms per master.
• You should not go beyond three generations of metal shims because there is a little image degradation with each step.

• The image on the final shim that you use for stamping should look just like the image you want the customer to see.

This means that you have to plan ahead. First, you have to decide how many holograms you are going to have. If it is 20,000 or less, then you make second generation shims and use them as stamping shims. If you plan for a very large production run, you use the third generation shims as your stamping shims. So if you intend to use third generation shims, you may have to reverse the image when exposing the photoresist. Saxby points out that this can be accomplished by making a silver-halide transfer and then contact printing the projected slit image, so that when viewed through the photoresist the image is reversed left to right. Another method for reversing the image is to use a negative photoresist and expose it through the back of the plate.

After your first generation shim has been washed with solvent to remove residual photoresist, it is necessary to passivate the first generation shim so that the second generation shim does not permanently bond to the first shim. To do this, the first shim is immersed in a 2.5% potassium dichromate solution for 1-10 minutes and then is rinsed in distilled water[17]. It is then mounted in a jig and you proceed to plate the second generation shim as you did the first. The third generation shims are made in the same manner.

6) Embossing

Now we have our stamping plate and we are ready to emboss. There are three methods of producing embossed holograms.
• Flat bed embossing
• Roll embossing
• Hot stamping (also called hot foil blocking)

The most commonly used material in both flat bed embossing and roll embossing is thermoplastic, typically PVC (polyvinylchloride), and it is usually .075mm to .20mm (.003 to .008 inches) thick. Polyester is sometimes used but is more difficult to control. The unembossed vinyl base is the viewing side because the embossed hologram surface is only a few thousand angstroms deep and very vulnerable to scratches. This is the reason the stamped hologram has its image reversed left to right.

Flat Bed Embossing: There are three main reasons this method is used.
• The machines are inexpensive (approximately US$3,000.+).
• Some companies that do a lot of embossing want to have a proof press.
• You are able to do large format holograms that cannot fit on roll or hot stamping machines.

The way flat bed embossing works is that two flat metal platens are each internally heated, and pressure is applied as the shim is pressed into the plastic. The platens are then cooled until the temperature is below the plastic flow point. Then pressure is released. The entire cycle is measured in seconds to minutes depending on the type of plastic and machine. This process is obviously slower than roll embossing but it can yield better quality holograms.

Production rates can be increased by stacking the shim and plastic so you have a plastic-shim-plasticshim-plastic-shim sandwich. If you have a large platen, you can also spread the shims out and make several holograms at a time. If the plastic is thin enough and on rolls, you can figure a way to advance the roll through the machine between impressions.

The metal shims for platen work have to be rigid and very flat. It is also desirable to have the shims a little thicker since they will be subjected to continual heating and cooling and they sometimes have to be pried apart from the plastic. The shim's metal thickness varies with hologram size from about a .25 mm (.01 inches) shim for a 5 x 5 cm (2 x 2 inches) hologram to a 1mm (.04 inches) shim for a 30 x 30 cm (12 x 12 inches) hologram or larger.

Roll Embossing: Here are a few main points about the machinery.
• The machinery is expensive (costs are around US$25,000.+).
• Very fast production rates (30 meters/minute+). Millions of holograms are made using this method.
• Because of the relatively low heat and pressure subjected to the shim at anyone time, the shims last considerably longer than with platen presses.
• This is the most ideal method for any quantity of holographic stickers like the ones you see on cereal boxes, etc. This method is also ideal for any short run hologram that cannot justify the expense of hot stamping.

These machines can emboss with masters that are as large as 30 x 60 cm (12 x 24 inches). Much larger thermoplastic roll embossing machinery exists but it has not yet been used for embossing holograms.

Roll embossing machines use a multiple pressureroller system. The metal master is attached to an internally heated roller. A roll of plastic is passed through this roller and an adjacent pressure roller. The metal master should be about .05 mm (.002 inches) thick and should be flexible enough to conform exactly to the shape of the roller. A new roller is used for each job because the metal masters are permanently attached to the roller. The deviation between masters should be kept to a minimum (±.005 mm or .0002 inches). Variations create dark areas or image voids.

In some systems, a roll of plastic is mounted on the machine. In other systems, plastic is extruded from pellets, calendered, and embossed in one continuous operation. The thickness of the typical embossed hologram is .076 mm (.003 inches) to .20 mm (.008 inches).

After embossing, the embossed roll is aluminized on the embossed side to create a mirror backing. This protects the hologram from scratching and creates the mirror-like backdrop we are all familiar with in embossed holograms. This aluminized backing is also subject to degradation and it is usually covered with a self adhesive backing and a plastic or paper peel-off: called a liner if your desired final roll is a sticker.

Hot Stamping (or Hot Foil Backing): A few of the major points of hot stamping are:
• This is just about the only method used for security applications such as banknotes, etc. because the hologram cannot be removed without destroying the product.
• The final hologram is very attractive since it is al'" lost flush with the product and appears to be a part of the product.
• This is cost effective for the bindery of a printing company because they can use their hot-foil equipment, with some adjustments, to apply the hologram directly to the printed product.
• Once committed to the hot stamping process, you cannot use the hologram as a sticker so you should consider your objective carefully.

Hot foil stamping has been around for a long time. If we had a roll of hot stamping foil in our hands and we started at the top and peeled each layer off we would go through the following layers.
• Plastic carrier sheet that comes off when hot stamped (Approximately 25 llm meters thick)
• Wax release layer that melts and releases the plastic carrier sheet when heated (Approximately .05 μm meters thick)
• Lacquer layer (Approximately 1.6 μm meters thick)
• Aluminum foil (Approximately .03 μm meters thick)
• Hot metal adhesive to bond aluminum to printed product. (Approximately 1.0 μm meters thick)

The lacquer layer, although very thin, is three times thicker than the indentation made by our embossing master. Therefore, the lacquer layer is embossed with the hologram before the aluminum foil and hot metal adhesive layers are put on. You obviously have to coordinate this with your foil provider.

In hot stamping, a heated platen presses on top of the above sandwich. The wax release layer and the hot metal adhesive both soften and the plastic carrier comes off while the hot metal adhesive bonds to the product.

Our hologram is safe since it is under the lacquer layer. It is also bonded to the product and cannot be removed without destroying it.

7) Bindery (applying the hologram)
In a great many of the cases the company that does the embossing is not the company that affixes the hologram to the product. In both roll embossing and hot stamping the end product is a roll of holograms ready to go to your printer's bindery. The bindery then applies the holograms to your product either by a label machine, by hand, or by hot stamping the hologram onto the product. This last step should not be taken for granted. You have a considerable investment in your product at this stage and there are numerous cases where projects with very large investments have met defeat at the bindery.

There is also a little confusion of terminology in the industry because printers that do hot stamping usually do embossing of printed products as well because their machines are designed to do both. Obviously embossing holograms is not the same as embossing paper but people can get confused and think that the printer is actually embossing the holograms.

Careful thought should be given to which product is best for your purpose. Labels can be applied in the bindery by hand or by label applicators. Any extra labels can be hand applied to other short run advertising products. Hot stamping's benefit is that for long runs it is very cost-effective. The hot stamping machine also registers your product extremely accurately.

There are a number of hot stamping machines on the market. Bose and Kluge are among the more famous and most widely used.

Potential Problems At The Bindery: One of the most important areas to supervise your project is at the bindery. Some common problems occur with:
• *Stickers:* Hand application always leads to register problems and needs to be watched carefully. There do exist automatic labeling machines, such as LabelAire, that can apply 250 to 1,000 stickers per minute. Machines like this help reduce the possibility of human error.
• *Hot stamp and stickers:* If the product you are putting your hologram on is uneven, the hologram will not play back well because the ripples in the surface cause ripples in the hologram and this causes the image to interfere with itself. This is especially critical for hot stamping since the hologram is so thin.
• *Hot stamp and stickers:* If the product on which you apply your hologram is not receptive to your adhesive, the hologram simply does not adhere to the product. Watch out for printed products that have been coated before applying your hologram.

• **Hot stamping:** The heat from the hot stamping machine, when overdone, can cause ripples in the surface of the product and destroy your hologram.

• **After application packaging problems:** An example of this would be making a hardbound book that has no dust jacket or a printed ad that will be affixed to a box. Your hologram, in this case, is hot stamped onto smooth plastic or paper that will cover the product. The hot stamping may look fine but later, when the bindery applies the stock (with hologram hot stamped on it) to the product, it may be subjected to a heating process that could warp the stock or product. The bindery people will feel the small amount of warping is well within their normal quality control limits but the heating and warping may be enough to destroy your hologram.

Buying An Embossed Hologram On The Commercial Market

A recent issue of a trade publication for magazine purchasing agents, was devoted to embossed holography and its costs[18]. The editors commissioned a hologram and had it hot stamped on the cover of the magazine. They reported the following costs involved:

• 3D models cost in the range of $1,000. to $4,000. depending on the amount of detail.

• Masters cost between $2,000. and $5,000. depending
on the complexity .

• The cover of the magazine had a 2.75 x 2.00 cm (7 x 5 inches) hot stamped hologram of a 3D model. Modeling and mastering, they reported, costs about $5,500. Embossing the hot stamp foil cost approximately $0.10 per image. Hot stamping cost $0.03 to $0.06 per image.

• Embossed foil costs $7.20 for 1,000 inches. Non-embossed foil sold to embossers costs $1.46 for 1,000 inches.

In general, the magazine editors found that the big appeal of embossed holography is that it definitely increases the percentage of returns on direct mail pieces and increases circulation for magazines and books. On the down side, they noted that the mastering cost is the one item that keeps this from being very cost-effective in short runs.

References and Footnotes:

1. Schell, K.J. (1985). White Light Holograms for Credit Cards, SPIE Proceedings, 523, 331-5

2. Light Impressions, Inc.

3. Burns, J. R. (1985). Large Format Embossed Holograms, Proceedings of the SPIE , 523, '7-14.

4. Saxby, G. (1988). Practical Holography, Prentice Hall International. Bartolini, R.A., Feldstein, N. & Ryan, R.J., (Oct. 1973). Replication of Relief-Phase Holograms for Prerecorded Video, Journ Electrochemical Society:Solid-State Science & Technology, 120, 1408-13. Clay, B.R. & Gore, DA (1974). SPIE Proceedings 45,149-55.

5. Towne Laboratories, Sommerville, NJ.

6. Norman, S.L. & Singh, M.P., (1975). Spectral Sensitivity and Linearity of Shipley AZ1350J Photoresist, Applied Optics, 14,818-20.

7. Peacock, Inc., Philadelphia, PA.

8. Graham, K.A.,Electroplating Engineering Handbook, Van Nostrand Reinhold Company, Third Edition,(1971) p. 479.

9. Electroplating Engineering, op cit. pp. 482-3.

10. Electroplating Engineering, op cit. pp. 212-3.

11. Elnic, Inc., Nashvilie, TN.

12. Plating Products, Inc., Roselie, NJ.

13. M&T Technical Data Sheet #P-N1-Sn, M&T Chemicals, Inc., Rahway, NJ, Nov. 1979.

14. DiBari, GA, "Electroforming", Metal Finishing - Guidebook & Directory Issue 1982, Metals and Plastics Publications, Inc. pp. 52~-24.

15. Electroplating Engineering, op cit. pp. 212-3.

16. Harstan Technical Bulliten, "Nickel Sulfamate", Harstan Chemical Corp. Brooklyn, NY, May 1973.

17. Fraunhofer, J.A., Basic Metal Finishing, Chemical Publishing Co., NY, 1976, pp. 149-50.

18. Publishing Technology (July 1989), North American Publishing Co., Philadelphia, PA. pp26-33.

HOLOGRAPHIC OPTICAL ELEMENTS (HOEs)

One of the most financially rewarding fields of holography has been the making of Holographic Optical Elements, commonly referred to as HOEs.

As the name implies, HOEs are optical elements such as lenses, mirrors, etc. that are made holographically. Although the fabrication can be quite involved technically, the concept of the HOE is relatively simple. The best way to explain the concept is in the following way.

We know from our earlier discussion that in making a transmission hologram the light from the object beam reflects off of our object at innumerable points and goes on to strike the photosensitive plate at the same time that the reference beam does. We then develop the plate to get our hologram. Whenever we shine a reference beam on our hologram at the same angle used to create the hologram, the light from the reference beam is diffracted by the hologram and forced to focus in space to recreate our original object so we can see it.

Although it sounds awkward, we could say that there are innumerable points of light coming from the object which together form the object beam. After the hologram is developed and we illuminate it, the transmission hologram diffracts the reference beam in such a way that it focuses the reference beam to recreate the points of light that were reflected off the object. So

what we actually see is a compilation of countless points of light, each with its own focal length, being refocused in space by the hologram.

Let's consider a much simpler transmission hologram. Instead of an object which reflects many points of light, suppose our object has just one point of light. If you stop to think about it, the transmission hologram of a single point source of light functions in the same way as as a traditional concave lens. Look at the similarities in Figure 15.

You will also find that a reflection hologram functions in much the same way as a convex mirror, as shown in Figure 16.

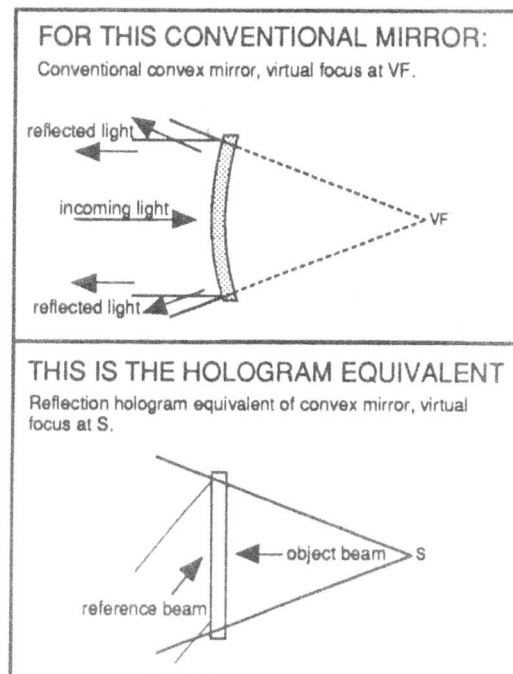

FOR THIS CONVENTIONAL MIRROR:
Conventional convex mirror, virtual focus at VF.

reflected light

incoming light

VF

reflected light

THIS IS THE HOLOGRAM EQUIVALENT
Reflection hologram equivalent of convex mirror, virtual focus at S.

object beam   S

reference beam

Figure 16. Conventional mirror and holographic equivalent.

These types of holograms are Holographic Optical Elements (HOEs). They are generally made without objects and with the intention of mimicking a conventional optical lens or a variation of conventional optics. With an HOE, the method for producing the image differs from conventional optics but HOEs do obey basic principles of conventional optics. These include such fundamentals as the equation for determining the focal length of the lens.

As you can imagine, there are differences between conventional lenses and HOEs. Here are a few of them:

• The angle of light used by holographic lenses is very selective since it is the reference angle.

• The frequency of light used by holographic lenses is very selective since it is the frequency used to make the hologram.

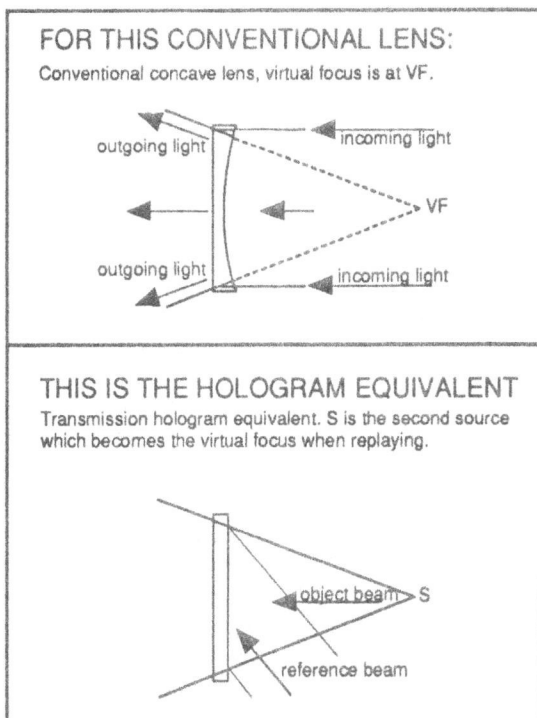

FOR THIS CONVENTIONAL LENS:
Conventional concave lens, virtual focus is at VF.

outgoing light   incoming light

outgoing light   VF

incoming light

THIS IS THE HOLOGRAM EQUIVALENT
Transmission hologram equivalent. S is the second source which becomes the virtual focus when replaying.

object beam   S

reference beam

Figure 15. Conventional lens, and holographic equivalent.

• Unlike a conventional mirror, a holographic mirror can be rotated 180 degrees to form a real image with a positive focal length.

Advantages of HOEs

• Since you can create a HOE lens that, unlike conventional lenses, is very angle and frequency specific it makes possible the construction of unique and very selective optical system configurations. One example is an HOE mirror that reflects light coming in from an angle greater than 45 degrees, or a mirror that reflects a given percentage of a specific wavelength you designate.
• Unlike conventional optical elements, the function of HOEs is relatively independent of the substrate geometry. Consequently HOEs can be produced on thin substrates that are relatively light even for large apertures.
• Since HOEs are holograms, spatially overlapping elements are possible because several holograms can be recorded in the same layer.
• HOEs can correct system aberrations, so that separate corrector elements are not required.
• HOEs offer the possibility of mass production at a significantly cheaper unit cost than conventional optics.
• Some HOEs can be produced from computer generated holograms. When the waveforms needed to form the HOE are calculated and generated by a computer, an HOE that has little noise and imperfection can be produced.

There are many different types of HOEs. In general, they are either gratings, diffraction lenses, beam splitters or combiners.

Diffraction Gratings

Diffraction gratings are among the simplest HOEs to construct. For a very simple diffraction grating, you only need the interference of two or more beams of laser light.

When white light illuminates the completed diffraction grating, the light of the exposing laser diffracts at the reference angle, and as you turn the hologram you see other wavelengths--thus producing a spectrum of colors. You can obviously do a large number of creative things in the lab such as multiple exposures, etc. to produce a variety of finished HOE products.

Holographic Lenses And Mirrors

The number of ways you can make holographic lenses and mirrors is limited only by your imagination. There are many ways to combine exposures or select specific wavelengths and angles to fit your needs. Variable beamsplitters, for example, can be produced by varying the exposure across the hologram. Sometimes you can use existing traditional optical devices to make your HOE. One technique for making holographic collimating mirrors, for example, is simply to make a hologram of a collimating mirror. Computer generated holograms are often used in making holographic lenses because the wavefronts, in many cases, are relatively simple and this allows you to compute an idealized wavefront, with almost no interference noise, for the holographic lens or mirror.

In summary, there is enough demand for holographic optical elements that they are now established as a permanent discipline. They have worked their way into airplanes, cars, supermarket scanners and laser discs for home entertainment to name a few of the very big markets that assure them of continued existence. Computer generated holograms, from which many HOEs originate, is covered next in this section.

COMPUTER GENERATED HOLOGRAMS (CGH)

As we explained in the introduction, some types of holograms sacrifice their depth so they may be seen in a wide variety of commonly encountered lighting conditions. For the sake of convenience, authors sometimes divide holograms into two broad categories: "thin" holograms and "volume" holograms. Basically, holograms are considered "thin" or 20 when the thickness of the recording material is small in relation to

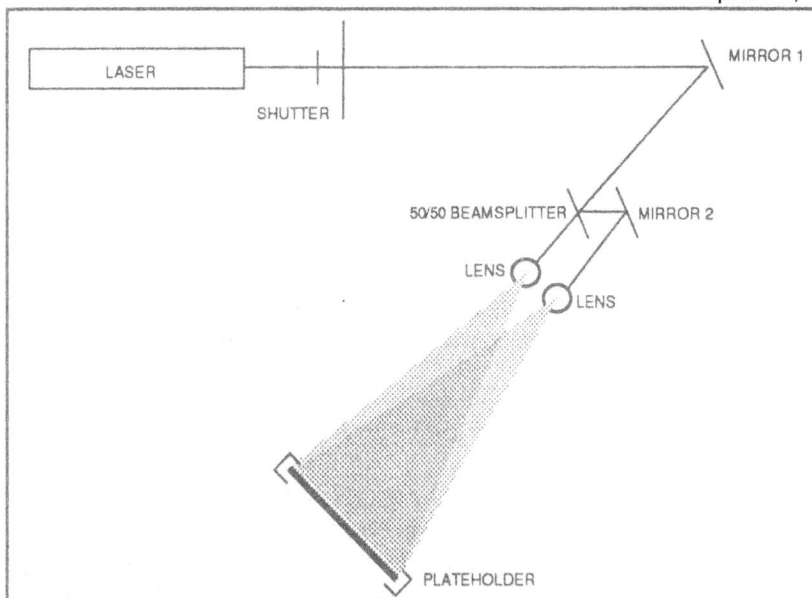

Figure 17. Simple diffraction grating

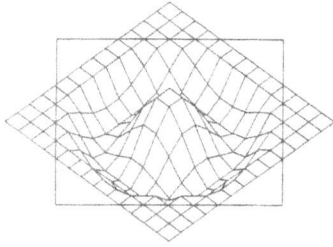

the average spacing of the interference fringes (less than 1:1) and volume holograms have a ratio greater than 1 :11. "Thin" holograms have little or no depth. "Volume" holograms can have great depth. Although volume holograms have all the glamorous 3D qualities, thin holograms have their own reasons for being popular. Thin holograms are:

• Easy to view even under harsh lighting conditions
• Easy to make since we are only dealing with 20 subjects
• Marketable holograms because of embossing and many are HOEs like gratings, mirrors, and lenses

Let us consider only the thin holograms, since CGH are thin holograms. With a fairly thin hologram, we could look at the hologram surface and can think of it as a flat, two dimensional object like a piece of paper. We could store all the information for this flat hologram in a computer by simply recording, on an (x,Y) coordinate graph, the size and location of all the holes in the hologram.

Following is an overview of how a Computer Generated Hologram (CGH) is made.

Making A Computer Generated Hologram

The general procedure used to make a computer generated hologram follows.

1) Object
We need to start with an object but the object does not need to exist physically. It can be imaginary.

2) Input Hologram.
We either have a real physical object or an imaginary, digitized object:

Real Object: If we have a real object, we make a "thin" hologram at this point. Then we have the holographic patterns on the plate digitized by scanning the plate using a microdensitometer. A microdensitometer measures the percentage of light that goes through a material in an extremely small area.

Imaginary Object: If we have an imaginary object, we calculate the wave patterns at the plate surface by using theories appropriate for Fraunhofer, Fresnel, or near-field diffraction. It should be pointed out that the more complex our object is, the more difficult this task will be.

3) Edit Hologram
If we need to edit or change our hologram in any way, we do so in the computer at this point.

4) Output Computer Generated Hologram
We now output or reproduce our hologram onto some media. Eventually we want to wind up with a transparency that has our holographic pattern on it. There are several programming techniques used to map the hologram pattern and we cover the techniques later in this section. The most common hardware used to output CGHs are:

• Computer driven plotter
• Computer driven laser beam
• Computer driven electron beam
• Computer monitor displays hologram and a photograph is made of the monitor

5) Reconstruct and Detect Hologram
Once the hologram is output, we copy the hologram and mass produce it or illuminate the hologram and record it using means that suit our purpose. Detectors may include:

• Eye-simply view the hologram if that is the purpose
• Charged couple devices
• Photographic film
• Small antennae and receivers have been used for detecting microwaves

Advantages of the Computer Generated Hologram
The advantages of the computer generated hologram are clear.

Objects: No objects are necessary in the simple cases where we can calculate the wave patterns by computer.

HOEs: Since we can use no objects, idealized waveforms can be made. This eliminates unwanted extraneous light and "noise" one would get from a normal hologram. This is of great value if we are making HOEs.

Editing: We can digitally edit our hologram in the computer.

Storage: We store our holograms electronically instead of on time sensitive emulsions.

Transmittal: We can send our holograms electronically worldwide in a matter of moments.

Translation: As we look into a Fourier Transform hologram we see an unique feature. We see a central spot of light and two images, one on each side of the spot. One object is an inverted and pseudoscopic image of the other. If we rotate the plate clockwise, the images rotate with the plate. If we simply turn the plate so back becomes front and vice versa, like we turn the pages of a book, the images stay fixed in space. This unique feature is used in some specialized cases.

Disadvantages of Computer Generated Holograms

Hologram Thickness: The major disadvantage to computer generated holograms is that they currently only work for "thin" holograms. One of the great attractions to holography is the "volume" or three dimensional qualities of a hologram. It is true, however, that there are clever ways to create depth using thin holograms such as stereograms. One of the most notable current items is the Computer Generated alcove hologram being done by Dr. Stephen Benton at MIT's Media Lab.

Computer Time And Storage: Generating the holographic fringe structure of a simple object such as a toy car can require tremendous amounts of memory and computer time, making it difficult and not cost effective for many applications.

Programming Techniques for Outputting Computer Generated Holograms

As we mentioned earlier, there are several programming techniques for plotting or mapping the holographic pattern we send to our output hardware. Four of the most commonly used methods are:
- Binary Detour Phase Holograms
- Kinoform
- ROACH (Referenceless On-Axis Complex Hologram)
- Computer Generated Interferograms

Binary Detour Phase Holograms

This method was first reported in 1966. The final product is an opaque mask with transparent holes or apertures that represent the hologram.

To output this, the computer calculates the image of the hologram mathematically using a Fourier transform. The paper or media on which the computer prints our hologram is subdivided into miniature cells. In each cell, the computer prints a dot which later becomes an aperture. The magnitude of the Fourier transform at the center of the cell is calculated and that calculation determines the height and width of the aperture. The lateral position of the aperture within each cell is proportional to the transform's phase at the center of the cell.

On photoreduction, the black dot becomes a clear aperture on a black background. Representing the phase by using a lateral shift of the aperture within the cell led to the name "Detour Phase Hologram", an analogy to diffraction gratings with unequally spaced rulings. Another term for these holograms is "binary hologram" because any point on the hologram we create has a transmittance value of zero or unity.

There are several variations of the Detour Phase Hologram which allow for a more refined output.

Advantage: It is possible to use a simple pen-and-ink plotter to prepare the binary master and problems of linearity do not arise in the photographic reduction process.

Disadvantage: This method is very wasteful of plotter resolution, since the number of addressable plotter points in each cell must be large to minimize the noise.

The Kinoform

If the object is diffusely illuminated, the magnitudes of the Fourier coefficients are relatively unimportant, and the object can be reconstructed using only the value of their phases. This gave rise to the Kinoform. To record a Kinoform, "the computed values of the phase are recorded on a multilevel gray scale which is used to control a photographic plotter that exposes a piece of film. The master is then photographed again, to reduce it to final size, and bleached to convert the gray levels to corresponding changes in optical thickness.

Advantage: Kinoforms can diffract all the incident light into the final image.

Disadvantage: Less information is available since only the values of the phases are used. Also, If there is any error in the recorded phase shift, light is diffracted into the zero order which can spoil the image with an extremely bright light in the center of the hologram.

The ROACH (Referenceless On-Axis Complex Hologram)

This method uses multilayer colored film as a recording medium to obtain most of the advantages of the Kinoform without · its major disadvantages. Both the phase and amplitude are recorded. Using Kodachrome film, the intensity-variation pattern is exposed through a red filter and the phase-variation pattern is exposed through cyan filter. If the processed film is illuminated with a HeNe laser, the cyan layer modulates the amplitude and the other two layers modulate the phase.

Advantage: Since all the light is diffracted into a single image, the diffraction efficiency of the ROACH is very high. In addition, because both the amplitude and phase information is recorded, the image quality is superior to that of the kinoform. The ROACH is superior to the Detour Phase Hologram because only one display spot is required for each Fourier coefficient and quantization noise is negligible.

Disadvantage: The steps required to produce this hologram are much more involved than for the Kinoform.

Computer Generated Interferograms

Problems were encountered with the Detour Phase Holograms when encoding wavefronts with large phase variations since the apertures in the cells overlapped in cases where the phase of the wavefront moved through a multiple of 21t radians or more. To solve this, an alternative approach was taken by noting that the case of a wavefront which has no amplitude variations is essentially similar to an interferogram. The non-linearity of the emulsion can then be exploited to produce a hologram that is approximately binary. There are methods that can then be used to record the amplitude variations in the binary fringe pattern.

Advantage: Computer generated interferograms are an improvement to the Detour Phase Hologram where large phase variations are encountered.

Disadvantage: Since amplitude variation is not recorded initially, it has to be calculated or derived later by one of several means.

Three Dimensional Computer Generated Holograms

Computer generated holograms were first generalized to a three dimensional object in 1968 by Waters. The process involved approximating the 3D object by making a number of equally spaced cross sections perpendicular to the z axis. Then a number of holograms from different angles are produced showing the resulting changes in parallax. Due to the fact that there are lines "hidden" from the front view of our object that we encounter as we make successive slices though the object, one must add the contributions to the object wave arising from the hidden lines as we go along.

In 1970, King, Knoll and Berry took a different approach. Their technique makes a tall, thin, holographic stereogram and then "multiplexes" or joins large quantities of these stereograms together on a single plate. The computer produces a series of perspective projections of an object by either programming the holograms or by filming the subject from a number of different perspectives along a horizontal axis, and inputting the holograms to the computer by microdensitometer readings. This is a nice theory, but reports are that it is difficult in practice.

The computer then outputs a series of thin vertical strips, each of which are holograms, on a single plate.

Since this is a thin hologram, it can be illuminated with white light to construct a bright, almost achromatic image. When the hologram is illuminated by the reference beam, or any white light, we see the real image, which is two dimensional and located in the plane of the final hologram. However, since our eyes see a number of different holograms and each one is from a slightly different perspective, the viewer has a stereoscopic view and sees a three dimensional image. If the plate is large and there are numerous images, we can tilt the plate from side to side and see the image move about.

The alcove hologram is the latest (Benton, 1987) in the evolution of 3D computer generated holograms. This is a multiplexed reflection stereogram except the multiplexed holograms are arranged in a semi-circular alcove into which the viewer looks. The advantage is that as the viewer walks from side to side, the three dimensional hologram rotates left to right allowing the viewer to see around the object. To date, a viewing angle of about 30 degrees has been accomplished but Benton suggests enlarging and extending the arrangement so as to have close to 180 degrees of viewing using 900 slit holograms that are 300 mm high and 1 mm wide (the interior of the concave would be 600 mm across).

Advantage

Although a volume hologram is the only true three dimensional recreation of an object, it has a problem with the restricted angle of view and subject matter. Holographic stereograms use any movie film that is filmed with the proper perspectives and, recently, color has been used in holographic stereograms. Industry benefits include being able to view stereoscopical-

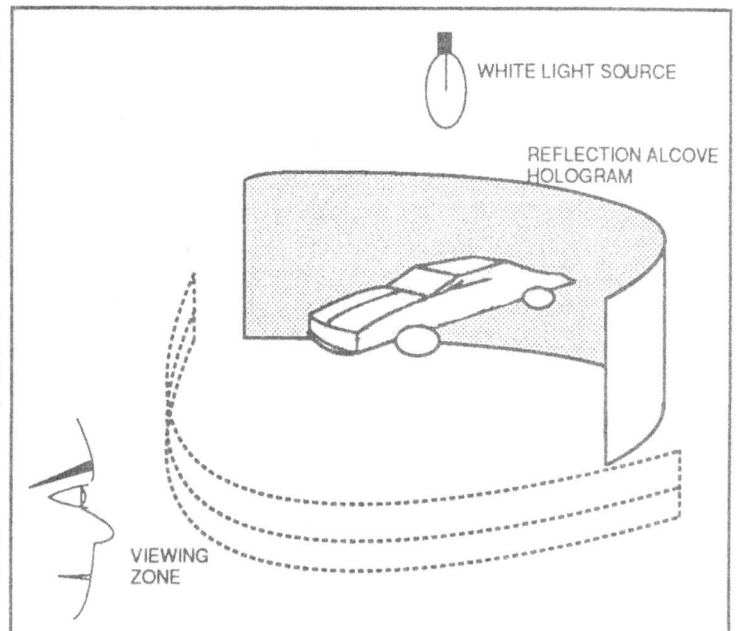

Figure 18. Oblique view of reflection alcove hologram and its viewing zone.

ly and move around any desired object at will without having the object present.

Disadvantage

One of the big problems with the CGH is getting the holographic image into the computer. Using a micro-densitometer to read 900 holograms into the computer is more of a challenge than anyone wants to take on. Generating the hologram internally by programming the wave patterns is very difficult for all but the simplest Objects. Progress is being made, however, and developing technology is on the side of the multiplexed CGH. It should be pointed out that curved, muiltiplexed holograms made without the aid of the computer are commercially available but the computer offers enormous benefits when the system is perfected.

Uses of Computer Generated Holograms

The uses for CGHs and HOEs are very similar since one of the main values of the CGH is its ability to produce almost noise free HOEs.

Some Applications for CGHs

• Commercial Embossing: Used to make simple artistic holograms for later use embossing, etc.

• Multiplexed Displays: 3D artistic displays using holographic stereograms such as the alcove hologram.

• Medicine: Currently under research at MIT is a process that takes CAT scans and MRI (magnetic resonance imaging), both already

---

Footnotes:

1. There are two popular mathematical models used to describe the pattern of light waves coming through a hologram. One model, the Raman-Nath diffraction pattern, is based on the assumption that the thickness of the emulsion is small compared to the average spacing of the interference fringes. The other model, called the Bragg diffraction pattern , is used to describe results when the thickness of the emulsion is large compared to the average spacing of the interference fringes. There is a "boundry" between the two models where neither model holds up perfectly. Therefore, one has to say that the Raman-Nath "thin hologram" model works when the emulsion thickness is considerably less than the fringe spacing and the "volume hologram" Bragg model works when the thinkness of emulsion is consderably greater than the fringe spacing.

References for Binary Detour Phase Holograms
Brown, R.B. & Lohmann, A.w. (1966). Complex Spatial filtering with binary masks. Applied Optics, 5,967-9.
___ . (1969). Computer Generated Binary Holograms. IBM Journal of Research & Development, 13, 160-7.

Burckhardt, CB. (1970). A simplification of Lee's method of generating holograms by computer. Applied Optics, 9, 1949.

Dallas, W.J. (1971 a.) Phase quantization - a compact derivation. Applied Optics, 10, 674-6.
___ . (1971 b.) Phase quantization in holograms - a few illustrations, Applied Optics, 10, 674-6.

Goodman, J.W. & Silvestri, A.M. (1970). Some effects of Fourier domain phase quantization. IBM Journal of Research & Development, 14, 478-84.

Haskell, R.E. & Culver, B.C. (1972). New coding technique for computer generated holograms. Applied Optics, 11, 271 2-14.
___ . (1973). Computer generated binary holograms with minimum quantization errors. Journal of the Optical Society 01 America., 63, 504.

Lee , W.H. (1970). Sampled Fourier transform hologram generated by computer. Applied Optics, 9, 639-43.

Lohmann, A.W. & Paris, D.P. (1967). Binary Fraunhofer holograms generated by computer. Applied Optics, 6, 1739-48.

References for The Kinoform:
Kermisch, D. (1970). Image reonstruction from phase information only. Journal of the Optical Society of America, 60, 15-7.

Lesem, L.B., Hirch, P.B., & Jordan, JA (1969). The Kinoform: A new wavefront reconstruction device. IBM Journal of Research & Development, 13, 150-5.

Lohmann, et alia. Binary Fraunhofer holograms generated by computer. op.cit., pp.1739-48.

References for the R.O.A.C.H.:
Chu, D.C., Fienup, J.R. & Goodman, J.w. (1973). Multiemulsion, on-axis, computer generated holograms. Applied Optics, 12, 1386-8.

References for Computer Generated Interferograms:
Bryngdahl O. and Lohmann, A.w. (1968) Interferograms are image holograms. Journal of the Optical Society of America, 58,141-2.

Lee, W.H. (1974). Binary synthetic holograms. Applied Optics, 13, 1677-82.

___ . (1979). Binary computer generated holograms. Applied Optics, 18, 3661-9 .

References for Three Dimensional Computer Generated Holograms:
Benton, SA (1982), Survey of holographic stereograms, SPIE Proc., 367, 15-19.

___ .(1987), Alcove Holograms for Computer-Aided Design, SPIE Proc, 761, 53-61 .

Brown, et alia. Computer generated binary holograms. op.cit. pp. 160-7.

Holzbach, M., (Sept. 1986), Three Dimensional Image Precessing for Synthetic Holographic Stereograms, M. Sc. thesis, Massachusetts Institute of Technology.

King, M.C., Noll, A.M. & Berry, D.H. (1970). A new approach to computer generated holography. Applied Optics, 7,1641-2.

Krantz, E .. (Sept. 1987), Optics for Reflection Holographic Stereogram Systems, M. Sc. thesis, Massachusetts Institute of Technology.

Lesem, et alia. The kinoform: a new wavefront reconstruction device. op. cit. pp.150-5.

Teitel, M. (Sept. 1986), Anamorphic Ray Tracing for Synthetic Alcove Holographic Stereograms, M. Sc. thesis, Massachusetts Institute of Technology,

Waters, J.P. (1968). Three-Dimensional Fourier transform method for synthesizing binary holograms. Journal of the Optical Society of America, 58, 1284-8.

HOLOGRAPHIC NON-DESTRUCTIVE TESTING (NDT) ALSO CALLED HOLOGRAPHIC INTERFEROMETRY

NDT is a fast growing field of holography. Although the techniques and analysis can be very involved, the general concept is not difficult to understand.

Suppose you shoot and develop a transmission hologram of a still object. Then, without disturbing the setup, you make another transmission hologram of the object. If you were to take the two holograms and put them on top of each other, you simply see one hologram pattern reinforced by the other since both holograms are identical. If, however, there is a small amount of movement in your object between exposures, the two patterns would be slightly different at the site of the dislocation. When you put the two patterns on top of each other you then see a Moire pattern at the site where the two hologram patterns differ. In holographic interferometry, these Moires are called interference patterns even though the hologram itself is an interference pattern. Since the interference patterns occur only where the holograms do not match at the site of the dislocation, the interference patterns show us where and by how much the object is dislocated. With investigations that started in 1965 it was discovered that in some cases, very exact measurements can be obtained (measurements down to the order of light waves in some cases) by studying the interference patterns.

The immediate applications are clear. You can take one holographic exposure of an object, subject the object to stress and then take a second holographic exposure. Analyzing the interference patterns that result when the two patterns are superimposed tells you where the object deformed under stress. This analysis has become a whole industry in itself and is variously called "Holometry", "Holographic Interferometry", or by its much more sales-oriented name of "Holographic Non Destructive Testing", abbreviated to NOT.

Some applications of holographic interferometry include:
• Locating the presence of a structural weakness: the object is stressed by the application of a load or change in pressure or temperature
• Detecting cracks and the location of areas of poor bonding in composite structures
• Medical and dental research
• Aerodynamics, heat transfer, and plasma diagnostics
• Solid mechanics such as measuring the changes in shape due to absorption of water and corrosion

There are a number of techniques for holographic interferometry tests and each has its advantages and disadvantages. Some of the major classifications are Real Time Holographic Interferometry, Double Exposure Holographic Interferometry, Sandwich Holographic Interferometry, Time Averaged Holographic Interferometry, and Strobed Holographic Interferometry. We discuss each of these methods in general terms.

Real Time Holographic Interferometry

This procedure is done to obtain immediate results. A hologram is shot, developed and then put back in the original holder it was in for the exposure. When the object is illuminated as if to expose it for another shot, you have two holographic patterns on the plate. One pattern will be the one you developed and the second pattern is created by the light you just turned on reflecting from the object still in its original position. They should match exactly if nothing moved and the plate is replaced exactly in its original spot. Therefore, if you look through the plate at the illuminated object, you should see just one, reinforced, holographic pattern. Suppose, however, that there was some small movement in your object while you were developing the plate. If there was movement, then after the plate is put back in the holder and the laser is turned on the observer viewing the object sees it covered with a pattern of interference fringes around the areas where the object has changed shape. If you move the object or put it under stress while viewing it, you see interference fringes change to map the area where the object is being deformed.

It is awkward to have one person at a time viewing the interference fringes. Capturing the image being viewed is also a problem. These problems may be solved by using a closed circuit television camera. A

Polaroid camera can photograph any desired scene. It is then possible to color code the fringes to identify the direction of the displacement by using filters and a double exposure. Another method is to have the video signals read digitally and input directly to a computer for analysis.

Since pulsed lasers are not subject to vibration restrictions, it is also possible to mount a pulsed laser on a truck, drive to an object that cannot be moved, such as a water tower, and from the truck perform a real time NOT stress analysis. This also applies to buildings, bridges, etc. for applications like earthquake analysis.

Although the above procedure for real time holographic interferometry sounds very simple and straightforward, there are many problems with it. Precise positioning of the hologram after processing is necessary and uniform drying of the emulsion to avoid deformations in the hologram is another potential problem. One method of solving these problems is to develop the hologram on the spot using a "liquid gate" arrangement.

Another alternative for processing the hologram, which eliminates the need for wet processing altogether, is to use thermoplastic recording. There are commercial units available in which a hologram is recorded and viewed in less than a minute.

Advantage
The advantage of real time NOT is its instant results. There are numerous cases where you must have instant analysis so that work on a project can continue. You are also able to view the interference patterns as the object is undergoing stress. In other words, you see a real time "movie" of the stress points developing as the experiment is conducted, and you may photograph or videotape what you see.

Disadvantage
One disadvantage is that there is a serious drop in the visibility of fringes because the light diffracted by the hologram remains linearly polarized while the light scattered by the object is largely depolarized. To avoid the decline in fringe visibility, it is necessary to use a polarizer when viewing or photographing the fringes. A second disadvantage is the difficulty of exact registration of a developed hologram with the original object. A third disadvantage is that you do not wind up with a precise, permanent record of the interference fringes except for what was on the television monitor. Recording on photographic emulsion offers more precision.

Double Exposure Holographic Interferometry
In this case you simply make a double exposure of the hologram on the same plate. The first exposure is usually made with the object in an unstressed condition and the second exposure is made with stress applied to the object.

Advantage
Double exposure interferometry is much easier than real time holography because the two holograms are in exact register. Any distortions to the emulsion affect both holograms equally. Both holograms have the same polarization and therefore the fringes are much clearer and no special care needs to be taken when illuminating the hologram. It is possible to use rainbow holograms for double exposure interferograms. Fringe patterns due to different effects can be displayed in different colors. .

Disadvantage
The double exposure tends to brighten the image of the object which makes it difficult to see small displacements. You can help correct this by shifting the phase of the reference beam or tilting the object beam between exposures. Another disadvantage is that with a double exposure you have the "before" and "after" snapshot but you do not see what happens in-between. This can be overcome, to some extent, by multiplexing techniques. Another problem is that you cannot compensate for and control the fringes if the object tilts between exposures. This problem can be solved by using two holograms, recorded on the same plate, with two, different, angularly separated reference waves.

Sandwich Holographic Interferometry
The sandwich hologram takes advantage of the fact that multiple exposures can be made on the same holographic plate. This procedure provides a much more elegant solution to the problem of viewing varying stages of stress on an object between exposures.

Two plates (with no anti-halation backing) are set in the same plate holder with their emulsion surface toward the object. An exposure is made of the object under no stress, then the object is tilted and/or stressed, and a second exposure is made. You can tilt the object again, apply more stress, make another exposure, etc.

To view the hologram, you put the plate that recorded the object under no stress in its original plateholder and add to the plateholder one of the holograms of the object under stress. You see interference patterns at the points of deformation. You can see the incremental deformations of your object by going through all the plates in sequence. Using any combination of two plates you can see the incremental changes between any two stages of deformation. By tilting the plates you can compensate for object tilt and also determine in which direction the object tilted by analysis of the interference patterns.

Advantage

This technique allows you to see a wide latitude or combination of stress loads. You could, for example, select the the plate B1 and F2 to view what happens when the first stress load is applied.

Disadvantage

You are looking through two different plates to see the image and there may be some parallax problems. It is also time and material consuming to catalogue and handle all these different plates.

Holographic Interferometry through Dense Media

One example of this procedure begins when a hologram is taken of an undisturbed chamber filled with a gas. The chamber is disturbed in some way and a second hologram is taken. Any variation in density of the gas alters the path length of the laser light and hence creates a hologram which produces interference patterns when compared to the undisturbed chamber's hologram.

Another example is plasma diagnostics where measurements of the light's refraction at two wavelengths make it possible to determine the electron density directly.

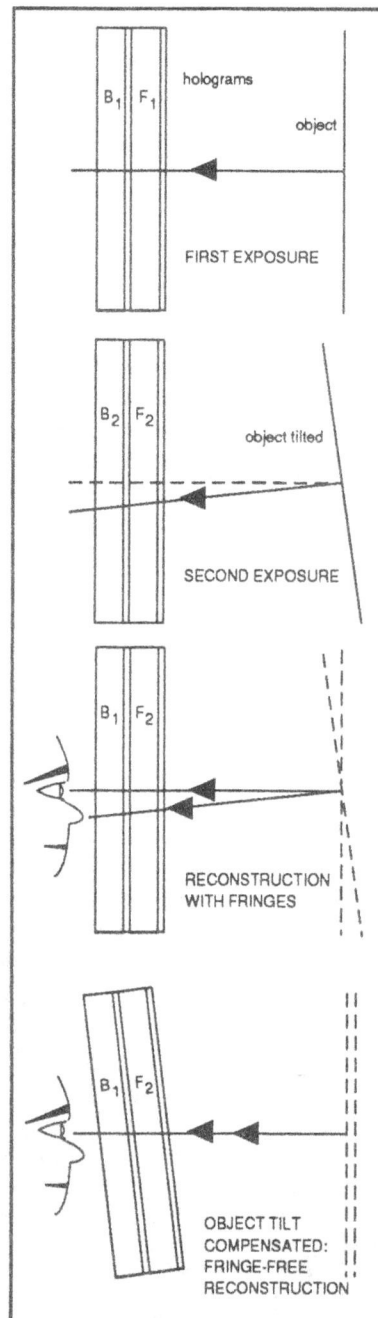

Figure 19. Sandwich hologram interferometry (Hariharan, p. 213)

Time Averaged Holographic Interferometry

There are numerous instances where the object can only be properly tested while operating. To see clearly all the uneven stress in a hi-fi speaker or a rotating fan, for example, you need to test it while it is in operation. In order to make interferometer tests on these subjects a number of clever methods have been developed. Time averaged holographic interferometry is one of them.

If you take a flexible metal ruler and whip it back and forth in the air while holding the end of it tightly in your fist you see the image of the ruler at the two most extreme points of the flex but the image is a blur inbetween. If you were to calculate it, you would find that the amount of time spent at the two extremes of the cycle is a substantial percentage of the time in the entire cycle.

Suppose we look directly into a hi-fi speaker. When it is operating, the speaker membrane or cone spends most of its time at the two ends of its vibration (Le. the cone is either fully extended toward us or fully extended away from us). We simply make a hologram of the speaker while it is operating and extend the exposure so that the exposure time is long compared to the cycle of the speaker membrane moving in and out. What is the result? Since the time spent between the two extremes is a blur, the result is a hologram of the speaker membrane fully extended toward us superimposed on a hologram of the speaker membrane fully extended away from us. In other words, we seem to obtain a "double exposure hologram" even though it is only one exposure.

If the speaker is not in perfect resonance and the distance the speaker membrane extends towards us is not exactly equal to the distance the speaker membrane moves away from us, the two superimposed holographic patterns create an interference pattern. You can calculate how out of resonance the speaker is by counting the fringes. The distance between fringes is slightly greater than one quarter of a wavelength.

The above technique of time averaged holographic interferometry works well for studying objects vibrating in a stable manner. It was first reported in 1965 and is usually used to identify the resonances of a test object by monitoring its response while varying the excitation frequency and the point of excitation on the object.

Typical applications of the process use musical instruments, electromechanical devices such as loudspeakers, turbine blades and aircraft structures.

Advantage

In time averaged holographic interferometry you are allowed to view and adjust your object to obtain an almost perfect resonance with little effort.

Disadvantage

Time averaged holographic interferometry is good only for objects vibrating in a stable manner and compares only the two extremes of the vibration cycle.

Also, this procedure does not show the relative phases of the various modes of vibration. Also, the contrast of the fringes falls off rapidly as the fringe order increases.

Strobed Holographic Interferometry

Suppose that you have a object that vibrates in a stable manner but the point you want to inspect in the vibrating cycle and not at one of the two extremes of the cycle. In 1968, articles were written describing a solution to this problem using a sort of strobe light effect. In this process, a sequence of laser light pulses are triggered at desired times during the cycle . The resulting hologram is equivalent to a double exposure hologram (or sandwich hologram if you use that method) recorded while the object is in the desired state of deformation at any time in the cycle.

Advantage

Allows you to study the object at any time in the vibration to the object at any other time in the vibration.

Disadvantage

This is a much more involved set up than the time averaged
technique.

References:
Hariharan, P. (1984) Optical Holography. Cambridge: Cambridge University Press.

References for applications of holographic interferometry Greguss, P. (1975). Holography in Medicine. London: IPC Press

___ . (1976). Holographic interferometry in biomedical sciences. Optics & Laser Technology, 8, 153-9.

Sciammarella, CA, (1987). Advances in the application of holography for NDE, SPIE Proceedings, 523, 137-44.

Von Bally, G. ed. (1979). Holography in Medicine and Biology. Berlin: Springer-Verlag.

Willenbourg, G.C., (1987). Holography and Holometry Apolictions in Dental Research, SPIE Proceedings, 747, 51-6.

References for Interference Fringes
Powel, L. & Stetson, K.A. (1965). Interferometric Vibration Analysis by Wavefront reconstruction, Journal of the Optical Society of America, 55, 1543-8.

References for real time holographic interferometry
Dandliker, R., Ineichen, B. & Mottier, F.M. (1973). High resolution hologram interferometry by electronic phase measurement. Optics Communictions, 9, 412-16.

Hariharan, Oreb & Brown, (1982). A digital phaseeasurement system for real time holographic interferometry Optics Communications, 41, 393-6.

___ . (1982). Real-time holographic interferometry: a microcomputer system for the measurement of vector diso, acements. Applied Optics, 22, 876-80.

References for liquid gate recording
Hariharan, P. (1977). Hologram interferometry: identification of the sign of surface displacements. Optica Acta, 24, 989-90.

Hariharan, P. & Ramprasad, B.S. (1973). Rapid in situ processing for real-time holographic interferometry. Journal of Physics E: Scientific Instruments, 6, 699-701.

Van Deelen, W. & Nisenson, P. (1969). Mirror blank testing by real time holographic interferometry. Applied Optics, 8, 951 -5.

References for thermoplastic recording
Hariharan, P. & Hegedus, Z.S. (1975). Relative phase shift of images reconstructed by phase and amplitude holograms. Applied Optics, 14,273-4.

Saito, T., Imamura, T. & Tsujiuchi, J. (1980). Solvent vapor method in thermoplastic photoconductor media. Journal of Optics (Paris), 11, 285-92.

Thinh, V.N. & Tanaka, S. (1973). Real time interferometry using thermoplastic hologram. Japanese Journal of Applied Physics, 12, 1954-5.

References for double exposure holographic interferometry
Yu, F.T.S. & Chen, H. (1978). Rainbow Holographic interferometry. Optics communications, 25, 173-5.

Yu, F.T.S., Tai, A. & Chen, H. (1979). Multiwavelength rainbow holographic interferometry. Applied Optics, 18, 212-18.

___ . (1979). Multislit one-step rainbow holographic interferometry. Applied Optics, 18, 6-7.

References for shifting reference beams
Collins, L.F. (1968). Difference holography. Applied Optics, 7, 203-5.

Hariharan, P. & Ramprasad, B.S. (1972). Simplified optical system for holographic subtraction. Journal of Physics E: Scientific Instruments, 5, 976-8.

___ . (1973). Wavefront tilter for double-exposure holographic interferometry. Journal of Physics E: Scientific Instruments, 6, 173-5.

Jahoda, F.C., Jeffries, A.A. & Sswyer, GA (1967). Fractional fringe holographic plasma interferometry. Applied Optics, 6,107-10.

References for multiplexing techniques
Caulfield, H.J. (1972). Multiplexing double-exposure holographic interferograms. Applied Optics, 11, 2711-12.

Hariharan, P. & Hegedus, Z.S. (1973). Simple multiplexing technique for double-exposure hologram interferometry. Optics Communications, 9,152-5.

Parker, R.J. (1978). A new method of frozen-fringe holographic interferometry using thermoplastic recording media. Optica Acta, 25, 787-92.

Two holograms on the same plate
Ballard, G.S. (1968). Double exposure interferometry with separate reference beams. Journal of Applied Physics, 39, 4846-8.

Gates, J.W.C. (1968). Holographic phase recording by interference between reconstructed wavefronts from separate holograms. Nature, 220, 473-4.

Tsuruta, T., Shiotake, N. & Itol, Y. (1968). Hologram interferometry using two reference beams. Japanese Journal of Applied Physics, 7, 1092-100.

References for sandwich hologram interferometry
Abramson, N. (1974). Sandwich hologram interferometry: a new dimension in holographic comparison. Applied Optics, 13,2019-25.

___ . (1975). Sandwich hologram interferometry: 2. Some practical calculations. Applied Optics, 14, 981-4.

___ .. (1976). Sandwich hologram interferometry: 3: Contouring Applied Optics, 15,200-5.

___ . (1976). Holographic contouring by translation. Applied Optics, 15, 1018-22.

___ . (1977). Sandwich hologram interferometry: 4: Holographic studies of two milling machines. Applied Optics, 16, 2521-31.

Abramson, N. & Bjelkhagen, H. (1978). Pulsed sandwich holography. 2. Pratical application. Applied Optics, 17, 187-91.

Hariharan, P. & Hedgedus, Z.S. (1976). Two-hologram interferometry: a simplified sandwich technique. Applied Optics, 15, 848-9.

References for holographic interferometry through dense media
Ostrovskaya, G. V. & Ostrovskii, Yu. I. (1971). Twowavelength hologram method for studying the dispersion properties of phase objects. Soviet Physics - Technical Physics, 15, 1890-2.

Radley Jr, R.J. (1975). Two-wavelength holography for measuring plasma electron density. Physics of Fluids, 18, 175-9.

Zaidel, A.N., Ostrovskaya, G. V. & Ostrovskil, Yu. 1.(1969). Plasma diagnostics by holography. Soviet Physics - Technical Physics, 13, 1153-64.

References for time averaged holographic interferometry
Agren, C.H. & Stetson, K.A. (1972). Measuring the resonances of treble viol plates by hologram interferometry and designing an improved instrument. Journal of the Acoustical Society of America, 51, 1971-83.

Bjelkhagen, H. (1974). Holographic time average vibration study of a structure dynamic model of an airplane fin. Optics & Laser Technology, 6,117-23.

Chomat, M. & Miller, M. (1973). Application of holography to the analysis of mechanical vibration in electronic components. TESLA Eletronics, 3, 83-93.

Saxby, G., (1988). Practical Holography, Prentice Hall, 321.
Stetson K.A. & Powell, A.L. (1965). Interferometric hologram evaluation and real-time vibration analysis of diffuse objects. Journal of the Optical Society of America, 55, 1694-5.

References for strobed holographic interferometry
Archbold, E. & Ennos, A. E. (1968). Observation of surface vibration modes by stroboscopic hologram interferometry. Nature, 217, 942-3.

Shajenko, P. & Johnson, C.D. (1968). Stroboscopic holographic interferometry. Applied Physics Letters, 13,44-6.

Watrasiewicz, B.M. & Spicer, P. (1968). Vibration analysis by stroboscopic analysis. Nature, 217, 1142-3.

ARTISTIC HOLOGRAPHY

Work done in-house = ●

Broker = ○

COMPANY NAME	H-1 Maker	Full Color H-1 Maker	Model Maker	Animator/Computer Animator	Silver Halide Trans. Maker	Silver Halide Reflection Maker	Fine Art Ltd. Edition Maker	Stereogram Maker	Dichromate Maker	Pulsed Laser Portrait Maker	Photopolymer H-1 Maker	Photoresist Master Maker	Lighting Consultant	Framing Consultant	Marketing Consultant	Educational Consultant	Other
			ARTISTS										CONSULTANTS				
Acme Holography	○	○	○	○	○	○	○	○			○		●			●	
Adel Rootstein, Inc			●														
Advanced Environ. Research	●				●	●											
Advanced Holographics Corp.	●		●						●								
Advanced Holographics, Ltd	●				●	●	●						●				●
AG Prismatic	○				○										○		
Aites Lightworks					●	●	●										
AKS Holographie-Galerie Gmb	●		●	●	●	●	●	●									
Amazing World Of Holograms													●	●	●	●	
Amazon							●	●									
American Bank Note Hologr...	●	●	●				●	●		●		●					
Amherst Media							●										
Anait Studio						●				●							
Angstrom Industries Inc.	●	●		●	●			●				●					
A.N.Sevchenko Research Inst.	●																
Applied Holographics Corp.	●	●		●	●	●						●	●	●			
Aptec Engineering Limited	○																
Arbeitskreis Holografie B.V	●																
Architectural Glass & Holog.															●		
Armstrong World Industries	●																
Art Freund Holography						●	●										
Artigliography Co.	○	○	○	○	○	○	○	○	○	○	○	○	●	●	●	●	
ArtKitek	●		●		●							●	●			●	
Artplay Holographic Studio	○	○	○	○	○	○	○	○	○			○	●			●	

ARTISTIC HOLOGRAPHY

Work done in-house = ●

Broker = ○

COMPANY NAME	ARTISTS												CONSULTANTS				
	H-1 Maker	Full Color H-1 Maker	Model Maker	Animator/Computer Animator	Silver Halide Trans. Maker	Silver Halide Reflection Maker	Fine Art Ltd. Edition Maker	Stereogram Maker	Dichromate Maker	Pulsed Laser Portrait Maker	Photopolymer H-1 Maker	Photoresist Master Maker	Lighting Consultant	Framing Consultant	Marketing Consultant	Educational Consultant	Other
Art,Science & Technology Inst	●	●	●		●	●	●		●	●	●	●	●	●	●	●	●
Ascot Laser Picture Studio					●	●	●										
Asociacion Española de Holog.	●															●	
Assoc. of Science & Technol.													●		●	●	●
Atelier Holographique de Paris	●	●	●	●	●	●	●						●	●	●	●	●
Atomika Technische Physik	●																
Barr & Stroud, Ltd	●																
Bob Mader Photography										●					●		
Brighton Imagecraft													●				
Brodel Holograms								●	●	●							
Burns Holographics Ltd.	●	●			●		●	●				●	●		●	●	●
Burton Holmes International	●						●										
Cambridge Consultants Ltd															●		
Cambridge Stereographics Grp															●		●
Casdin-Silver Holography	●	●	○	○	●	●	●	●		●			●	●		●	●
Center For Applied Rsch	●												●		●		
Central Electricity Generat. Brd										●							
Cherry Optical Company	●				●	●	●						●	●	●		
Chimeric Images, Inc													●	●	●		●
Chromagem Inc	●	●			●	●	●				●	●					
CISE SpA Technologie Innov.	●				●	●	●		●			●	●	●	●	●	
Coburn Corporation												●					
Darkroom Eight Ltd							●										
David Schmidt Holography	●		●		●	●		●					●	●	●	●	●

ARTISTIC HOLOGRAPHY

Work done in-house = ●

Broker = ○

COMPANY NAME	ARTISTS												CONSULTANTS				
	H-1 Maker	Full Color H-1 Maker	Model Maker	Animator/Computer Animator	Silver Halide Trans. Maker	Silver Halide Reflection Maker	Fine Art Ltd. Edition Maker	Stereogram Maker	Dichromate Maker	Pulsed Laser Portrait Maker	Photopolymer H-1 Maker	Photoresist Master Maker	Lighting Consultant	Framing Consultant	Marketing Consultant	Educational Consultant	Other
Deep Space Holographics	○																
Deutsche Gesellschaft Für Hol																●	●
Deutscher HolographieVertrieb	●																
Dialectica AB	●																
The Diffraction Company Inc.													●		●	●	
Dimensional Imaging Technol.	○																
Dovecote Studio	●																
Dream Images															●		
Duston Holographic Services	●	●			●	●	●	●									●
Dutch Holographic Laboratory	●	●		●	●	●	●	●				●					
E.C. Schultz & Co.																	●
E/F Productions	●																
ERBA	●																
Focal Image Ltd		●															
The Foreign Dimension			○												●		
Fringe Research Holographics					●	●				●							
FTI Joffe	●																
Gardener Promotion Marketing	○				○	○			○				●	●	●		
General Imaging Corporation										●					●		
Gerald Marks Studio							●										●
Gray Scale Studios Ltd.			●														
Hickmott & Austin Holograms	●																
Hoechst Celanese Corporation	●														●		
Holage	●					●	●										

ARTISTIC HOLOGRAPHY

Work done in-house = ●

Broker = ○

COMPANY NAME	ARTISTS												CONSULTANTS				
	H-1 Maker	Full Color H-1 Maker	Model Maker	Animator/Computer Animator	Silver Halide Trans. Maker	Silver Halide Reflection Maker	Fine Art Ltd. Edition Maker	Stereogram Maker	Dichromate Maker	Pulsed Laser Portrait Maker	Photopolymer H-1 Maker	Photoresist Master Maker	Lighting Consultant	Framing Consultant	Marketing Consultant	Educational Consultant	Other
Holar Seele KG							●										
Holaxis Corporation	●										●		●		●		
Holicon Corporation	●				●	●				●							
Holo 3													●	●	●	●	●
Holo ARP							●									●	
Holocom							●										
Holocom Holographie							●										
Holocrafts of Long Island							●										
Holodesign							●										
Holodesign Studies															●		
Holo-Dimensions Inc							●										
Holofar Lab (SRL).	●																
Holofax Limited						●											
Holoflex Company					●	●											
Holo Gmbh Holografielab. Osn.	●														●	●	
Holografica							●										
Holografie - Hofmann Labor							●									●	
Holograma Lab. Holographique										●							
Hologramm Werkstatt+ Galerie	○		○		○	○	○						●	●		●	●
Holographic Applications	○	○	○	○	○	○	○	○	○	○	○	○	●	●	●	●	●
Holographic Art			●												●		
Holographic Concepts-MA															●		
Holographic Creations	●																
Holographic Design Incorporat	○	○	○			○	○			○	○	○	●	●	●	●	●

Holographic T-Shirts and Sweatshirts

FRONT OF SHIRT

BACK OF SHIRT

Golden Gate San Francisco

THE BOOKSHELF

Below is a selection of unique or hard-to-find holography items. Some of the books are not found in your local bookshop and can be ordered directly from us, using the order form on the next page. The sweat- and T-shirts pictured opposite use an unique application process that has just been patented. The soft washable fabric is 50% cotton/poly and is imprinted with non-lead, permanent plastisol ink. Fully machine washable

All prices include shipping and handling. To order, use form on the other side. <u>Overseas shipping and insurance</u>: please add (US)$9.00 for ONE item, $6.00 <u>per item</u> for 2 or more items.

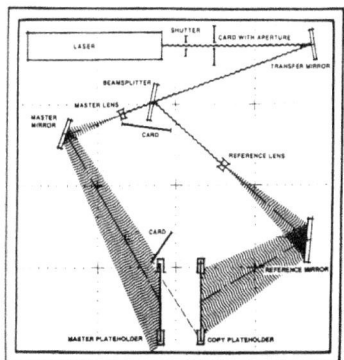

pictured on opposite page

HOLOGRAPHY HANDBOOK
by Unterseher, et alia
407pp, $21.00(US), paperback, illustrated, with a hologram in every copy!

World's best selling how-to handbook on holography. This book guides you with easy, step-by-step instructions in how to make your own holograms at home. Everything you need to know is in here.

PRACTICAL HOLOGRAPHY
by Graham Saxby.
488pp, $48.00(US), hardback, illustrated, indexed, bibliography. Includes a silver halide hologram in each copy!

For the "technologist or professional", this book covers the practical aspect of holography at all levels.

OPTICAL HOLOGRAPHY
by P. Hariharan
320pp, $28.00(US), paperback, illustrated, indexed, bibliography

Self-contained treatment of principles, techniques and applications of optical holography. For those who want to learn more about optical holography, plus further detailed treatment for advanced readers.

ORDER INFORMATION

CABLE CAR T-SHIRT
$19.00(US), £12, (UK)

CABLE CAR SWEATSHIRT
$35.00(US), £22, (UK)

EAGLE T-SHIRT
$17.00(US), £11, (UK)

EAGLE SWEATSHIRT
$30.00(US), £19, (UK)

These colorful and machine washable T-shirts and sweatshirts demonstrate the newest successful application of holography integrated with clothing. Textile Graphics patented process permanently bonds the hologram to the clothing, producing an unique product that you won't find in your local T-shirt shop. Eagle T-shirt also comes in white on white. Care instructions are included.

HOLOGRAPHY HANDBOOK
by Unterseher, et alia. 407pp, $21.00(US), £14,(UK)
The world's best selling book on holography. Designed as a do-it-yourself manual of practical holography, this book is laid out in an easy-to-read manner and gives the reader absolutely all the information he needs to buy and construct his own home holography studio. Complete step-by-step instructions with hundreds of illustrations are given on making beginning and advanced holograms. Holography Handbook is currently used by several universities as a lab manual.

OPTICAL HOLOGRAPHY
by P. Hariharan.320pp, $28.00(US), £18, (UK)
The leading book on the theory of optical holography. Haraharan covers in considerable depth all the major procedures used in holography. Thorough footnotes and references list over 700 works on holography. If you need more depth of understanding and research material, this is unquestionably the book for you.

PRACTICAL HOLOGRAPHY
by Graham Saxby. 488pp, $48.00(US), £30, (UK)
For the "technologist or professional". this book covers the practical aspect of holography at all levels. Saxby gives wide coverage of all aspects of the field. This book gives a detailed technical understanding of the practical techniques used in holography and comes with an impressive silver halide hologram in each copy.

ORDER FORM: 3 or more items=10% discount!
Phone Order: (415) 841 2474; Toll Free In US: (800) 367 0930; FAX Order: (415) 841 2695

Quantity	Description	Price	Total

Mail your completed order form and payment to:
Order Department
Ross Books
P.O. Box 4340
Berkeley, CA 94704 USA

Subtract 10% discount if 3 or more items ordered_____
Add overseas shipping $9.00 (US) £6 (UK) for one item ordered;
$6.00(US), £4,(UK) per item for two or more items
TOTAL_____

CHEQUES WRITTEN ON ANY BANK IN THE WORLD ARE ACCEPTED
CREDIT CARD USERS: (Check One):

☐ Master Card ☐ VISA ☐ American Express

Card Number:_____

Expiration Date:_____

Signature:_____

SHIP TO ADDRESS:

Name:_____

Address:_____

City/State/(Province):_____

Country/Postal Code:_____

ARTISTIC HOLOGRAPHY

Work done in-house = ●

Broker = ○

COMPANY NAME	H-1 Maker	Full Color H-1 Maker	Model Maker	Animator/Computer Animator	Silver Halide Trans. Maker	Silver Halide Reflection Maker	Fine Art Ltd. Edition Maker	Stereogram Maker	Dichromate Maker	Pulsed Laser Portrait Maker	Photopolymer H-1 Maker	Photoresist Master Maker	Lighting Consultant	Framing Consultant	Marketing Consultant	Educational Consultant	Other
					ARTISTS								CONSULTANTS				
Holographic Design Systems					●	●	●	●							●		
Holographic Dimensions, Inc							●										
Holographic Images Inc	●		●		●		●										
Holographic Industries, Inc	○	○			○		○			○			●		●		
Holographic Products Inc	●	●	●				●		●				●	●	●	●	
Holographics Australia	●																
Holographic Service															●		
Holographics North Inc	●		●		●	●	●			●	●		●			●	
Holographic Studios													●		●		
The Holographic Studio, Ltd	●				●	●	●						●	●		●	
Holographics (UK) Ltd							●		●	●							
Holographic Visions													●		●		
Holographie Design							●										
Holographie Konzept							●										
Holography Institute	○	○	○	○	○	○	○	○		○	○	○	●	●	●	●	●
Holography Ltd															●	●	
Holo-Images, Inc	●																
Holo-Laser	●	●			●	●	●					●					
Holo Laser Tech Ltd	○	○	○	○	○	○		○		○		○	●				
Holomagic Inc							●										
Holomedia AB/Hologram Mus.							○										
Holomedia Inc.	●		●		●	●	●		●								
Holomorph Visuals, Inc							●										
Holoproduction													●	●	●	●	●

ARTISTIC HOLOGRAPHY

Work done in-house = ●

Broker = ○

COMPANY NAME	H-1 Maker	Full Color H-1 Maker	Model Maker	Animator/Computer Animator	Silver Halide Trans. Maker	Silver Halide Reflection Maker	Fine Art Ltd. Edition Maker	Stereogram Maker	Dichromate Maker	Pulsed Laser Portrait Maker	Photopolymer H-1 Maker	Photoresist Master Maker	Lighting Consultant	Framing Consultant	Marketing Consultant	Educational Consultant	Other
	ARTISTS												CONSULTANTS				
Holopublic																●	
Holo-Service							●										
Holo-Service.Fries							●										
Holos Gallery													●	●	●		
Holos-Holos															●		
Holo/Source Corporation								●									
Holo-Spectra													●	●			
Holosystems Inc							●										
Holotec CC							●										
Holotec Plc															●		
Holotek, S.A	●		●		●	●	●										
Holotron SRL															●		
Holtronic										●							
Hyperspace Studio					●	●	●										
Ian M. Lancaster Holographics		○			○	○	○	○	○	○	○		●		●	●	●
Ibero Gestão - Gestão Integra					●										●	●	
Ibou Inc	●		●		●		●						●	●	●		
Icon Holographic							●										
Idhol					●	●	●						●	●			
Imac International, Inc															●		
Images Company	●				●	●		●					●			●	
Imaging & Design	●	●	●		●	●	●	●				●	●		●		
Ingenieur Büro Geiger	●		●		●			●				●			●	●	
Institute of Art & Design	●				●		●			●						●	

ARTISTIC HOLOGRAPHY

Work done in-house = ●

Broker = ○

COMPANY NAME	H-1 Maker	Full Color H-1 Maker	Model Maker	Animator/Computer Animator	Silver Halide Trans. Maker	Silver Halide Reflection Maker	Fine Art Ltd. Edition Maker	Stereogram Maker	Dichromate Maker	Pulsed Laser Portrait Maker	Photopolymer H-1 Maker	Photoresist Master Maker	Lighting Consultant	Framing Consultant	Marketing Consultant	Educational Consultant	Other
				ARTISTS									CONSULTANTS				
Institute of Physics, Ukraine					●	●											
K.C. Brown Holographics							●			●							
Kyoto Technical University	●	●			●	●											
Labor für Holografie							●										
Laminex/High Tech UK Ltd	●																
Lasart Ltd	○		○				○		○								
Laser Affiliates					○	○							●	●			
Laser Arts													●	●	●	●	
L.a.s.e.r. Co							●						●				
Laser Fare Ltd											●				●		
Laserfilm Eckhard Knuth							●										
Lasergrafics							●										
Laser Graphics	●				●	●	●										
Lasergruppen Holovision AB							●										
Laser Holographics, Inc.															●		
Laser Image Design							○										
Laser Innovations Inc							●										
Laserion Handels GmbH							●										
Laserlabbet	●																
Laser Light Expressions Pty.	●	●	●	●	●	●	●					●	●	●	●		
Laser Light Ltd							●										
Laser Lightworks							●										
The Lasersmith	●																
Laserworks	●																

ARTISTIC HOLOGRAPHY

Work done in-house = ●

Broker = ○

COMPANY NAME	ARTISTS												CONSULTANTS				
	H-1 Maker	Full Color H-1 Maker	Model Maker	Animator/Computer Animator	Silver Halide Trans. Maker	Silver Halide Reflection Maker	Fine Art Ltd. Edition Maker	Stereogram Maker	Dichromate Maker	Pulsed Laser Portrait Maker	Photopolymer H-1 Maker	Photoresist Master Maker	Lighting Consultant	Framing Consultant	Marketing Consultant	Educational Consultant	Other
Lasiris Inc	●				●	●						●					
Laza Holograms	●	●	●	●	●	●	●	●		●							
Lazart Holographics					●	●											
Les Productions Hololab!							●										
Light Construction							●										
Light Fantastic Plc								●				●					●
Light Impressions, Inc.	●	●	●									●			●	●	
Linda Law Holographics							●										
MacShane Holography	●		●		●	●	●										
Man Environment, Inc							●										
Markem Systems Ltd	●				●	●	●					●					●
Martinsson Elektronik AB												●					
Marwell AB							●										
Matt Hannifin Co.							●										
McCain Marketing														●	●	●	●
Media Interface, Ltd							●									●	
Messerschmitt-Boelkow-Blohm		●															
Metamorfosi Olografia Italia Srl									●						●		
Metaplast Electrochemicals																	●
MGD Modulations							●										
Mind's Eye: Hologr. Consultant															●		
Mirage Holograms Ltd					●	●											
Moeller Wedel Optische Werk							●										
Museum für Holographie															●	●	●

ARTISTIC HOLOGRAPHY

Work done in-house = ●

Broker = ○

COMPANY NAME	ARTISTS												CONSULTANTS				
	H-1 Maker	Full Color H-1 Maker	Model Maker	Animator/Computer Animator	Silver Halide Trans. Maker	Silver Halide Reflection Maker	Fine Art Ltd. Edition Maker	Stereogram Maker	Dichromate Maker	Pulsed Laser Portrait Maker	Photopolymer H-1 Maker	Photoresist Master Maker	Lighting Consultant	Framing Consultant	Marketing Consultant	Educational Consultant	Other
Nancy Gorglione/fine Art Holog	●	●	●		●	●											
Neovision Productions							●										
Newbold Wells Company					●	●											
Newcastle Up.Tyne Polytechn.																●	
New Dimension Holographics																●	
New York Hall of Science							●										
New York Holographic Labs							●										
North American Holographics										●					●		
Northwestern Univer.-Biomed										●							
Odhner Holographics	●				●	●											
OP-Graphics (Holography) Ltd							●										
Optical Images	○	○	○	○	○	○	○	○	○			○				●	●
Optical Laboratory							●										
Optical Surfaces Ltd							●										
Optical Works Ltd							●										
Oriel Scientific Ltd							●										
Pacific Holographics, Inc	●				●	●	●					●	●	●			
Pennsylvania State University		●															
Perception Holography							●										
Phoenix Holograms	○		○		○	○	○		○	○			●		●	●	●
Photon League	●				●		●										
Point of View Dimensions, Ltd								●					●		●		
Portson, Inc	●		●		●	●	●			●		●					
Ralcon									●								

ARTISTIC HOLOGRAPHY

Work done in-house = ●

Broker = ○

COMPANY NAME	ARTISTS												CONSULTANTS				
	H-1 Maker	Full Color H-1 Maker	Model Maker	Animator/Computer Animator	Silver Halide Trans. Maker	Silver Halide Reflection Maker	Fine Art Ltd. Edition Maker	Stereogram Maker	Dichromate Maker	Pulsed Laser Portrait Maker	Photopolymer H-1 Maker	Photoresist Master Maker	Lighting Consultant	Framing Consultant	Marketing Consultant	Educational Consultant	Other
Ralph Cullen Holographics	○	○			○	○	○		○	○	○	○	●				
Randy James/Holography	●				●	●	●				●						
Raven Holo Ltd							●										
Real Image Holographics (Ack							●										
Reel Image							●										
Revelation Holography							●										
Richard Bruck Holography	●				●	●	●						●			●	
Richmond Holographic Studios	●	●			●	●	●			●							●
Ritcher Holograms							●										
Robert Sherwood Holographic	○	○	○			○	○		○	○	○		●	●	●	●	●
Robinson Hologram Lighting													●				
Royal College of Art										●							
Royal Photographic Society																●	
Scope Optics Ltd							●										
Soviskusstvo V/o Mezhdunaro	●		●		●	●	●		●				●	●			●
Space Age Designs Inc.							●		●						●		
Spindeler & Hoyer GmbH	●														●		
Spot															●		
Starlight Holographic Inc.		○	○	○		○	○	○		○	○	○	●				
Studio für Holographie	●				●	●	●					●					
Studio Weil-Alvaron	●						●										
Superior Technology Implem.	●				●	●		●	●								
Swede Holoprint							●								●		
Synchronicity Holograms						●										●	

ARTISTIC HOLOGRAPHY

Work done in-house = ●

Broker = ○

COMPANY NAME	ARTISTS												CONSULTANTS				
	H-1 Maker	Full Color H-1 Maker	Model Maker	Animator/Computer Animator	Silver Halide Trans. Maker	Silver Halide Reflection Maker	Fine Art Ltd. Edition Maker	Stereogram Maker	Dichromate Maker	Pulsed Laser Portrait Maker	Photopolymer H-1 Maker	Photoresist Master Maker	Lighting Consultant	Framing Consultant	Marketing Consultant	Educational Consultant	Other
Technoexan/Semicon, Inc.	●						●			●							
TechNorth							●										
Textile Graphics, Inc																	●
Third Dimension Ltd	●			●	●	●				●							
Three Dimensional Imagery Lt							●										
Tjing Ling Industrial Research							●										
Touchwood Holographics	●				●	●			●								
Tridimensionale Hologramas							●			●							
Trilone Holographie Corp	●		●				●			●		●			●		
U.K. Gold Purchasers															●		
UK Optical Supplies	○	○			○	○	○		○	○	○	○	○				
Univ Gen t / Workshop Hologr.	●		●			●	●			●			●	●	●	●	●
Vniktk Kultura															●		
Wavefronts															●	●	
Wave Guides Inc							●										
Wave Mechanics					●	●											●
Wenyon & Gamble							●										
White Tiger Holograms	●							●	○								
Whole Hography					○	○	○										
Wöber Design Hololab Austria	●				●	●	●						●				
Wonders of Holography Galler															●		
World Art Project	●																
Wotan Lamps Ltd	●																
3D Gallery	●	●		●	●	●	●	●					●		●	●	

ARTISTIC HOLOGRAPHY

Work done in-house = ●

Broker = ○

COMPANY NAME	H-1 Maker	Full Color H-1 Maker	Model Maker	Animator/Computer Animator	Silver Halide Trans. Maker	Silver Halide Reflection Maker	Fine Art Ltd. Edition Maker	Stereogram Maker	Dichromate Maker	Pulsed Laser Portrait Maker	Photopolymer H-1 Maker	Photoresist Master Maker	Lighting Consultant	Framing Consultant	Marketing Consultant	Educational Consultant	Other
			ARTISTS												CONSULTANTS		
3D Media																	●
INDIVIDUALS																	
Abrams, Claudette							●										
Alexander							●										
Allen, Jeffrey		○	○	○	○		○	○		○	○	○			●	●	●
Andrews, Matthew	○	○			○	○	○			○			●	●	●	●	●
Benyon, Margaret	●				●	●	●			●							
Berkhout, Rudie	●			●	●		●										
Boyd, Patrick	○	○	○	○	○	○	○	○		○							
Bunts, Frank/Flatiron Studio							●										
Byrne, Kenneth G							●										
Capucci, Pier Luigi															●	●	●
Carlsson, Torgny E							●										
Carlton, David							●										
Conner, Arlie							●										
Cossette, Marie-Andrée							●										
Cowles, Susan Ann							●										
Cross, Lloyd																	●
Deem, Rebecca	●				●	●	●									●	●
Defreitas, Frank	●				●		●						●	●	●	●	●
Dewar, David	●																
Didier, Leduc	●	●	●		●	●	●										
Dietrich, Edward																	●
Dinsmore, Sydney	●	●		●	●	●			●								

ARTISTIC HOLOGRAPHY

Work done in-house = ●

Broker = ○

COMPANY NAME	ARTISTS												CONSULTANTS				
	H-1 Maker	Full Color H-1 Maker	Model Maker	Animator/Computer Animator	Silver Halide Trans. Maker	Silver Halide Reflection Maker	Fine Art Ltd. Edition Maker	Stereogram Maker	Dichromate Maker	Pulsed Laser Portrait Maker	Photopolymer H-1 Maker	Photoresist Master Maker	Lighting Consultant	Framing Consultant	Marketing Consultant	Educational Consultant	Other
Dominguez, Carmenza															●		
Dubov, Philip							●										
Dyens, Georges M							●										
Feferman, Bennett J.							●										
Feroe, James							●										
Fischer, Julian							●										
Foltz, Susannah D.							●										
Fornari, Arthur David					●	●											
Frigon, Jacques							●										
Gilles, Jean							●										
Gittoes, George	●																
Goldberg, Bruce																●	
Hansen, Matthew E							●										
Harris, Nick					●	●		●								●	
Heaton, Lorna							●										
Howlett, Glenn							●										
Ishii, Setsuko					●		●		●								
Jackson-Smith, Rosemary							●										
Jung, Dieter							●										
Kac, Eduardo	●				●	●	●									●	
Kaufman, Andreas							●										
Lafreniere, Anik							●										
LeSar, Christopher J.							●										
Levine, Chris							●										

ARTISTIC HOLOGRAPHY

Work done in-house = ●

Broker = ○

COMPANY NAME	H-1 Maker	Full Color H-1 Maker	Model Maker	Animator/Computer Animator	Silver Halide Trans. Maker	Silver Halide Reflection Maker	Fine Art Ltd. Edition Maker	Stereogram Maker	Dichromate Maker	Pulsed Laser Portrait Maker	Photopolymer H-1 Maker	Photoresist Master Maker	Lighting Consultant	Framing Consultant	Marketing Consultant	Educational Consultant	Other
Light, Gail							●										
Lijn, Liliane							●										
Lloyd, S																●	
Lysogorski, Charles	●																●
McClean, James	●		●		●		●										
McIntyre, Jim	●		●		●	●	●			●							
Menning, Melinda							●										
Merlin	●																
Merrick, Michael G.					●	●	●										
Merrill, Mark C							●										
Miller, Mr. Peter	●																●
Molteni, William															●		
Mulhem, Dominique	●	●	●	●	●	●	●	●	●	●							
Muskovitz, Aaron							●										
Myre, Robert							●										
Naimark, Michael							●										
Nemtzow, Scott							●										
Newman, Paul	●				●	●	●			●				●		●	
Padnos, W.R	●				●	●	●						●	●			
Palmer, Caroline							●										
Pepper, Andrew																●	
Petersen, Joel							●										
Peticov, Antonio							●										
Pizzanelli, David	●																

ARTISTIC HOLOGRAPHY

Work done in-house = ●

Broker = ○

COMPANY NAME	H-1 Maker	Full Color H-1 Maker	Model Maker	Animator/Computer Animator	Silver Halide Trans. Maker	Silver Halide Reflection Maker	Fine Art Ltd. Edition Maker	Stereogram Maker	Dichromate Maker	Pulsed Laser Portrait Maker	Photopolymer H-1 Maker	Photoresist Master Maker	Lighting Consultant	Framing Consultant	Marketing Consultant	Educational Consultant	Other
Pye, Tim							●										
Quinlan, Denis															●		
Reilly, John							●										
Reutersward, Carl Fredrick							●										
Roa, Warna J															●		
Rosewell, Michael P							●										
Ross, Jonathan															●		
Roy, Joanne							●										
Senechal, Jacques							●										
Stallard, Penn							●										
Stockler, Len	●																
Tunnadine, Graham							●										
Unterseher, Fred	●	●	●		●	●	●		●	●	●	●	●	●	●	●	●
Vila, Doris	●														●		
Volker, Mirau							●										
Weber, Sally	●																
Weil-Alvaron, Hans	●						●										
Welby, Mr. Dominic							●										●
Wells, Sandrajean							●										
Zeman, Lee							●										

ARTISTIC HOLOGRAPHY

Work done in-house = ●
Broker = ○

COMPANY NAME	MASS-MANUFACTURING (WORK DONE IN HOUSE)									BUYING & SELLING ARTISTIC HOLOGRAMS & SUPPLIES										
	Embossed (see embossed chart also)	Stereogram	Silver halide Transmission	Silver Halide Reflection	Photoresist Masters for Embossing	Photopolymer	Dichromate	Jewelry/Novelties	Portrait	Retail Shop	Wholesaler/Distributor	Catalogue/Mail Order	Gallery/Museum Shop	Private Collection	Hologram Display Units	Lighting Fixtures	Jewelry/Novelties	Dichromate Holograms	Embossed Holograms	Silver Halide Holograms
Acme Holography			○	○		○														
Active Image											●								●	
Advaned Holog. Cor.							●				●				●			●		
AD 2000 / Another ...											●		●							
A.H. Prismatic Inc.								●		●	●	●	●		●		●		●	
A.H. Prismatic Ltd.										●	●		●		●		●		●	
Aites Lightworks										●										
AKS Holographie-Gal		●	●	●	●														●	●
Alpha Foils, Inc	●																			
Amazing World of Hol										●	●		●	●	●	●	●	●	●	●
Amazon														●						
Amblehurst	●																			
American Bank Note	●	●																		
Applied Holographics	●	●	●	●	●						●				●				●	●
Art Freund Holog.														●						●
Artigliography Co.			○	○	○	○	○													
Artigliography Gallery										●	●	●	●	●	●	●	●	●	●	●
Artplay Holog. Studio	○	○	○	○	○															
Art,Science & Techn.	●											●	●			●	●		●	
Assc.Science & Tech.										●			●	●			●	●	●	●
Atelier Holographique														●	●					●
Beddis Kenley Engin.	●																			
Bobst Group	●																			

ARTISTIC HOLOGRAPHY

	MASS-MANUFACTURING (WORK DONE IN HOUSE)									BUYING & SELLING ARTISTIC HOLOGRAMS & SUPPLIES										
COMPANY NAME	Embossed (see embossed chart also)	Stereogram	Silver halide Transmission	Silver Halide Reflection	Photoresist Masters for Embossing	Photopolymer	Dichromate	Jewelry/Novelties	Portrait	Retail Shop	Wholesaler/Distributor	Catalogue/Mail Order	Gallery/Museum Shop	Private Collection	Hologram Display Units	Lighting Fixtures	Jewelry/Novelties	Dichromate Holograms	Embossed Holograms	Silver Halide Holograms
Brandtjen & Kluge Inc	●																			
Burns Holographics		●	●		●			●	●					●	●				●	●
Casdin-Silver Holog	○	○	○	○	○	○			○		●			●				●	●	●
Cherry Optical Co.											●	●								
Chimeric Images, Inc-			○	○			○	○	○											
Chromagem Inc	●		●	●	●					●	●		●				●	●	●	●
CISE SpA Technolog.	●		●	●	●		●				●	●		●	●	●		●	●	●
Coburn Corporation	●				●															
Crown Roll Leaf, Inc.	●																			
Dai Nippon Printing	●																			
David Schmidt Holog.		●	●	●				●			●		●		●	●				●
Diaures S.A.	●																			
Die Dritte Dimension										●			●				●	●	●	●
The Diffraction Co Inc	●																			
Dream Images													●							
Duston Holog. Svcs																				●
Dutch Holog. Laborat.	●	●	●	●	●															
The DZ Company								●			●		●				●			
E.C. Schultz & Co.														●						
Edmund Scientific Co.									●	●	●									
Elusive Image													●							
Fasion Moda													●							
Foreign Dimension							●	●			●				●	●	●	●		

Work done in-house = ●

Broker = ○

ARTISTIC HOLOGRAPHY

	MASS-MANUFACTURING (WORK DONE IN HOUSE)									BUYING & SELLING ARTISTIC HOLOGRAMS & SUPPLIES										
COMPANY NAME	Embossed (see embossed chart also)	Stereogram	Silver halide Transmission	Silver Halide Reflection	Photoresist Masters for Embossing	Photopolymer	Dichromate	Jewelry/Novelties	Portrait	Retail Shop	Wholesaler/Distributor	Catalogue/Mail Order	Gallery/Museum Shop	Private Collection	Hologram Display Units	Lighting Fixtures	Jewelry/Novelties	Dichromate Holograms	Embossed Holograms	Silver Halide Holograms
Foundation Ideecentr.													●							
Fringe Resch Hologr.													●							
Gallerie Illusoria													●							
Galvanoart	O	O	O	O	O															
General Hologr. Corp											●				●				●	●
General Hologr. Inc.				O			O	O			●				●		●		●	●
General Imaging Corp						●														
Halo Power-Track																●				
High Tech Network																			●	
HOL 3 Galerie Holog.										●			●							
Holaxis Corporation-	●														●				●	
Holicon Corporation									●											
Holo 3		●	●	●					●											
Holo ARP											●		●							
Holocom																			●	
Holocrafts		●					●	●										●		
Holografia Galleria										●			●							
Holograf. Galler.-Oulu										●			●							
Holografie - Hofmann											●									
Holo Lab Holographiq	●																			
Hologram Europe sprl										●										
Hologram. Industries	●	●	●																	
Hologramm Werkstatt											●			●	●	●				●

Work done in-house = ●

Broker = O

ARTISTIC HOLOGRAPHY

Work done in-house = ●
Broker = ○

COMPANY NAME	MASS-MANUFACTURING (WORK DONE IN HOUSE)									BUYING & SELLING ARTISTIC HOLOGRAMS & SUPPLIES										
	Embossed (see embossed chart also)	Stereogram	Silver halide Transmission	Silver Halide Reflection	Photoresist Masters for Embossing	Photopolymer	Dichromate	Jewelry/Novelties	Portrait	Retail Shop	Wholesaler/Distributor	Catalogue/Mail Order	Gallery/Museum Shop	Private Collection	Hologram Display Units	Lighting Fixtures	Jewelry/Novelties	Dichromate Holograms	Embossed Holograms	Silver Halide Holograms
Hologram One											●									●
Hologram Roadshow										●										
Holograms & Other St										●		●					●			
Hologram Schoppe										●										
Hologram World										●										
Holographic Applic.		○	○	○	○	○					●				●	●	●	●	●	●
Holographic Art								●		●	●	●		●	●	●	●	●	●	●
Holographic Design	○		○	○	○	○	○	○		●	●	●		●			●	●	●	●
Holographic Images		●	●	●		●			●											●
Holographic Industr.			○	○						●	●		●	●			●	●	●	●
Holographic Products						●	●			●	●	●	●	●	●			●	●	
Holo Shop Milwaukee										●										●
Holographics North			●	●																
The Hologr Studio Ltd														●	●					
Holographic Visions										●	●		●							
Holographie Labor										●								●		●
Holography Institute			○	○	○	○					●	●		●	●	●			●	●
Holography Ltd				○						●	●		●	●	●				●	
Holo-Laser	●		●	●	●									●	●				●	●
Holo Laser Tech Ltd.										●	●							●		●
HoloMart--Prem Tech-										●									●	●
Holomedia AB/Museu											●		●							●
Holomedia Inc.			●	●						●	●				●			●		●

ARTISTIC HOLOGRAPHY

Work done in-house = ●

Broker = ○

COMPANY NAME	Embossed (see embossed chart also)	Stereogram	Silver halide Transmission	Silver Halide Reflection	Photoresist Masters for Embossing	Photopolymer	Dichromate	Jewelry/Novelties	Portrait	Retail Shop	Wholesaler/Distributor	Catalogue/Mail Order	Gallery/Museum Shop	Private Collection	Hologram Display Units	Lighting Fixtures	Jewelry/Novelties	Dichromate Holograms	Embossed Holograms	Silver Halide Holograms
Holoprint Rosowski											●								●	
Holoproduction			●	●																
Holopublic														●					●	
Holos Art Galerie										●			●							
Holos Gallery			○	○		○	○			●	●	●	●	●	●	●	●	●	●	●
Holo/Source Corp.	●	●																		
Holo-Spectra											●								●	●
Holotek, S.A								●		●	●	●	●	●	●	●	●	●	●	●
Holovision										●										●
Ian M. Lancaster Holo	○	○	○	○		○	○		○											
Ibero Gestão - Gestão											●									●
Ibou Inc										●	●	●	●	●	●		●		●	●
Images Company	○	○	○	○				○			●	●					●			
Imaging & Design	●	●	●	●	●	●					●						●	●	●	●
Ingenieur Büro Geiger		○	○	○	○				○	●			●	●	●				●	●
Institute. Art & Design													●	●						●
Integraf										●	●									
Kolbe-Druck, Coloco	●										●	●							●	
Lasart Ltd						○	○				●						●	●	●	
Laser Fare Ltd						●														
Laser Holographics	○										●									
Laser International													●							
Laser Light Designs										●	●		●							

ARTISTIC HOLOGRAPHY

Work done in-house = ●

Broker = O

COMPANY NAME	MASS-MANUFACTURING (WORK DONE IN HOUSE)									BUYING & SELLING ARTISTIC HOLOGRAMS & SUPPLIES										
	Embossed (see embossed chart also)	Stereogram	Silver halide Transmission	Silver Halide Reflection	Photoresist Masters for Embossing	Photopolymer	Dichromate	Jewelry/Novelties	Portrait	Retail Shop	Wholesaler/Distributor	Catalogue/Mail Order	Gallery/Museum Shop	Private Collection	Hologram Display Units	Lighting Fixtures	Jewelry/Novelties	Dichromate Holograms	Embossed Holograms	Silver Halide Holograms
Laser Light Expressio.	●		●	●	●			●			●		●	●	●	●	●	●	●	●
Laser Light Image										●	●		●	●	●	●	●	●	●	●
Lasiris Inc	●			●															●	
Laza Holograms			●	●																
Lazart Holographics										●	●	●	●				●	●		●
Lazer Wizardry											●	●					●	●	●	
Leonard Kurz GmbH	O																			
Let There Be Neon										●			●							
Licht-Blicke-Buro													●							
Light Angels													●							
Light Engineering										●			●							
Light Fantastic-CO										●		●	●				●	●	●	●
Light Fantastic Plc	●	●			●					●	●		●	●			●		●	●
Light Impressions, Inc	●				●										●	●	●		●	
Light Wave Gallery										●			●							
MacShane Holog.			●	●											●	●				●
Magic Laser										●	●	●			●	●	●	●	●	●
Magic Light Holograf.													●							
Markem Systems	●				●															
McCain Marketing										●	●		●	●					●	
Metamorfosi Olografia											●				●			●	●	
Munday Spatial Imag.										●										●
Musée de L'holograph											●	●								

ARTISTIC HOLOGRAPHY

Work done in-house = ●
Broker = ○

COMPANY NAME	MASS-MANUFACTURING (WORK DONE IN HOUSE)									BUYING & SELLING ARTISTIC HOLOGRAMS & SUPPLIES										
	Embossed (see embossed chart also)	Stereogram	Silver halide Transmission	Silver Halide Reflection	Photoresist Masters for Embossing	Photopolymer	Dichromate	Jewelry/Novelties	Portrait	Retail Shop	Wholesaler/Distributor	Catalogue/Mail Order	Gallery/Museum Shop	Private Collection	Hologram Display Units	Lighting Fixtures	Jewelry/Novelties	Dichromate Holograms	Embossed Holograms	Silver Halide Holograms
Museum für Hologr.										●	●	●	●	●	●		●	●	●	●
Museum Hol/Chicago										●	●	●	●							
Museum of Hologr-NY													●							
Mus.Fine Arts Rsch													●							
N.Gorglione/Fine Art			○	○							●	●	○	●	●	●				●
New Clear Imports										●			●							
New Dimension Holo										●										
Newport Holograms										●			●				●	●		
New York Hall Of Sci.													●							
Odhner Holographics			●	●																
Ontario Science Cen													●							
Optical Images		○	○	○	○		○	○					●		●		●	●		●
Oxford Holographics										●	●						●	●	●	●
Pacific Holographics	●	●			●															
Photon League														●						●
Pilkington P.E. Ltd.						●														
Portson, Inc	●		●	●	●	●	●										●	●	●	●
Qué Sera Sera											●							●		
Ralcon						●	●	●	●								●	●		
Ralph Cullen Hologr.			○	○	○	○	○			●	●			●	●	●		●	●	●
Regal Press Inc	●																			
Richard Bruck Hologr.				●							●									●
Richmond Holographi			●	●					●		●				●	●				●

ARTISTIC HOLOGRAPHY

Work done in-house = ●
Broker = ○

COMPANY NAME	Embossed (see embossed chart also)	Stereogram	Silver halide Transmission	Silver Halide Reflection	Photoresist Masters for Embossing	Photopolymer	Dichromate	Jewelry/Novelties	Portrait	Retail Shop	Wholesaler/Distributor	Catalogue/Mail Order	Gallery/Museum Shop	Private Collection	Hologram Display Units	Lighting Fixtures	Jewelry/Novelties	Dichromate Holograms	Embossed Holograms	Silver Halide Holograms
	MASS-MANUFACTURING (WORK DONE IN HOUSE)									**BUYING & SELLING ARTISTIC HOLOGRAMS & SUPPLIES**										
Robert Sherwood Hol	○	○	○	○		○	○	○	○											
Soviskusstvo V/o.				●				●												
Space Age Designs							●	●			●	●					●			
Spycatcher Ltd											●								●	
Starcke, KY										●	●		●						●	●
Starlight Holographic			○	○	○	○	○	○		●	●	●	●	●		●	●	●	●	●
Studio für Holograph.					●														●	●
Superior Technology	●	●			●															
Technoexan/Semicon											●	●								●
Third Dimension Arts							●	●												
Third Dimension Ltd			●	●					●						●	●				●
Three-D Light Gallery													●							
Trend													●							
Trilone Holographie					●				●		●					●	●			
U.K. Gold Purchasers											●	●	●		●	●	●	●	●	●
UK Optical Supplies			○	○	○	○	○			●	●		●	●	●		●	●	●	●
Univ Gent Workshop			●	●					●					●						●
Wave Mechanics														●						●
White Light Works Inc	●																			
Whole Hography															●					
Whole Picture Gallery										●			●				●	●	●	●
Wöber Design Holo											●		●			●			●	
Wonders of Holograp										●	●		●	●			●	●	●	

ARTISTIC HOLOGRAPHY

COMPANY NAME	MASS-MANUFACTURING (WORK DONE IN HOUSE)									BUYING & SELLING ARTISTIC HOLOGRAMS & SUPPLIES										
	Embossed (see embossed chart also)	Stereogram	Silver halide Transmission	Silver Halide Reflection	Photoresist Masters for Embossing	Photopolymer	Dichromate	Jewelry/Novelties	Portrait	Retail Shop	Wholesaler/Distributor	Catalogue/Mail Order	Gallery/Museum Shop	Private Collection	Hologram Display Units	Lighting Fixtures	Jewelry/Novelties	Dichromate Holograms	Embossed Holograms	Silver Halide Holograms
3D Gallery			●	●				●		●	●		●	●	●	●	●	●	●	●
INDIVIDUALS																				
Allen, Jeffrey											●	●		●	●		●	●	●	
Andrews, Matthew			○	○			○				●			●						●
Benyon, Margaret																				●
Berkhout, Rudie				●										●						
Capucci, Pier Luigi														●						
Didier, Leduc														●						●
Dinsmore, Sydney				●										●						●
McClean, James														●						●
Merrick, Michael G.														●						
Mulhem, Dominique	●	●	●					●						●						
Unterseher, Fred		●	●	●	●	●		●							●			●	●	●

Work done in-house = ●

Broker = ○

COMPANY NAME	Museum/Curator	College Level Instruction	Independ. Educat. Facility	Hands-On Workshop	Holography Teacher Training	Curriculum Development	Educational Materials	Academic Research	Industrial Research	Medical Research	Time Sequence Movie of Holograms	Microscopic Holography	Endoscopic Holography	Plasma Research	Dental Research
Acme Holography		○		○	○	○	○								
Aerospatiale									●						
Alcan International Ltd												●			
Angstrom Industries, Inc			●	●	●				●						
A.N. Sevchenko Research Inst.									●						
Applied Holographics Corp.									●						
Artigliography Gallery		○		○	○	○	○	○							
ArtKitek		●													
Art, Science and Technology	●	●	●	●	●	●	●	●	●						
Ascot Laser Picture Studio				●											
Associates of Science & Techn	○			○											
Atelier Holographique de Paris			●			●									
AT&T Bell Labs-Materials Grou									●						
Beijing Institute Of Posts & Tele									●						
Beijing Normal University								●	●						
Bremer Institute für Angewandt									●						
Brookhaven National Laborator									●						
Burns Holographics Ltd	●	●	●	●	●	●		●	●						
Casdin-Silver Holography	●	○	○	●	●	●		●							
Center For Advanced Visual								●							
Central Electricity Generating									●						
Central Michigan University		●		●											
Cherry Optical Company	●				●	●									

Work done in-house = ●

Broker = ○

COMPANY NAME	Museum/Curator	College Level Instruction	Independ. Educat. Facility	Hands-On Workshop	Holography Teacher Training	Curriculum Development	Educational Materials	Academic Research	Industrial Research	Medical Research	Time Sequence Movie of Holograms	Microscopic Holography	Endoscopic Holography	Plasma Research	Dental Research
Cinema & Photo Research Inst									●						
CISE SpA Technologie Innovat-		●	●	●	●		●	●	●						
Citröen Industrie									●						
College of Manufacturing		●						●							
Columbia University													●		●
David Schmidt Holography				●	●			●	●		●	●			
Deutsche Gesellschaft Für Holo			●												
Duston Holographic Services	●	●		●				●	●						
Dutch Holographic Laboratory								●	●						
École Nationale Superieure									●						
Edmund Scientific Company							●								
Elan Bio-Medical														●	
Environmental Research Institut								●	●				●		
ERBA		●	●												
Eve Ritscher Associates Ltd		●	●												
F.A.S.T. Electronic Bulletin							●								
Focal Image Ltd													●		
Ford Scientific Labs									●						
Free University Of Brussels								●							
Fuji Photo Optical Co., Ltd									●						
Holo 3	●	●	●	●	●	●	●	●	●	●	●	●	●	●	●
Holo Arp			●	●											
Holo GmbH Hologr. Osnabrück			●	●	●		●	●	●						

	HOLOGRAPHY EDUCATION — ARTISTIC HOLOGRAPHY EDUCATION										MEDICAL HOLOGRAPHY — BIOL. SAMPLES, POLYMERS, CONTAMINANTS, TEST./ MEASUREMENT				
COMPANY NAME	Museum/Curator	College Level Instruction	Independ. Educat. Facility	Hands-On Workshop	Holography Teacher Training	Curriculum Development	Educational Materials	Academic Research	Industrial Research	Medical Research	Time Sequence Movie of Holograms	Microscopic Holography	Endoscopic Holography	Plasma Research	Dental Research
Hologramm Werkstatt & Galerie	●		●	●	●	●									
Holographic Applications		○		○		○			○						
Holographics International							●								
Holographics North Inc		●		●			●								
Holographic Studio, Ltd	●		●	●	●										
Holographic Studios				●											
Holography Institute	●	●	●	●	●	●	●	●	●	●	●				
Holography Ltd	●	●													
Holography News						●	●	●	●	●					
Holography Workshop-Goldsmit				●											
Holography Yearbook							●								
Holo-Laser		●		●				●	●	●		●	●		
Holo Laser Tech Ltd		○		○	○	○	○	○	○						
Holomedia AB / Hologram Mus	●														
Holoproduction	●	●	●	●	●	●	●	●	●	●	●	●	●	●	●
Holopublic						●	●		●						
Holo/Source Corporation										●					
Ian M. Lancaster Hologr	●														
Ibero Gestão - Gestão Int		●		●	●			●	●	●					
Illinois Institute of Technology									●						
ILlinois Valley Magnetic Resona								●							
Images Company				●											
Imperial College of Science								●							

Work done in-house = ●

Broker = ○

Work done in-house = ●

Broker = ○

COMPANY NAME	HOLOGRAPHY EDUCATION — ARTISTIC HOLOGRAPHY EDUCATION										MEDICAL HOLOGRAPHY — BIOL. SAMPLES, POLYMERS, CONTAMINANTS, TEST./ MEASUREMENT				
	Museum/Curator	College Level Instruction	Independ. Educat. Facility	Hands-On Workshop	Holography Teacher Training	Curriculum Development	Educational Materials	Academic Research	Industrial Research	Medical Research	Time Sequence Movie of Holograms	Microscopic Holography	Endoscopic Holography	Plasma Research	Dental Research
Ing.-Agentur fur Neue Technol.									●						
Ingenieur Büro Geiger					●										
Institute of Art & Design-Tsukub		●	●		●			●							
Institute of Electronics. BSSR								●							
Institute Nuclear Physics-Lenin.								●							
Institute of Opthalomology										●			●		
Institute of Optical Research								●				●			
Institute Optical Science-ROC										●					
Institute of Physics-Ukrainian A								●							
Interchange Studios				●											
Isast/Leonardo							●								
The Johns Hopkins University								●		●					
Karolinska Institutet										●					●
Kraftwerk Union AG								●							
Labor für Holografie			●	●											
Lake Forest College Holograph		●	●	●											
Laser Affiliates	●	●			●	●									
Laser Focus World							●								
Laser Institute of America			●												
Laser Labs, Inc													●		
Laser Light Expressions Pty Ltd									●						
Laser Lightworks		●													
Lasermetrics, Inc									●						

Work done in-house = ●

Broker = ○

COMPANY NAME	HOLOGRAPHY EDUCATION — ARTISTIC HOLOGRAPHY EDUCATION										MEDICAL HOLOGRAPHY — BIOL. SAMPLES, POLYMERS, CONTAMINANTS, TEST./ MEASUREMENT				
	Museum/Curator	College Level Instruction	Independ. Educat. Facility	Hands-On Workshop	Holography Teacher Training	Curriculum Development	Educational Materials	Academic Research	Industrial Research	Medical Research	Time Sequence Movie of Holograms	Microscopic Holography	Endoscopic Holography	Plasma Research	Dental Research
Lasiris Inc									●						
Lawrence Berkeley Laboratory									●						
LCPC--Lab Central des Ponts									●						
Leningrad Subsidiary in Machin									●						
Light Construction, Inc			●												
L.I.R. E.R.A.									●						
Los Angeles School of Hologr.			●	●	●										
Lulea University of Technology									●						
Lund Institute of Technology		●						●							
LURE Institute D'Optique								●							
MacShane Holography		●	●	●	●	●			●						
Markem Systems Ltd									●						
Massachusetts Institute of Tech		●				●									
McCain Marketing			●			●	●	●							
Medical University South Caroli										●					
Messerschmitt-Boelkow-Blohm									●						
Mitsubishi Heavy Industries Ltd									●						
Moscow Physical Engineering															
Musée de L'holographie	●			●	●		●								
Museum für Holographie & neu	●			●											
Museum of Holography/Chicag	●		●												
Museum of the Fine Arts Resea															
Nancy Gorglione/fine Art Hologr	●	●			●	●									

Work done in-house = ●

Broker = ○

COMPANY NAME	HOLOGRAPHY EDUCATION — ARTISTIC HOLOGRAPHY EDUCATION										MEDICAL HOLOGRAPHY — BIOL. SAMPLES, POLYMERS, CONTAMINANTS, TEST./ MEASUREMENT				
	Museum/Curator	College Level Instruction	Independ. Educat. Facility	Hands-On Workshop	Holography Teacher Training	Curriculum Development	Educational Materials	Academic Research	Industrial Research	Medical Research	Time Sequence Movie of Holograms	Microscopic Holography	Endoscopic Holography	Plasma Research	Dental Research
Nasa Marshall Space Flight									●						
National Physical Laboratory									●						
Newcastle Upon Tyne Polytech.		●	●	●				●	●						
Newport Corporation															
New York Hall Of Science	●														
New York Holographic Labs				●											
New York Institute of Technolog									●						
Norges Tekniske Hogskole									●						
Northern Illinois University									●						
Northwestern University		●						●	●	●					
Novator Research Center										●					
Odhner Holographics		●	●	●	●		●								
Ontario College of Art		●													
Ontario Hydro									●						
Ontario Science Centre				●											
Optische Fenomenon							●								
Oxford University		●		●					●						
Pennsylvania State University									●						
Phoenix Holograms	●				●		●								
Photon League				●											
Physics Institute. Latvian SSR									●						
Qué Sera Sera				●											
Rainbow Symphony Inc.							●								

Work done in-house = ●

Broker = O

COMPANY NAME	HOLOGRAPHY EDUCATION — ARTISTIC HOLOGRAPHY EDUCATION										MEDICAL HOLOGRAPHY — BIOL. SAMPLES, POLYMERS, CONTAMINANTS, TEST./MEASUREMENT				
	Museum/Curator	College Level Instruction	Independ. Educat. Facility	Hands-On Workshop	Holography Teacher Training	Curriculum Development	Educational Materials	Academic Research	Industrial Research	Medical Research	Time Sequence Movie of Holograms	Microscopic Holography	Endoscopic Holography	Plasma Research	Dental Research
Ralcon		●							●						
Ralph Cullen Holographics									●						
Reconnaissance, Ltd	●			●		●	●	●	●						
Richard Bruck Holography		O	O	O	O		O								
Rochester Institute of Technol.		●	●	●				●							
Rolls-Royce PLC--Advanced									●						
Ross Books							●								
Rottenkolber Holo-System									●						
Rowland Institute for Science									●						
Royal Institute of Technology			●						●						
Royal Photographic Society		●	●												
Royal Sussex Hospital										●					
Saab-Scania									●						
Sandia National Laboratories									●						
School-Art Institute of Chicago		●													
School of Holography				●											
School of Holography/Chicago	●		●	●	●	●	●								
Science Kit & Boreal Laborator.							●								
Scientific Council on Exhibitions	●														
S.I. Vavilov State Optics Institut									●						
Society for Photo-Optical Instru							●								
SOI - Society Olografica Italia			●												
Spectral Images										●					

COMPANY NAME	HOLOGRAPHY EDUCATION / ARTISTIC HOLOGRAPHY EDUCATION										MEDICAL HOLOGRAPHY / BIOL. SAMPLES, POLYMERS, CONTAMINANTS, TEST./ MEASUREMENT				
	Museum/Curator	College Level Instruction	Independ. Educat. Facility	Hands-On Workshop	Holography Teacher Training	Curriculum Development	Educational Materials	Academic Research	Industrial Research	Medical Research	Time Sequence Movie of Holograms	Microscopic Holography	Endoscopic Holography	Plasma Research	Dental Research
Spectron Development Laborat.								●							
State Research and Project Inst															
Studio für Holographie				●											
Superior Technology Implement									●						
Sydney College of the Arts		●													
Systems Group of TRW Inc									●						
Synchronicity Holograms			●	●		●		●							
Technical University of Budape										●					
Technical University of Eindhov								●							
Technical University of Wroclaw								●							
TNO Institute-Applied Physics									●						
Tokai University									●						
Tokyo University										●					
Tri-Ess Sciences:Student Scien							●								
Trilone Holographie Corp									●						
Ultrafine									●						
UK Optical Supplies							●	●	●						●
Universidade do Porto									●						
Universita di Roma-La Sapienz									●						
Universitat Erlangen - Nurnberg								●							
Université de Franche-Comte									●						
Universite de Neuchatel									●						
Université de Paris-Sud									●						

Work done in-house = ●

Broker = ○

| | HOLOGRAPHY EDUCATION | | | | | | | | | | MEDICAL HOLOGRAPHY | | | | |
| | ARTISTIC HOLOGRAPHY EDUCATION | | | | | | | | | | BIOL. SAMPLES, POLYMERS, CONTAMINANTS, TEST./ MEASUREMENT | | | | |
COMPANY NAME	Museum/Curator	College Level Instruction	Independ. Educat. Facility	Hands-On Workshop	Holography Teacher Training	Curriculum Development	Educational Materials	Academic Research	Industrial Research	Medical Research	Time Sequence Movie of Holograms	Microscopic Holography	Endoscopic Holography	Plasma Research	Dental Research
Université de Paris-Sud									●						
Université Laval				●											
Université Louis Pasteur								●							
University Essen									●						
University Gent, Workshop Hol				●	●	●	●	●	●			●	●		●
University of Aberdeen									●						
University of Alabama in Huntsv									●						
University of Alicante				●											
University of Arizona									●						
University of Bologna			●												
University of California			●						●						
University of Dayton, Research		●						●							
University of Michigan				●				●							
University of Munich										●					
University of Munster										●					
University of Oxford--Holograph				●				●							
University of Rochester								●							
University of Southern Californi								●							
University of Stratchclyde									●						
University of Stuttgart									●						
University of Tokyo								●		●					
University of Wisconsin		●	●												
University of Zagreb									●						

Work done in-house = ●

Broker = ○

Work done in-house = ●

Broker = ○

COMPANY NAME	HOLOGRAPHY EDUCATION — ARTISTIC HOLOGRAPHY EDUCATION										MEDICAL HOLOGRAPHY — BIOL. SAMPLES, POLYMERS, CONTAMINANTS, TEST./ MEASUREMENT				
	Museum/Curator	College Level Instruction	Independ. Educat. Facility	Hands-On Workshop	Holography Teacher Training	Curriculum Development	Educational Materials	Academic Research	Industrial Research	Medical Research	Time Sequence Movie of Holograms	Microscopic Holography	Endoscopic Holography	Plasma Research	Dental Research
Vincennes University		●		●	●										
Volkswagen AG									●						
Waseda University										●					
Wenyon & Gamble						●									
Whole Hography										●					
Wöber Design Hololab Austria	●														
Worcester Polytechnic Institute									●	●					
WYKO Corporation									●						
York University															
INDIVIDUALS															
Andrews, Matthew		●	●	●	●										
Capucci, Pier Luigi					●		●	●							
Deem, Rebecca					●										
DeFreitas, Frank		●	●	●	●	●	●								
Didier, Leduc				●	●			●							
Dinsmore, Sydney	●														
Harris, Nick				●	●										
Kac, Eduardo				●											
Means, Marcia	●														
Newman, Paul		●		●	●	●	●	●							
Pepper, Andrew		●													
Stockler, Len				●											
Unterseher, Fred	●	●	●	●	●	●	●	●	●	●					

EMBOSSED HOLOGRAPHY

Work done in-house = ●
Broker = ○

COMPANY NAME	ARTISTIC: CREATION OF IMAGE — H-1 Maker	Full Color H-1 Maker	Stereogram H-1 Maker	Model Maker	Artistic Consultant	OWNER HOLOG. EMBOSSING MACHINE — Photoresist Maker	Shim Maker	Holographic Embossing	EQUIPMENT & SUPPLIES — Manufacturer of Holographic Embossing Machine	Manufacturer of Label Applying Machine	Manufacturer of Hot Stamp Machine	Photoresist Maker	Foil Maker	Shim Maker	BINDERY — Application to clothes/toys	Stickers applied	Application to paper products	Hot-Stamping / Foil applied
Advanced Environmental	○	○		○														
AKS Holographie-Galerie	●		●	●		●												
Alpha Foils, Inc								●					●					
Amblehurst Ltd	●	●		●	●	●	●	●										
American Bank Note Hol.	●	●	●	●	●	●	●	●				●			●	●	●	
Angstrom Industries, Inc	●	●	●		●							●						
Applied Holographics Cor	●	●	●		●	●	●	●				●	●	●				
Artigliography Co	○		○	○	○	○		○				○	○	○		○	○	○
Artplay Holographic Studi	○	○	○	○	○	○	○					○		○	○	○	○	○
Art, Science and Technol.	○	○		○	○											○	○	○
Beddis Kenley Engineerin												●						
Bobst Group										●								
Brandtjen & Kluge, Inc											●							
Burns Holographics Ltd	●	●	●		●							●		●		○	○	○
Casdin-Silver Holography	●	●	○	○	●			○										
Chimeric Images, Inc						○		○								○	○	○
Chromagem Inc	○	○	○	○	○	●	●	●			○	○	○	○		○	○	○
CISE SpA Technologie	●					●	●					●		●				
Coburn Corporation						●		●										
Creative Label															●	●	●	●
Crown Roll Leaf, Inc							●	●					●	●				
Dai Nippon Printing Co.						●	●	●					●	●	●	●	●	●
David Schmidt Holograph	●		●	●	●	○	○	○							○	○	○	○
The Diffraction Company								●										

EMBOSSED HOLOGRAPHY

Work done in-house = ●
Broker = ○

COMPANY NAME	H-1 Maker	Full Color H-1 Maker	Stereogram H-1 Maker	Model Maker	Artistic Consultant	Photoresist Maker	Shim Maker	Holographic Embossing	Mfr. of Holographic Embossing Machine	Mfr. of Label Applying Machine	Mfr. of Hot Stamp Machine	Photoresist Maker	Foil Maker	Shim Maker	Application to clothes/toys	Stickers applied	Application to paper products	Hot-Stamping / Foil applied
Diversified Graphics, Ltd		●													●			
Dutch Holographic Lab	●	●	●		●							●			○	○	○	○
E.C. Schultz & Company					●	●												
The Foreign Dimension				○				○					○			○	○	○
George M. Whiley Limited												●						
Global Images, Inc									●									
Gray Scale Studios Ltd				●	●													
Holaxis Corporation	●	●		●		●	●	●				●		●		○	○	○
Holocor I.B.F. Printing Inc				●				●						●				
Hologramas De Mexic																●	●	●
Hologram. Industries	●	●	●			●	●	●										
Holographic Applications		○		●	●		○	○				○		○		○	○	○
Holographic Design Inc.	○	○		○	○							○	○	○	○	○	○	○
Holography Institute	●	●	●	●	●							○	○	○		○	○	○
Holo-Laser	●	●		●		●	●	●				●		●				
Holo Laser Tech Ltd		○	○	○								●		●				○
Holomedia Inc	●			●	●													
Holo/Source Corporation	●	●	●			●	●	●				●	●	●				
Ian M. Lancaster Hologr.			○		○													
Ibero Gestão - Gestão Int			○													○	○	○
Images Company	●			●	●	○	○	○										
Imaging & Design	●	●	●	●	●							○	○	○		●	●	●
Ingenieur Büro Geiger	●	●	●	●			●	●						●				
Jaeger Graphic Technol.													●				●	●

EMBOSSED HOLOGRAPHY

Work done in-house = ●

Broker = ○

COMPANY NAME	ARTISTIC: CREATION OF IMAGE					OWNER HOLOG. EMBOSSING MACHINE			EQUIPMENT & SUPPLIES FOR EMBOSSED HOLOGRAPHY						BINDERY: APPLICATION IN HOUSE			
	H-1 Maker	Full Color H-1 Maker	Stereogram H-1 Maker	Model Maker	Artistic Consultant	Photoresist Maker	Shim Maker	Holographic Embossing	Manufacturer of Holographic Embossing Machine	Manufacturer of Label Applying Machine	Manufacturer of Hot Stamp Machine	Photoresist Maker	Foil Maker	Shim Maker	Application to clothes/toys	Stickers applied	Application to paper products	Hot-Stamping / Foil applied
James River Products									●									
Jayco Holographics							●	●	●									
Kolbe-Druck,Coloco Gmb							●	●							●	●	●	●
Laser Light Expressions	●	●	●	●	●	●	●	●					○				●	●
Lasiris Inc	●					●	●	●				●		●		●	●	●
Letterhead Press Inc																	●	●
Light Fantastic Plc	●	●	●		●	●	●	●				●		●		●	●	●
Light Impressions, Inc	●	●		●	●	●	●	●						●				
Loughborough University	●							●						●				
Markem Systems Ltd	●		●			●	●	●			●							
Metaplast Electrochemic.							●							●				
Polaroid Corporation	○		○	○	○											○	○	
Portson, Inc	●		●	●		●		●				●	●			○	○	○
Ralph Cullen Holographic												○	○	○				
Regal Press Inc																	●	●
Reynolds Metals Co																	●	●
Robert Sherwood Hologr.	○	○		○	○							○	○	○	○	○	○	○
Sillcocks Plastics Internat			○	○	○							○		○			●	●
Smith & McKay Printing																	●	●
Starlight Holographic Inc			○	○	○		○	○								○	○	○
Studio für Holographie	●					●												
Superior Technology Impl	●					●						●		●				
Textile Graphics, Inc	○	○			●										○			
Toppan Printing Co. Ltd	●	●				●	●	●									●	●

EMBOSSED HOLOGRAPHY

Work done in-house = ●
Broker = ○

COMPANY NAME	ARTISTIC: CREATION OF IMAGE					OWNER HOLOG. EMBOSSING MACHINE			EQUIPMENT & SUPPLIES FOR EMBOSSED HOLOGRAPHY						BINDERY: APPLICATION IN HOUSE			
	H-1 Maker	Full Color H-1 Maker	Stereogram H-1 Maker	Model Maker	Artistic Consultant	Photoresist Maker	Shim Maker	Holographic Embossing	Manufacturer of Holographic Embossing Machine	Manufacturer of Label Applying Machine	Manufacturer of Hot Stamp Machine	Photoresist Maker	Foil Maker	Shim Maker	Application to clothes/toys	Stickers applied	Application to paper products	Hot-Stamping / Foil applied
Transfer Print Foils Inc													●					●
UK Optical Supplies												○	○	○				
White Light Works, Inc	●											●						
Wöber Design Hololab	●																	
Wonderlight Gallery	○	○		○	○											○	○	○
X-IAL			●															
INDIVIDUALS																		
Allen, Jeffrey			○	○	○							○	○	○	○	○	○	
Capucci, Pier Luigi					●													
Unterseher, Fred	●	●		●	●													

EQUIPMENT & SUPPLIES

Manufacturer = ●
Distributor = ⇨
Both = ⊃

COMPANY NAME	LASER EQUIPMENT							LAB EQUIPMENT					OPTICS & SET UP				SERVICE / OTHER			
	Rods	Tubes	Diodes	Collimators	Etalons / Etalon Filters	Phase Conjugators	Cleaning Equipment	Chemicals	Film / Recording Material	Lamps / Isolation Tables	Plates / Emulsions	Safety Equipment	Spatial Filters	Beamsplitters / steerers	Mirrors / Lenses	Wavelength Shifters	Holography Kits	Used Laser Sales	Laser Repair	Free Catalogue
Aerotech Inc.		●		●																
AG Electro-Optics Ltd									●											
Agfa Corporation									●		●									
Agfa Gevaert N.V.									●		●									
American Holographic													●							
Alpha Photo Products								⇨	⇨		⇨	⇨								●
Apollo Lasers, Inc	⊃	⊃		⊃	⊃	⊃	⊃		⊃	⊃	⊃		⊃	⊃	⊃	⊃	⊃			●
Beddis Kenley Engine									⇨											
Brighton Imagecr.									●								⊃			
Burns Holographics																	●			
CISE SpA Technolog.																	●			
C.Itoh & Company									●		●									
City Chemical								●			●									
Corion Corp.														●	●					●
Coulter Optical Co.															●					●
CSI															●					
CVI Laser Corp.				●										●	●		⇨			●
Die Dritte Dimension		⇨													⇨					
Ealing Electro-Optics				●									●	●	●					●
Eastman Kodak Co.									●		●						⇨			
Edmund Scientific Co.																				
E.I. Dupont Co.									●		●				●					●
Electro Optical Indust.				●																
Electro Optics Devel.														●	●					
Excitek Inc.		⊃																●	●	

EQUIPMENT & SUPPLIES

Manufacturer = ●
Distributor = ⇨
Both = ↻

COMPANY NAME	LASER EQUIPMENT							LAB EQUIPMENT					OPTICS & SET UP				SERVICE / OTHER			
	Rods	Tubes	Diodes	Collimators	Etalons / Etalon Filters	Phase Conjugators	Cleaning Equipment	Chemicals	Film / Recording Material	Lamps / Isolation Tables	Plates / Emulsions	Safety Equipment	Spatial Filters	Beamsplitters / steerers	Mirrors / Lenses	Wavelength Shifters	Holography Kits	Used Laser Sales	Laser Repair	Free Catalogue
Fisher Scientific EMD			⇨	⇨										⇨	⇨		⇨			●
Fresnel Technologies															●					
Fuji Photo Optical Co.														●	●					
Galvoptics Ltd.															●					
General Imaging Corp									●		●									
G.M. Vacuum Coating											●									
Halo Power-Track										↻										
Holofax Limited								●		●										
Holo Laser Tech Ltd									⇨		⇨		↻	↻						●
Holomex Ltd.																	●			
Holoproduction	⇨	⇨		⇨	⇨	⇨	⇨		⇨	⇨	⇨	⇨	⇨	⇨	⇨		⇨			
Holo-Spectra									⇨				⇨		⇨			●	●	
Howard Smith Precis.															●					
ICI Americas															●					
Ilford Limited								●	●		●									
Images Company		⇨						↻	⇨	↻		●	↻	⇨	↻		↻			
Imaging & Design								⇨	⇨	⇨	⇨	⇨								
Institute of Electronics	↻				↻	↻	↻	↻						↻	↻					
Integraf									⇨		⇨									
Jodon Inc		●		●			●	●	●	●	●		●	●	●					●
Kendall Hyde Ltd														●	●					
Keystone Scientific								⇨	⇨		⇨						⇨			●
Lambda/Ten Optics															●					
Laser Applications Inc	●			●	●		●						●	●	●					
Laser Electronics Pty.	⇨	↻	⇨	↻			↻			⇨		⇨	↻	↻	↻		↻			

EQUIPMENT & SUPPLIES

Manufacturer = ●
Distributor = ⇨
Both = ⊃

COMPANY NAME	LASER EQUIPMENT							LAB EQUIPMENT					OPTICS & SET UP				SERVICE / OTHER			
	Rods	Tubes	Diodes	Collimators	Etalons / Etalon Filters	Phase Conjugators	Cleaning Equipment	Chemicals	Film / Recording Material	Lamps / Isolation Tables	Plates / Emulsions	Safety Equipment	Spatial Filters	Beamsplitters / steerers	Mirrors / Lenses	Wavelength Shifters	Holography Kits	Used Laser Sales	Laser Repair	Free Catalogue
Laser Fare Ltd.									●		●									
Laser Ionics Inc		●			●															
Laser Resale Inc																		●	●	
Laser Science LSI	●			●																
LiCONix		●																		
Litton Systems Can.															●					
MacShane Hologr.																	⇨			
Melles Griot		●	●	●						●			●		●					●
Meredith Instruments					●														●	
Metrologic Instrument.															●					
Mitutoyo Measuring																				●
Newport Corporation			⇨	⊃	⇨				⇨	⊃	⇨		⊃	⇨	⊃		⊃			
Norland Products, Inc								⊃												
Optical Images				⊃					⊃		⊃			⊃	⊃	⊃			●	
Optics Plus Inc															●					
Optilas B.V.	⇨	⊃	⇨	⊃	⊃	⊃	⊃	⇨	⊃	⊃	⊃	⊃	⊃	⊃	⊃	⇨	⇨		●	●
Optimation Inc.													⊃							
Phase-R Corporation		●		●	●															
Photographer's Form								●	●		●									
Plasma Technology																				●
Polaroid Corporation									●											
Portson, Inc.									●		●									
Ralcon														●						
Ralph Cullen Hologr.	⇨	⊃	⇨	⊃	⊃	⊃	⊃	⊃	⊃	⊃	⊃	⊃	⊃	⊃	⊃	⇨	⊃		●	●
Rochester Photonics				⊃											⊃					

EQUIPMENT & SUPPLIES

Manufacturer = ●
Distributor = ⇨
Both = ↺

COMPANY NAME	LASER EQUIPMENT							LAB EQUIPMENT					OPTICS & SET UP				SERVICE / OTHER			
	Rods	Tubes	Diodes	Collimators	Etalons / Etalon Filters	Phase Conjugators	Cleaning Equipment	Chemicals	Film / Recording Material	Lamps / Isolation Tables	Plates / Emulsions	Safety Equipment	Spatial Filters	Beamsplitters / steerers	Mirrors / Lenses	Wavelength Shifters	Holography Kits	Used Laser Sales	Laser Repair	Free Catalogue
Science & Mechanics																				●
Shipley Chemical Co								↺	↺		↺									
Spectra-Physics Inc		↺		↺			↺						↺	↺	●				●	●
Spectrogon				●									●							
Spectrolab Ltd	↺	↺	↺	↺	↺	↺	↺	↺	↺	↺	↺	↺	↺	↺	↺	↺	↺			●
Steinbichler Optotech.	⇨	⇨	⇨	↺	⇨				⇨	⇨	⇨		●	●	↺		●			
Tri-Ess Sciences																	●			●
UK Optical Supplies	⇨	⇨	⇨	↺	↺	⇨	↺	⇨	↺	↺	↺	↺	↺	↺	↺	⇨	⇨			●
Vinten Electro Optics															↺					
Wentworth Laborator.										●										
Wise Instruments				●													●			
Wonders of Hologr.		⇨															⇨			
Wotan Lamps Ltd.										●										
X-IAL											●									
3M--Optics Technol.																	●			

HOLOGRAPHIC NON-DESTRUCTIVE TESTING

COMPANY NAME	Continuous Wave	Pulsed	Real Time	Time Average	Double Exposure	Flow Visualization	Vibration Visualization	Impact Pulse Probing	Load/Thermal Strain	Pressure Cycling / Effects	Particle Density	Turbulent Velocity	Particle Size Distribution	Thermoplastic Material	Videocamera / tape	DCG	Photopolymer ("monobath")	Silver Halide	Videocamera/tape convertible to personal computer	Free Catalogue	Self Contained Holography Unit
Abbott Laborator		●	●						●												
Aerospatiale		●		●		●						●		●							
Amer Bank Note	●	●	●	●	●															●	
ArtKitek	●		●				●		●						●			●			
Artplay Hologr.	●			●			●		●									●			
Beijing Normal U			●			●					●							●			
Bremer Institute				●								●	●					●			
British Aerospac	●	●	●	●	●	●	●		●						●	●		●			
Central Electric		●	●						●									●			
Chernovtsy Univ	●	●	●	●	●	●					●	●	●		●			●	●		
Chiba University	●			●																	
CISE SpA Tech	●	●	●	●	●	●	●		●	●	●	●	●		●	●		●	●	●	●
Citröen Industrie	●		●					●	●						●						
College of Manu		●	●																		
Daimler Benz	●		●					●	●												
Dutch Holog Lab	●		●	●	●		●		●									●			
École NationSup	●			●								●		●				●			
Eurolaunch Ltd	●					●															
Ford Scien Labs	●					●				●		●									
HOLO 3	●	●	●	●	●	●	●	●	●	●	●	●	●	●	●		●	●	●	●	
Holo-Laser	●	●	●	●	●		●								●			●			
Holo Laser Tech	●		●	●	●									●						●	
Holometric AB	●			●																	
Holoproduction	●	●	●	●	●	●	●	●	●	●	●	●	●	●	●			●	●	●	

HOLOGRAPHIC NON-DESTRUCTIVE TESTING

COMPANY NAME	TYPE OF LASER AVAIL.		SENSING METHOD			EQUIPMENT & PARTICLE TESTING DONE IN HOUSE								RECORDING / ANALYSIS ON:						PRODUCTS & SERVICES	
	Continuous Wave	Pulsed	Real Time	Time Average	Double Exposure	Flow Visualization	Vibration Visualization	Impact Pulse Probing	Load/Thermal Strain	Pressure Cycling / Effects	Particle Density	Turbulent Velocity	Particle Size Distribution	Thermoplastic Material	Videocamera / tape	DCG	Photopolymer ("monobath")	Silver Halide	Videocamera/tape convertible to personal computer	Free Catalogue	Self Contained Holography Unit
Ibero Gestão	●		●	●	●		●		●					●	●			●			
Illinois Institute		●	●						●												
Imperial College		●									●		●		●						
Ing.-Agentur fur		●				●			●												
Institute of Elect		●		●	●	●	●	●	●	●											●
Jodon Inc.	●		●	●	●		●								●			●			
Johns Hopkins		●	●								●		●	●	●						
Keystone Sci.	●		●	●	●													●		●	●
Kraftwerk Union	●											●						●			
Labor Dr. Steinb	●	●	●	●	●	●	●	●	●	●	●	●	●	●	●			●	●		●
Lasiris Inc.	●		●	●	●		●		●	●								●			
Litton Systems		●								●											
Lulea University		●	●																		
Lumonics Inc.																					●
MacShane Holo	●		●		●				●									●			
Micraudel	●	●	●	●	●	●	●	●	●	●				●	●						
Mitsubishi Ltd		●							●												
Nat PhysicalLab		●	●																		
Newport Corp.	●	●	●	●	●	●	●		●	●				●	●	●	●	●	●	●	●
New York Inst		●		●																	
Norges Teknisk	●		●																		
Northwestern U		●	●																		
Odhner Hologr	●		●	●	●		●		●	●											
Optical Images	●	●	●	●	●		●		●	●	●	●	●				●	●	●		

HOLOGRAPHIC NON-DESTRUCTIVE TESTING

COMPANY NAME	Continuous Wave	Pulsed	Real Time	Time Average	Double Exposure	Flow Visualization	Vibration Visualization	Impact Pulse Probing	Load/Thermal Strain	Pressure Cycling / Effects	Particle Density	Turbulent Velocity	Particle Size Distribution	Thermoplastic Material	Videocamera / tape	DCG	Photopolymer ("monobath")	Silver Halide	Videocamera/tape convertible to personal computer	Free Catalogue	Self Contained Holography Unit
Optilas B.V.	●	●	●	●	●																
Portson, Inc	●				●											●		●		●	
Ralcon	●	●	●						●									●			
R Cullen Hologr	●	●	●	●	●		●		●								●	●		●	●
Spectrolab Ltd	●	●	●	●	●		●		●												●
Spectron Devel		●		●	●									●							
Steinbichler Opt	●	●	●	●	●	●	●	●	●	●	●	●	●	●	●			●	●		●
Stoltz AG		●		●										●							
Swiss Federal In		●													●						
Systems Group		●									●		●								
Ultrafine		●																		●	
UK Optical Supp	●	●	●	●	●		●		●									●		●	●
UTRC		●	●	●	●	●	●		●	●					●			●			●
University Gent	●	●	●	●	●		●	●	●	●				●	●			●	●	●	●

OPTICS & HOLOGRAPHIC OPTICAL ELEMENTS

Work done in-house = ●

Broker = ○

COMPANY NAME	H.O.E.S MADE IN HOUSE				COMPONENTS MADE IN HOUSE								SERVICES / OTHER				
	Mirrors	Gratings	Prisms / Lenses	Head-Up Displays	Collimators	Beamsplitters	Filters	Head-Up Displays	Molds	Polarizers	Prisms / Lenses	Windows	Polishing	Cutting / Grinding	Coating	Optics Testing	Software for Designing Optical Systems
American Holographic		●														●	
APA Optics, Inc			●	●				●								●	
Architectural Glass & Hologr.												●					
Ashland Electric Products										●	●						
Burns Holographics Ltd		●			●												●
Cambridge Stereographics	○		○		○		○										
Grp	●		●														
Chromagem Inc.	●	●	●	●	●		●									●	
CISE SpA Technologie Innov						●				●							
Continental Optical	●		●			●	●				●	●			●		
Corion Corp.										●							
Coulter Optical Company						●				●							
CSI	●		●		●	●	●			●		●	●	●	●	●	
CVI Laser Corporation						●											
Datasights Ltd		●	●		●												
David Schmidt Holography						●											
Davin Optical Ltd	○		○														
Die Dritte Dimension			●														
Duston Holographic Services		○	●	○	○												
Dutch Holographic Laboratory						●				●							
Ealing Electro-Optics	●		●	●			●										
E.I. DuPont De Nemours & Co					●												
Electro Optical Industries Inc.							●									●	
Free University Of Brussels											●						
Fresnel Technologies Inc					●						●						

OPTICS & HOLOGRAPHIC OPTICAL ELEMENTS

Work done in-house = ●

Broker = ○

COMPANY NAME	H.O.E.S MADE IN HOUSE				COMPONENTS MADE IN HOUSE								SERVICES / OTHER				
	Mirrors	Gratings	Prisms / Lenses	Head-Up Displays	Collimators	Beamsplitters	Filters	Head-Up Displays	Molds	Polarizers	Prisms / Lenses	Windows	Polishing	Cutting / Grinding	Coating	Optics Testing	Software for Designing Optical Systems
General Imaging Corp.	●		●	●	●		●										
Holographic Applications							○										○
Holographic Design Incorporat	●	●	●	●		●	●										
Holographic Products Inc.	●	●	●	●			●										
Holo-Laser		●	●										●	●	●		
Holo Laser Tech			○		○					○							
Holo-Or Ltd		●			●		○	○								●	○
Holos Gallery					○												
Holotek Ltd	●		●														
Holtronic	●		●						●								
Howard Smith Precision Optics	●		●														
IBM Almaden Research Center		●			●												
ICI Americas		●	●														
Images Company	●		●			●	●		○	●							
Institute of Electronics, BSSR	●		●		●	●				●	●	●				●	
Institute of Optical Research			●												●		
Institute Of Optical Science	●						●										
Kaiser Optical Systems, Inc			●	●						●							
Kendall Hyde Ltd	●		●	●						●	●					●	
Lambda/Ten Optics	●				●		●									●	
Laser Applications	●		●		●	●	●			●		●					
Laser Electronics Pty., Ltd			○		○		○				○	○	●	●	●		
Laser Fare Ltd	●	●	●	●	●		●			●							
Lasiris Inc.	●	●	●				●										
L.I.R. E.R.A.	●																

OPTICS & HOLOGRAPHIC OPTICAL ELEMENTS

Work done in-house = ●
Broker = ○

COMPANY NAME	H.O.E.S MADE IN HOUSE — Mirrors	Gratings	Prisms / Lenses	Head-Up Displays	COMPONENTS MADE IN HOUSE — Collimators	Beamsplitters	Filters	Head-Up Displays	Molds	Polarizers	Prisms / Lenses	Windows	SERVICES / OTHER — Polishing	Cutting / Grinding	Coating	Optics Testing	Software for Designing Optical Systems
Markem Systems Ltd.			●		●												
Melles Griot	●		●		●	●	●			●		●	●	●	●	●	
Meredith Instruments	●		●		●		●										
Mitutoyo Measuring Instrument																●	
Odhner Holographics		○															○
Optical Images			●														
Optics Plus Inc			●	●	●	●								●	●	●	
Optilas B.V.	○	○	○		○		○			○	○		○	○			
Optimation Inc.							●										
Pilkington P.E. Ltd	●		●		●												
Portson, Inc		●															
Ralcon	●	●	●	●	●	●	○							●			
Ralph Cullen Holographics	●	●	●	●	●	●				●	●	●	○	○	○	○	○
Rochester Photonics Corp.		●	●	○	●		●								●	●	
Spectra-Physics Inc.	○																
Spectra-Phys Laser Products	●		●		●	●				●		●	●	●	●		
Spectrogon	●	●	●			●	●					●				●	
Spectrolab Ltd	●	●	●		●	●	●		●	●	●	●	●		●	●	●
Superior Technology Implemen		●								●							
men																●	
T.A.I. Incorporated																●	
Towne Laboratories, Inc	●	●	●	●	●	●	○	○	○	●	●	●	○	○	○	○	○
UK Optical Supplies	●	●	●		●	●							●	●			
University Gent											●						
Wise Instruments		●	●	●	●	●				●		●				●	○

LASERS

LASERS MANUFACTURED / DISTRIBUTED AT THIS SITE

Pulsed = ★
Continuous Wave = ☆
Both Pulsed and CW = ⊙

COMPANY NAME	Argon	Krypton	Helium Cadmium	Helium Neon	Ruby	Krypton Chloride	Krypton Fluoride	Xenon	Xenon Chloride	Carbon Monoxide	Carbon Dioxide	Nitrogen	Nitrogen Fluoride	Hydrogen
Aerotech Inc				☆							⊙			
AG Electro-Optics Ltd				☆										
Apollo Lasers, Inc					★						★			
Coherent	☆	☆												
Die Dritte Dimension	☆			☆										
Ealing Scientific Ltd	☆			☆										
Fisher Scientific E.M.D. Divisio	⊙		☆	☆										
Fuji Electric Co. Ltd											⊙			
Holoproduction	☆	☆		☆	☆									
Hughes Aircraft Co.-Laser Prod											⊙			
Images Company				☆										
Institute of Electronics, BSSR					⊙									
Ion Laser Technology Inc	☆													
Jodon Inc	☆	☆		☆										
Laser Applications, Inc					⊙									
Laser Electronics Pty., Ltd	⊙	☆	☆	☆			☆		☆	⊙	⊙	☆		
Laser Ionics Inc	☆	☆												
LiCONix	⊙	⊙	☆											
Lumonics					★	★	★	★	★		★			
Melles Griot				☆							⊙			
Meredith Instruments	☆			☆										
Newport Corporation	☆	☆	☆											
Optilas B.V.	⊙	⊙	⊙	☆		★	★		★	⊙	⊙		⊙	
Ralph Cullen Holographics	☆	☆	☆	☆	☆						☆			
Siemens Ltd	☆			☆										
Spectra-Physics Inc	⊙	⊙		⊙			★							
Spectrolab Ltd.	⊙	⊙	⊙	⊙	⊙	⊙	⊙	⊙	⊙	⊙	⊙	⊙	⊙	⊙

Pulsed = ★
Continuous Wave = ☆
Both Pulsed and CW = ◑

COMPANY NAME	Argon	Krypton	Helium Cadmium	Helium Neon	Ruby	Krypton Chloride	Krypton Fluoride	Xenon	Xenon Chloride	Carbon Monoxide	Carbon Dioxide	Nitrogen	Nitrogen Fluoride	Hydrogen
LASERS — LASERS MANUFACTURED / DISTRIBUTED AT THIS SITE														
UK Optical Supplies	☆	☆	☆	☆	☆						☆			
Uniphase Vetreibs-GmbH	☆			☆										
Wonders of Holography Gallery	☆	☆	☆	★										

NAMES AND ADDRESSES

BUSINESSES

A

ABBOTT LABORATORIES. Department 93F, (Building AP-9), Routes 43 and 137, Abbott Park, IL 60064 USA. Telephone : (312) 9374117. Contact: Dr. Gerald Cohn. Company description: Holographic NonDestructive testing.

ACME HOLOGRAPHY. Established in 1988. 2 Employees at this address. 12 Sunset Road, West Somerville, MA, Boston Area. USA. Telephone: (617) 623 0578. Contact: Betsy Connors. President, Betsy Connors; Vice-President, David Chen. Company description: Acme Holography is Boston's first private holography lab. We offer full service in reflection, transmission and computer generated holography, including design consultation and H1 mastering for Polaroid Mirage photopolymer holograms.

ACTIVE IMAGE. Established in 1989. 1 Employee at this address. P.O. Box 97, Boulder Creek, CA 95006 USA. Telephone: (408) 338 2405. FAX: (408) 338 9833. Contact: Nelson Poe. Subsidiary of Steve Provence Holography. Company description: Exclusive wholesale distributor of packaged embossed holograms by Steve Provence Holography.

ADEL ROOTSTEIN, INC. 205 West 19th Street, New York, NY 10011 USA. Telephone: (212) 645 0202. Company description: Manufacture mannequins for displays and/or holography models.

ADVANCED ENVIRONMENTAL RESEARCH. Route 1, Box 1830, Woolwich, ME 04579 USA. Telephone; (207) 443 6587. Contact: E.King, R. Ian, S. Weber. Company description: Embosed; Silver halide hologram makers.

ADVANCED HOLOGRAPHICS CORP. Established in 1987. 16 Employees at this address. 2469 East Fort Union Blvd. Suite 108, Salt Lake City, UT 84121 USA. Telephone: (801) 943 1809. FAX: (801) 942 7006. Contact: Gary Mangum, Vice-President. President, Jay Greenan; Vice-President, Gary Mangum; Sales Manager, Mark Cornelius; Customer Service, Keith Gulbransen. Branch office: Logan, UT USA. Company description: Advanced Holographics specializes in the mass production of quality dichromate holograms on both glass and continuous, roll-fed plastic. We service the novelty, security, packaging, and apparel industries.

ADVANCED HOLOGRAPHICS, LTD. 243 Lower Mortlake Rd.,Unit 11, Richmond, Surrey TW9 2LL, England, United Kingdom. Contact: John Andrews/ Don Tomkins. Company description: Artistic holography; presentation support

ADVANCE PHOTONICS, A-147 Ghatkopar Industrial Estate, Ghatkopar, Bombay 400 086, India. Telephone: (91)22 582 204. FAX: (91)22 202 4202. Branch office of Newport Corporation, Fountain Valley, CA USA.

AD 2000/ ANOTHER DIMENSION. 946 State Street, New Haven, CT 06511 USA. Telephone: (800) 334 4633. Contact: Jeffrey Levine. Company description: Gallery; wholesale.

AEROSPATIALE-DIVISION HELICOPTERES, 2 avenue Marchel-Cachin, F-93126 La Courneuve Cedex, France. Telephone: (33)(48) 389 178. Contact: M. Guignard. Description: Scientific research; industrial research; holographic non-destructive testing; interferometry.

AEROSPATIALE. Ets D'Aquitaine, Saint-Medard-en-Jalles, F-33165 Bordeaux, France. Telephone: (33) (56) 058405. Contact: H.C. LeFloch. Description: Scientific & industrial research; holographic nondestructive testing; interferometry.

AEROTECH INC., (MAIN HEADQUARTERS), 130 Employees at this address, Electro Optical Division, 101 Zeta Drive, Pittsburgh, PA 15238 USA, Telephone: (412) 963-7470, FAX: (412) 963-7459, Contact: Steve A. Botos, Marketing Manager. President, Stephen A. Botos; Vice-President, David Kincel; Sales Manager, Frank Armstrong; Customer Service, Wes Taylor. Subsidiary Companies: Aerotech UdEngland, Aerotech GMBH-Germany, Aerotech Australia- Australia.

Company description: Aerotech manufactures helium neon tubes, power supplies and complete systems for OEM and end users. Other product lines include optical table positioners and precision rotary and linear positioning systems.

AEROTECH NORTHEAST, Executive Suite 120, 270 Farmington Avenue, Farmington, CT 06032 USA Telephone: (203) 673-3330 Sales Office.

AEROTECH CENTRAL-WEST, 26791 Lake Vue Drive #8, Perrysburg, OH 43551 USA.Telephone: (419)874-3990. Sales Office.

AEROTECH WEST, Suite 217, 7002 Moody Street, La Palma, CA 90623 USA, Telephone: (213) 860-7470. Sales Office.

AEROTECH MIDWEST, P.O. Box 625, Dundee, IL 60118 USA, Telephone: (312) 428-5440. Sales Office.

AEROTECH SOUTH ATLANTIC, 8804 Lomas Court, Raleigh, NC 27615 USA, Telephone: (919) 848-1965. Sales Office.

AEROTECH CENTRAL-EAST, 856 Cottonwood Drive, Monroeville, PA 15146 USA, Telephone: (412) 373-4160. Sales Office.

AEROTECH MID-ATLANTIC, 521 Kingwood Road, King of Prussia, PA 19406 USA, Telephone: (215) 265-6446. Sales Office.

AEROTECH NORTHWEST, 444 Castro Street, Suite 32, Mountain View, CA 94041 USA, Telephone: (415) 967-4996. Sales Office.

AEROTECH SOUTHWEST, 6001 Village Glen Drive, #1301, Dallas, TX 75206 USA, Telephone : (214) 987-4556. Sales Office.

AG ELECTRO-OPTICS LTD. 29 Forest Road, Tarporley, Cheshire CW6 OHX, England, United Kingdom. Telephone: (44)(8293) 3305. Description: Manufacture lasers, optics, lab equipment.

AGFA CORPORATION. (Branch office). 100 Challenger Road, Ridgefield Park, NJ 07660 USA. Telephone: (201) 440 2500 ext. 4226. FAX: (201) 440-1512. Contact: Mark Redzikowski, Product Manager. President & CEO: Halge H. Wehmeier. Parent Compnay: Agfa Gevaert Ltd., England, United Kingdom. Main Headquarters: AGFA GEVAERT, Mortsel, Belgium. Branch offices: Atlanta GA USA; Des Plaines, IL USA; Burbank, CA USA; Brisbane, CA USA; Irving, TX USA.

Company description: Manufacturer of film, plates, emulsions and recording material

AGFA-GEVAERT LTD. (Subsidiary of AGFA GEVAERT N.V.). 27 Great West Road, Brentford, Middlesex, TW8 9AX England, United Kingdom. Main Headquarters: AGFA GEVAERT N.v., Mortsel, Belgium. Company description: Manufacturer of film, plates, emulsions and recording material

AGFA GEVAERT N.V. (MAIN HEADQUARTERS). Established 1894. 8000 Employees at this address. Holography Film Dept, Septestraat 27, B2510, Mortsel, Belgium. Telephone: (03) 444 2111. FAX: (03) 444 7094. Contact Person: R. De Winne, Product Manager-Holography. President, E. De Wolf; Sales Manager, H. Deschaumes; Customer Service: R. De Winne. Branch offices: Agfa-Gevaert LTD, England, United Kingdom. Subsidiary Company: Agfa Corporation, USA.

Company description: Manufacturer of film, plates, emulsions and recording material

AG PRISMATIC, Bimbolegge & Bimbogioca SRL, Via Borfuro 12, 1-24100 Bergamo, Italy. Telephone: (39)

(35) 213 015. Description: Artistic holography.

A.H. PRISMATIC INC. (SEE OUR ADVERTISEMENT ON PAGE 3) Established 1986.6 Employees 31 this address. 285 West Broadway, New York, NY 10013 USA Telephone: (212) 2190440. FAX: (212) 219 0443. Contact: Brian Szpakowski, Sales Manager.

Company description: Distributors of exclusive ranges of holographic gifts, toys and jewelry, and film holograms. Point of purchase displays available for all items. Hologram store in Museum of Holography, New York.

A.H. PRISMATIC, LTD. (SEE OUR ADVERTISE-MENT ON PAGE 3) Established 1982. 15 Employees at this address. New England House, New Engand Street, Brighton, BN1 4GH England, United Kingdom. Telephone: (44) 0273 686 966. FAX: (44) 273 676 692. Contact: Ian Dayus, Sales Manager. President, Barc Thompson; Sales Manager, Ian Days; Customer Service, Sheila Bagley. Company description: Manufacturers and distributors of holographic gifts, toys, and jewellery, together with exclusive range of film holograms. Point of purchase displays compliment all products

IMS OPTRONICS SNNV, Rue Ferd Kinnenstraat 30, B-1950 Kraainem, Belgium. Telephone: (32)(027) 310 440. FAX: (32)(027) 318 918. Branch office of Newport Corporation, Fountain Valley CA, USA.

AITES LIGHTWORKS. 2148 North 86th Street, Seatle, WA 98103 USA Telephone: (206) 526 5752. Contact: Edward Aites. Description: Artistic holographer; Buying & Selling holograms.

AKS HOLOGRAPHIE-GALERIE GmbH. Established in 1985. 3 Employees at this address. Potsdamer StraBe 10, 4300 Essen 1, Federal Republic of Germany. Telephone: (49)(0201) 704562. Contact person: Gudrun Sott, Administrative Director. President, Detlev Abendroth; Vice-President, Peter Kremer; Sales Manager, Gudrun Sott. Company description: Embossed hologram manufacturer, artistic hologram maker; Buying & selling holograms

ALCAN INTERNATIONAL LTD. Kingston Research & Development, P.O.Box 8400, Kingston, Ontario K7L 4Z4, Canada. Telephone: (613) 541 2245. Contact: Christine Gallerneault. Company description: Medical holography

ALPHA FOILS, INC. P.O. Box 152, Bernardsville, NJ 07924 USA Telephone: (201) 766 1500. Contact: Henry Ruschman. Company description: Manufacturers of embossed holography foil; company embosses foil and has other embossing services - please call for more information.

ALPHA PHOTO PRODUCTS, INC. Established in 1947. 50 Employees at this address. 985 Third Street, P.O. Box 23955, Oakland, CA 94623 USA Telephone: (415) 893 1436. Contact: Karl Mills, Marketing Manager. President, Alan W. Wuthnow; Customer Service, Spencer J. Umeki. Branch office: Elmwood Camera Shop, Berkeley, CA USA Company description: Since 1947, Alpha Photo has been privileged to serve the photo/graphic needs of a diverse and growing community Northern California, Western Nevada and points beyond.

AMAZING WORLD OF HOLOGRAMS. Established in 1984. 6 Employees at this address. Corrigan's Arcade, Foreshore Road, South Bay, Scarborough, North Yorkshire Y011 1 PB, England, United Kingdom. Telephone: (44)(0723) 354 090. FAX: (44) (0482) 492 286 ref holo. Contact: Carl Racey. Sales Manager, Carl Racey. Parent company: Laser Light Image, Hull, England. Company description: Exhibitors and retailers of film, glass, embossed, dichromate and related products. Permanent display of 200 holograms updated and changed regularly. Main season May-October. Distributors of film & glass.

AMAZON. c/o Ruggero Maggi, C. So Sempione 67, 20149-Milano, Italy. Telephone: (39)(02) 349 1947. Company description: Fine art holograms, Stereogram maker; Gallery sells fine art holograms.

AMBLEHURST LTD., Established in 1978. 40 Employees at this address. 52 Invincible Road, Farnborough, Hants GU14 7QU, England, United Kingdom. Telephone: (44) 0252 520052. FAX: (44) 0252 373871. Contact: Mr. Philip M.G. Hudson, Director. Director, PMG Hudson; Sales Manager, RE Kauffman; Customer Service, S. Doolan. Company description: Laminates, hot stamp foil and pressure sensitive labelling manufactured in-house in integrated plant.

AMERICAN BANK NOTE HOLOGRAPHICS, INC. (SEE OUR ADVERTISEMENT ON PAGE v) (MAIN HEADQUARTERS). Established in 1983. 50 Employees at this address. 500 Executive Boulevard, Elmsford, New York 10523 USA Telephone: (914) 592-

2355. FAX: (914) 592 3248. Contact: Russell R. LaCoste, Vice-President, Sales and Merketing.

President, Salvatore D'Amato; Vice-President, Russell LaCoste; Customer Service, Ellen Weiser. Parent company: International Bank Note Company. Branch offices: 400 Montgomery Street, Suite 810, San Francisco, CA 94104 USA; 999 Plaza Drive, Suite 400, Schaumburg, IL 60173 USA.

Company description: American Bank Note Holographies is the producer of the world's best known holograms; those seen on Visa, MasterCard and the three holographic covers of National Geographic Magazine.

AMERICAN BANK NOTE HOLOGRAPHICS, INC., (SEE OUR ADVERTISEMENT ON PAGE v) (Branch office) 999 Plaza Drive, Suite 400, Schaumburg, IL 60173 USA.

AMERICAN BANK NOTE HOLOGRAPHICS, INC. (SEE OUR ADVERTISEMENT ON PAGE v) (Branch office). 400 Montgomery Street, Suite 810, San Francisco, CA 94104 USA.

AMERICAN HOLOGRAPHIC INC. Established in 1976. 30 Employees at this address. P.O. Box 1310, 521 Great Road, Littleton, MA 01460 USA. Telephone: (508) 486 9621 . FAX: (508) 486 9080. Contact: Rick Dishman. Description: Design, develop & manufacture optical components & instruments for use in industrial & medical measurements. Using holographic diffraction grating design & maufacturing capability to produce components for unique measurement instruments.

AMHERST MEDIA. 418 Homecrest Drive, Amherst, NY 14226 USA. Telephone: (716) 883 9220. Description: Fine art hologram maker.

ANAIT STUDIO. 1685 Fernald Point Lane, Santa Barbara, CA 93108 USA. Telephone: (805) 969 5666. Contact: Anait Stevens. Description: Artist; reflection holography; portraiture; commissions.

ANGSTROM INDUSTRIES, INC. Established in 1988. 2 Employees at this address. 3202 Argonne, Houston, TX 77098 USA. Telephone: (713) 526 0006. Contact: Frank Davis, President. President, Frank Davis; Vice-President, Vikki Fruit. Description: Artistic holography; embossed holograms.

ANOTHER DIMENSION. (see AD 2000)

A.N. SEVCHENKO RESEARCH INSTITUTE OF APPLIED PHYSICAL PROBLEMS. 220106 Minsk, USSR. Contact: V.P. Mikhaylov. Description: Materials research; artistic holography.

ApA OPTICS, INC. 2950 Northeast 84th Lane, Blaine, MN 55432 USA. Telephone: (612) 784 4995. Contact: Anil Jain. Company description: Design and manufacture of head-up displays, HOEs.

APOLLO LASERS, INC. P.O. Box 2730, Chatsworth, CA 91311 USA. Telephone: (818) 4073000. Contact: Dexter Gocha, Sales and Marketing Manager. Description: Manufacturer of carbon dioxide & ruby lasers; equipment & supplies for holography.

APPLIED HOLOGRAPHICS CORPORATION, Established in 1989. 15 Employees at this location. 1721 Fiske Place, Oxnard, CA 93033 USA. Telephone: (805) 385-5670. FAX: (805) 385-5671. Contact: Mr. Chris Outwater, Executive Vice-President. President, Mr. O. Boxall; Vice-President, Mr. C. Outwater; Sales Manager, Mr. R. Gripp; Marketing, Mr. D. Buell. Parent Company: Applied Holographics, PLC., England. Branch office: Richmond, VA USA.

Company description: From embossed credit cards up to meter square holograms, using computer generated or real life graphics, Applied Holographics Corporation (formerly ADD) is the recognized leader in full color holographic stereography.

APPLIED HOLOGRAPHICS CORPORATION. (Branch office). Eastern Sales Office, 8508 Kenwin Road, Richmond, VA 23235 USA.

APPLIED HOLOGRAPHICS, PLC., (MAIN HEADQUARTERS), Braxted Park, Great Braxted, Malden, Essex CM8 3XB, England, United Kingdom. Subsidiary Companies: Applied Holographics Corporation, Oxnard, CA USA. Company description: Holography, stereograms, narrow and wide web embossing, hotstamp foil, pressure sensitive, polyester, OPP, acetate, large format holograms, holodisc.

APTEC ENGINEERING LIMITED. 4251 Steeles Avenue West, Downsview, Ontario, M3N 1 V7 Canada. Telephone: (416) 661 9722. Contact: Edward Zileba. Description: Broker for artistic holography.

ARBEITSKREIS HOLOGRAFIE B.v. Herman-Josef Bianchi, Boeckelter WEG 47, 4170 Geldern. Federal Republic of Germany. Telephone: (49)(2831) 3034

Contact: Christian Liegeois. Description: Artistic holography.

ARCHITECTURAL GLASS & HOLOGRAPHY. 1 South Street, Great Waltham, Chelmsford, Essex CM3 1 DF, England, United Kingdom. Contact: Graham Barker. Company description: Optics; consulting.

ARMSTRONG WORLD INDUSTRIES. P.O. Box 3511. Lancaster, PA 17604 USA. Telephone: (717)397 0611. Contact: Larrimore B. Emmons. Description: Artistic holography.

ART FREUND HOLOGRAPHY. Established in 1982. Employee at this address. 124 Brookwood Drive, Santa Cruz, CA 95065 USA. Telephone: (408) 426-4436. Description: Artistic holographer

ARTIGLIOGRAPHY CO. Established in 1987. 4 Emaloyees at this address. 7130 Mohawk West Drive_ Indianapolis, IN 46236 USA. Telephone: (317) 823-0069. Contact: Kerry J. Brown. Description: Broker for embossed holography & artistic holography.

ARTIGLIOGRAPHY GALLERY. (Branch of Artigliography Co.) 415 Massachusetts Ave, Indianapolis, IN 46204 USA. Description: Buying & Selling holograms; consulting; holography education.

ART INSTITUTE OF CHICAGO (See School of The Art Institute of Chicago)

ARTKITEK. Established in 1986. 1 Employee at this address. 122 Myrtle Avenue, Cotati, CA 94931 USA. Telephone: (707) 664 2330. Contact: Steve Anderson, Proprietor. Company description: Artkitek provides holographic services to the design community: holograms of architectural models, laser speckle interferometry, laser sculpture for lobby displays and special events, Hi mastering and purchasing consultation.

ARTPLAY HOLOGRAPHIC STUDIO. (MAIN HEADQUARTERS). Established in 1986. 12 Employees at this address. H-1191 Budapest, Ady Endre ut. 8, Hungary. Telephone: (36-1) 1270412. Contact: Tibor Balogh. President, Tibor Balogh; Vice-President, Zsuzsa Dobranyi; Sales Manager, Zsolt Szekely; Customer Service, Szilvia Bagdan. Subsidiary companies: Galvanoart, Budapest, Hungary. Company description:

Besides usual holographic services the ARTPLAY is the only Eastern European studio offering complex design work applying holograms, shim production and own galvanic services also for other studios and printers.

ART, SCIENCE AND TECHNOLOGY INSTITUTE-HOLOGRAPHY SOCIETY. Established in 1983. 6 Employees at this address. 2018 R Street, N.W.,Washington D.C. 20009 USA. Telephone: (202) 667 6322. FAX: (202) 861 0621 . Contact: L. Bussaut, Research Director. Company description: Research and educational organization for the advancement of the art, science and technology of holography/ Research focus: holographic motion picture, imagery/2 galleries for the artworks' promotion/travelling exhibit/Training

ASCOT LASER PICTURE STUDIO, 27 Upper Village Road, Sunninghill, Ascot, Berkshire SL5 7AJ, England, United Kingdom. Telephone : (44)(0990) 21789. Contact: Mr. Brode/. Description: Artistic holography; holography education; workshops

ASHLAND ELECTRIC PRODUCTS-VISUAL ELECTRONICS CORP., 80-39th Street, Brooklyn, NY 11232 USA. Telephone: (718) 855 3319. Contact: J. Conlan. Description: Manufacturer of optics.

ASOCIACION ESPAtiOLA DE HOLOGRAFIA. Avda. Filipinas 38-1A, Madrid 3, Spain. Description: Artistic holography.

ASSOCIATED PAPER INDUSTRIES. Parent Corp. of George M. Whiley Ltd. Silk House, Park Green, Macclesfield, Cheshire SK11 7NU, England, United Kingdom.

ASSOCIATES OF SCIENCE AND TECHNOLOGY (AST) INC., Established 1986. 2450 Lancaster Road, Suite 36, Ottawa K1 B 5N3, Canada. Telephone: (513) 521-2557, Contact: Majdeline A. Jaoude, Marketing & Promotion. President: Dr. J. William McGowan, Executive Director: Warna Roa. Company description: ASTI is a non-profit organization dedicated to promoting public awareness of Science and Technology. Producers of the largest travelling international exhibit of holograms entitled IMAGES IN TIME AND SPACE, AST act as consultants, producers of exhibits and galleries, making holography accessible to special interest groups as well as the general public

ATELIER HOLOGRAPHIQUE DE PARIS. 13, Passage Courtois, F-75011 Paris, France. Telephone: (33)(1) 437969 18. President, Pascal Gauchet; Vice-President, Jonathan Collins; Customer Service, Dominique Seuray. Company description: Artistic holography; Buying & Selling; Consulting

ATOMIKA TECHNISCHE PHYSIK GmbH. D-8000 Munchen 19, Kuglmuellerstrasse 6, Federal Republic of Germany. Telephone: (49)(89) 152 031 . Description: Artistic holography.

AT&T BELL LABS-MATERIALS GROUP. 600 Mountain Avenue, Murray Hill, NJ 07974-2070 USA. Telephone: (201) 582 3000 extension 3086. Contact: Thomas Dudderar. Description: Scientific & industrial research; holographic non-destructive testing.

B

BARR & STROUD, LTD. Caxton Street, Anniesland. Glasgow G13 1 HZ, Scotland,United Kingdom. Telephone: (44)(41) 954 9601. Company description : Artistic holography

BBT INSTRUMENTER APS., Dronning Olgasvej 6, DK-2000 Frederiksberg, Denmark. Telephone: (45)(01)198 208. FAX: (45)(01) 198747. Branch office of Newport Corporation, Fountain Valley, CA USA.

BEDDIS KENLEY ENGINEERING LTD. Lenton Drive, Dewsbury Road, Parkside Industrial Estate, Leeds, West Yorkshire LS11 7L, England, United Kingdom. Telephone; (44)(0532 709 874. Company description: Artistic holography; embossed holography; equipment & supplies.

BEIJING INSTITUTE OF POSTS AND TELECOMMUNICATIONS. Department of Applied Physics, Holography Laboratory, Beijing 10080, People's Republic of China. Telephone: ((86)(1) 668 1255. Contact: Hsu Da-Hsiung. Description: co lege courses in holography.

BEIJING NORMAL UNIVERSITY. Analysis and Testing Centre, Beijing 100875, Peoples Republic of China. Contact: Huang Wanyun. Description: Industrial and Scientific research, Non-Destructive testing

BOB MADER PHOTOGRAPHY. P.O.Box 796728, Richardson, TX 75080 USA. Telephone: (214) 690-5511 . Contact: Hope Hickman. Description: Artistic holography; pulsed laser portraits; marketing consultant.

BOBST GROUP. 146-T Harrison Avenue, Roseland, NJ 07068 USA. Telephone: (201) 226 8000. Contact: John Torsion. Company description: Manufacturer of hologram applicator machinery.

BRANDT JEN & KLUGE, INC, (MAIN HEADQUARTERS), Established in 1919.80 Employees at this address. 539 Blanding Woods Road, St. Croix Falls, WI 54024 USA. Telephone: (715) 483 3265. FAX: (715) 483-1640. Contact: Hank A. Brandtjen III, VicePresident. President, Henry A. Brandtjen, Jr. ; VicePresident,
Hank A. Brandtjen III, H.M. Williams. Company description: Manufacturer of Hot Stamp Machine

BREMER INSTITUTE FOR ANGEWANDTE. Strahltechnik, BIAS, ErmlandstraBe 59, D-2820 Bremen 71 , Federal Republic of Germany. Contact: Werner Juptnero Telephone: (49)(0421) 606 063. Description: Industrial research; holographic non-destructive testing.

BRIGHTON Imagecraft. 7 Bath Street, Brighton, East Sussex BN1 3TB England, United Kingdom. Telephone: (44)(0273) 202 069. Contact: Jeff Blyth. Description: Artistic holography consulting; recording material manufacturing.

BRITISH AEROSPACE PLC. Established in 1908. 8000 Employees at this address. Sowerby Research Centre, FPC: 267, P.O. Box 5, Filton, Bristol BS12 7QW, England, United Kingdom. Telephone: (44)(0272) 366 842. FAX: (44)(0272) 36 3733. Contact: Mr. S.C.J. Parker. Chairman, Prof. Smith; VicePresident, John Evans. Subsidiary company: Austin Rover, Royal Ordnance. Branch offices: Filton, Bristol; Stevenage ; Warton; Lostock; Bruff; Preston; Plymouth; Weybridge; Chadderton; Hatfield. Company description: Holographic Non-destructive testing.

BRITISH AEROSPACE PLC. (MAIN HEADQUARTERS). 11 Strand, London, England, United Kingdom.

BRODEL HOLOGRAMS. 15 School Road, Sunninghill, Ascot, Berkshire, England, SL5 7AE, United Kingdom. Telephone: (44) (990) 21789. Description: Artistic holography.

BROOKHAVEN NATIONAL LABORATORY. Building Side, Upton, NY 11973 USA. Telephone: (516) 282 3758. Description: Industrial research.

BURNS HOLOGRAPHICS LTD. Established in 1989. 3 Employees at this address. P.O. Box 377, Locust Valley, NY 11560 USA. Telephone: (516) 6743130. Contact: Joseph Burns. Company description: Since 1972, Holograms/Stereograms/Editions; Silver halide, Photoresist, Nickel, Embossed with Agam, Dali, Cossette, Nunez, Dieter Jung, Sam Moree, Others; 1979 - Injection-Molded Holograms; 1987 Design/Development NY Telephone Hologram CreditCard.

BURTON HOLMES INTERNATIONAL. 1004 Larrabee Street, West Hollywood, CA 90069 USA. Telephone: (213) 652 0970. Contact: Burton Holmes. Company description: Artistic holography; multiplex holograms.

C

CAMBRIDGE CONSULTANTS LTD. Science Park, Milton Road, Cambridge CB4 4DW, England United Kingdom. Telephone : (44)(223) 358 855. Company description: Marketing consultants.

CAMBRIDGE STEREOGRAPHICS GROUP. Established 1976. 4 Employees at this address. P.O. Box 159, Kendall Square Station, Cambridge, MA 02142-0002 USA. President: Stephen A. Benton, VicePresident, S. Lee Anthony. Company description: consultants in holographic optics and imaging, and producers of prototype holographic items.

CASDIN-SILVER HOLOGRAPHY. (MAIN HEADNames

QUARTERS). Established in 1968. 51 Melcher Street, #501, Boston, MA 02210 USA. Telephone: (617) 739 6868. Contact: Harriet Casdin-Silver, Owner. Company description: I have been creating holographic art and interactive holographic environments since 1968. Our company specializes in original holograms for advertising, architectrual and theatre settings, expositions. We are consultants, exhibition organizers/designers.

CASDIN-SILVER HOLOGRAPHY. (Branch office) 99 Pond Avenue, Suite D403, Brookline, MA 02146 USA. Telephone: (617) 423 4717. Contact: Harriet Casdin-Silver.

CENTER FOR ADVANCED VISUAL STUDIES. Massachusetts Institute of Technology, 40 Massachusetts Avenue, Cambridge. MA 02139 USA. Telephone: (617) 253 4478. Description: Scientific, academic research.

CENTER FOR APPLIED RESEARCH IN ART AND TECHNOLOGY. 72 Lange Boomgaardstraat, B 9000 Gent, Belgium. Subsidiary: University Gent, Workshop Holography, Gent, Belgium.

CENTRAL ELECTRICITY GENERATING BOARD. Marchwood Engineering Laboratories, Magazine Lane, Marchwood, Southampton S04 4ZB, England, United Kingdom. Telephone : (44)(0703) 865 711. Contact: John Webster. Company description: Artistic holography; pulsed portraiture; scientific & industrial research; holographic non-destructive testing

CENTRAL MICHIGAN UNIVERSITY, Art Department. Mt Pleasant, MI 48859 USA. (517) 774 3025. Contact: Richard Kline. Description: Artistic holography; holography education; courses/workshops.

CHERNOVTSY STATE UNIVERSITY, 2 Kotsyubinsky Str., 274012 Chernovtsy, USSR. Telephone: (700) Chernovtsy 44730. Contact: Oleg V. Angelsky, Correlation Optics Department head. Description: High-speed holographic and interference methods for non-destructive measurements of surfaces with surface roughness heights 0.001 - 0.5llm. Measurements of particle density, size and velocity distribution in disperse media.

CHERRY OPTICAL COMPANY. (SEE OUR ADVERTISEMENT ON PAGE 41). 2047 Blucher Valley

Road, Sebastopol, CA 95472 USA. Telephone: (707) 823 7171. FAX: (707) 823 8073. Contact: G. Cherry. Company description: Proprietors Greg Cherry and Nancy Gorglione produce art and display editions of silver halide reflection and transmission holograms in all size formats. Services include custom holograms and innovative exhibition installations.

CHIBA UNIVERSITY. Faculty of Engineering, 1-33 Yayoi-cho, Chiba 260, Japan. Telephone: (81)(0472) 511 111 ext. 2874. Contact: Jumpei Tsujiuchi. Description: Scientific research; holographic nondestructive testing.

CHIMERIC IMAGES, INC. Established in 1988. 1 Employee at this address. 713 1/2 Main Street, Lafayette, IN 47901 USA. Telephone: (317) 7420586. Contact: Ellen M. Shew, President. Company description: Chimeric Images, Inc., is a design and marketing firm specializing in Special EventlTrade Show & Promotional Item holography. Dichromate; Embossed; Photo Polymer Mediums.

CHROMAGEM INC. (Branch office). Established in 1981.4 Employees at this address. 573 South Schenley, Youngstown, OH 44509 USA. Telephone : (216) 799 0323. FAX: (216) 747 9371. Contact: Steve Lev. President, Thomas J. Cvetkovich; Vice-President, Steve Lev. Company description: A diversified holographic company with mUlti-service labs specializing in embossed mastering & full color display holography and is currently expanding into retail and distribution. 30 years combined staff experience in holography.

CHROMAGEM INC. (MAIN HEADQUARTERS). 25 Market Street, International Towers, Youngstown, OH 44503 USA.

CINEMA & PHOTO RESEARCH INSTITUTE. NIKFI, Leningradsky, Prospect 47, Moscow USSR. Telephone: (7)(1570 2923. Contact: I. Nalimov. Description: Scientific research.

CISE SPA TECHNOLOGIE INNOVATIVE. (SEE OUR ADVERTISEMENT ON PAGE 64c). Established in 1946. 600 Employees at this address. via Reggio Emilia, 39, 20090 Segrate, Milano, Italy. Mail address: P.O. Box 12081, 1-20134 Milano, Italy. Telephone: (39)(2) 2167 2634. FAX: (39)(2) 2167 2620. Contact: Mrs. M. Luciana Rizzi, Eng. President, F. Velona; Vice-President, S. Villani; Sales Manager, F. Banfanti; Customer Service, F. Bonfanti. Parent company: ENEL (Italian National Electric Power Authority) . Subsidiary company: Conphoebus, Siet, Co. Tim, Tim Tecnopolis CSATA, CNRSM.

Company description: CISE is a research company developing innovative technologies and transferring them to industry. In addition to research, CISE performs
technological services and produces equipment, instrumentation, and new technological items.

C.ITOH & COMPANY. Central P.O. Box 136, Tokyo 100-91, Japan. Telephon.e: (81)(3) 639 2946. Description: Holography lab equipment.

CITRoEN INDUSTRIE. 35, rue Grange Dame Rose, F-92360 Meudon-la-Foret, France. Contact: Thierry Manderscheid. Company description: Industrial research; holographic non-destructive testing.

CITY CHEMICAL. 132 West 22nd Street, Dept. H, New York, NY 10011 USA. Telephone: (212) 929 2723. Company description: Equipment & Supplies including photochemistry & emulsions.

COBURN CORPORATION. Marketing Department. 1650 Corporate Road West, Lakewood, NJ 08701 USA. Telephone : (201) 3675511. FAX: (201) 367 2908. Contact: Joe Coburn; John White. Company description: Embossing & shim making; training programs..

COHERENT. (MAIN HEADQUARTERS). Established in 1966. 350 Employees at this address. 3210 Porter Drive, Palo Alto, CA 94306 USA. Telephone: (415) 493 2111. FAX: (415) 858 7631. Contact: Christine Krieg, Marketing Assistant. President, Henry Gauthier; CEO, James Hobart; Sales and Marketing Director, Paul Crosby; Customer Service, Gustavo Pinto. Branch offices: Coherent (U.K.) Ltd., Cambridge, England; Coherent GmbH, Ober-Roden, Federal Republic of Germany ..
COHERENT (U.K.) LTD. Cambridge Science Park, Milton Road, Cambridge CB4 4RF, United Kingdom. Telephone: (44)(223) 420 501. FAX: (44)(223) 420 073.
COHERENT GMBH. SenefelderstraBe 10, 6074 Rodermark, Ober-Roden, Federal Republic of Germany.. Telephone: (49)(6074) 914/0. FAX: (49)(6074)95654.
Company description: Manufacturer of HighLight™ and PureLight™ ion laser systems for the entertainment industry. Both serve as economical sources of high-power laser light for light shows and other entertainment applications.

COLLEGE OF MANUFACTURING. Cranfield Institute of Technology, Cranfield, Bedford MK43 OAL, Engand, United Kingdom. Contact: J.M. Burch. Description: Scientific, industrial engineering; holographic non-destructive testing.

COLUMBIA UNIVERSITY. Deptartment of Otolaryngology, 630 West 168th Street, New York, NY 10032 SA. Telephone: (212) 305 3993. Contact: Shyam Khanna. Company description: Medical holography.

CONTINENTAL OPTICAL. 15 Power Drive, HaupJauge, NY 11788 USA. Telephone: (516)582 3388. Description: Optics and custom orders.

CORION CORP. (MAIN HEADQUARTERS), Estabshed 1967. 65 Employees at this address. 73 Jeffrey Ave., Holliston, MA 01746 USA, Telephone: (508) 429-5065. FAX: (508) 429-8983. Contact: Walter J. Lekki. President, Frank Mascis; Vice-President, Walter Lekki; Sales Manager, Michael Mascis. Company description: Corion Corp. manufactures; Volume and one-of-a-Kind/Custom and Stock, optical compoentes including coatings, filters, optics and optical assemblies for use in the UV-Visible-IR spectrum

COULTER OPTICAL COMPANY. P.O. Box K Dept. MP, 54121 Pinecrest Road, Idyllwild, CA 92349 USA. Telephone: (714) 659 2991 . Contact: Mary Braginton. Company description: Make telescope mirrors, parabolic mirrors and more. Send for free list of products.

CREATIVE LABEL. 2450 Estes Drive, Dept. M, Elk Grove Village, IL 60007 USA. Telephone: (312) 956 6960. Contact: Jerry Koril. Company description: Bindery application on Kluge (2 stream) and Bobst (4 stream) machines. Call for more information.

CROWN ROLL LEAF, INC., (MAIN HEADQUARTERS). Established in 1970.95 Employees at this address. 91 Illinois Ave., Paterson, NJ 07503 USA. Telephone: (201) 742-4000. FAX: (201) 742-0219. Contact: James Waitts, Holographic Mgr. President, Robert Waitts; Vice-President, Manuel Cueli; Sales Manager, Pamela Herforth; Customer Service, Maggie Carola. Branch offices: Crown Roll Leaf Inc .. , CA; Crown Roll Leaf Inc., IL. Company description: Crown Roll Leaf has been supplying embossing material internationally for 5 years. We have been manufacturing shims and embossed products for 4 years. Please call Jim Waitts with any questions.

CROWN ROLL LEAF, INC., 20705 South Western Avenue, Torrance, CA 90501 USA. Branch office.

CROWN ROLL LEAF, INC., 2456 Elmhurst Road, Elk Grove Village, IL 60007 USA. Branch Office. CSI. 7 Meadowfield Park South, Stocksfield, Northumberland, NE43 7QA, England, United Kingdom. Telephone: (44) (661) 842 741. Company description: Manufacture mirrors; optics.

CVI LASER CORPORATION, CVI WEST. Established in 1986. 470 Lindbergh Avenue, Livermore, CA 94550 USA. Telephone: (415) 440-1064. FAX: On request. Contact: Dr. Alex Jacobson. President, Dr. Yu H. Hahn; Vice-President, F.J. Brandiger. Parent company: CVI Laser Corporation. Branch offices: livermore, CA; Putnam, CT. Company description: Manufactures holographic quality single and multiple element lenses, mirrors, windows, and beamsplitters for all standard holographic laser sources. Free 104-page catalog available.

CVI LASER CORPORATION. (MAIN HEADQUARTERS). 200 Dorado Place SE, P.O. Box 11308, Albuquerque, NM 87192 USA.

D

DAIMLER BENZ AG. 0-7000 Stuttgart 60. Federal Republic of Germany. Contact: H.G.Leis. Description: Industrial Research; holographic non-destructive testing.

DAI NIPPON PRINTING CO. Ltd.Cental Research Institute 12, 1-Chome Ichigaya-Kagacho, Shinjuku -ku. Tokyo 162, Japan. Contact: Tokio Kodera. Telephone: (81)(03) 266 2310. Company description : Artistic holography; embossed holography; printing applications.

DARKROOM EIGHT LTD. Unit 8 - Impress house Vale Grove, Acton, London W3 7QH, United Kir.gdom. Telephone: (44)(1) 7492218. Company description: Artistic holography.

DATASIGHTS LTD. Alma Road, Ponders End, Enfield, Middlesex, EN3 7BB, England, United Kingdom. Telephone: (44)(1) 805 4157. Company description: Manufacture mirrors.

DAVID SCHMIDT HOLOGRAPHY. Established in 1985. 3 Employees at this address. 23962 Craftsman Road, Calabasas, CA 91302 USA. Telephone: (818) 992 1541. FAX: (818) 703 1182. Contact: David Schmidt, Owner. Company description: Holography courses offered. Company description: David Schmidt Holography is a full service mass production laboratory specializing in stereograms both cylindrical and image plane formats. We also mass produce reflection and transmission holograms for the trade.

DAVIN OPTICAL LTD. Reliant House, Oakmere Mews, Potters Bar, Hertfordshire, EN69XX, England, United Kingdom. Telephone: (44)(707) 44445. Company description: Manufacture mirrors; optics.

DB ELECTRONIC INSTRUMENTS S.R.L., Via Teano 2, 1-20161 Milano, Italy. Telephone: (39)(02) 646 9341. FAX: (39)(02) 645 6632. Branch office of Newport Corporation, Fountain Valley, CA USA.

DEEP SPACE HOLOGRAPHICS. 1328 Dunsterville Avenue, Victoria, British Columbia, Canada V8l 2X1. Telephone : (1)(604) 479 4357. Company description: artistic holography.

DEUTSCHE GESELLSCHAFT FOR HOLOGRAFIE E.V. (GERMAN HOLOGRAPHIC SOCIETY). Established in 1989. Lerchenstr. 142 a, 0-4500 OsnabrOck, Federal Republic of Germany. Telephone: (49)(0541) 186059. FAX: (49) (0541) 7102173. Contact: Dr. Peter Zec, President. President, Dr. Peter lec; VicePresident, Brigitte Burgmer. Company description: The society was founded to promote awareness of holography, and its members are mainly holographers and artists. To this end, the group intends to organise exhibitions.

DEUTSCHER HOLOGRAPHIE-VERTRIEB. Kolner Strasse 49-51, 0-4047 Dormagen 1, Federal Republic of Germany. Contact: Thomas Rost. Company description: Artistic holography.

D.GA MFG. CO. A SUBSIDIARY OF DIVERSIFIED GRAPHICS, Ltd., 3719 Joy Road, Columbus, GA 31906 USA.

DIALECTICA AB. Skanegatan 87, 6tr, S-116 37 Stockholm, Sweden. Contact: Ambjorn Naeve. Company description: Artistic holography.

DIAURES SA HOLOGRAPHIC DIVISION, Via 1 Maggio 262/A, 1-41019 Soliera (Modena) Italy. Telephone: (39) (059) 567 274. Company description: Artistic holography; embossed holography; equipment & supplies.

DIE DRITTE DIMENSION. Established in 1987. 2 Employees at this address. Frankfurter StraBe 132 -134, 0-6078 Neu Isenburg, Federal Republic of Germany .. Telephone: (49) (06102) 33367. FAX: (49) (06102) 36709. Contact: Mr. Carlo Westphal. President, Mr. Carlo Westphal; Sales Manager, Mrs. Elke Hein; Customer Service, Mrs. Elke Hein. Company description: Biggest special shop for holography in Federal Republic of Germany .. Always over 600 different holograms in stock. Comprehensive fine art section.

THE DIFFRACTION COMPANY, INC. (MAIN HEADQUARTERS). 38 Loveton Circle, Sparks, MD 21152 USA. Contact: Hugh Wynd.

THE DIFFRACTION COMPANY, INC. Established in 1957. 15 Employees at this address. P.O. Box 151, Riderwood, MD 21152 USA. Telephone: (301) 666 1144. FAX: (301) 472 4911 . Contact: Hugh C. Wynd, President. Vice-President, Christopher W. Wynd; Customer Service, Kim Price. Company description: We offer: A. 58 patterns available in 16 colors with a variety of adhesives. B. Color explosion graphics/microetching an alternative to 3D. C. Custom embossing of holograms. D. Dazzlers-Stickers.

DIMENSIONAL IMAGING TECHNOLOGY. 439 Walsingham Court, Dayton, OH 45429 USA. Telephone: (513) 4345818. Description: Artistic holography.

DIVERSIFIED GRAPHICS, LTD., (MAIN HEADQUARTERS). Established 1972. 120 Employees at this address. 5433 Eagle Industrial Court, Hazelwood, MO 63042 USA. Telephone: (314) 895-4600. FAX: (314) 895-4363. Contact: J.Jay Cassen, President! C.O.O. Chairman of the Board, Max R. Scharf. Subsidiary companies: K-Studio, D.G. Sportswear, Lazer Vision. Branch offices: D.GA Mfg. Co., GA. Company description: Diversified Graphics, Ltd., a Missouri Corporation, develops, manufactures and markets screen printed apparel. In 1988, it launched a Lazer Vision line of holographic apparel to major department and specialty stores.

DOVECOTE STUDIO. Witham Friary, Frome, Somerset, England, United Kingdom. Telephone: (44)(74) 985 691. Contact: Angela Coombes. Company description: Artistic holography; oversize format

DREAM IMAGES. Postfach 1602, Vermeerweg 15, D-5047 Wesseling, Federal Republic of Germany. Telehone: (49)(022) 364 3138. Contact: Klaus Thielker. Company description: Artistic holography; gallery; marketing consultant.

DUPONT COMPANY. (See E.1. DUPONT DE NEMOURS & CO.) DUSTON HOLOGRAPHIC SERVICES INC. Established 1988. 2 Employees at this address. 90 Sherbrooke Avenue, Ottawa, Ontario, Canada K1Y 1 R9. Telephone: (613) 722-9004. Contact: Deborah A. Duston, President. Company description: Duston Holographic Svcs consults corporate and government clients on HOEs, remotely sensed Holographic Stereograms and the educational and curatorial aspects of Holography. Deborah Duston is also a well known artist -holographer.

DUTCH HOLOGRAPHIC LABORATORY. Established in 1983. 6 Employees at this address. Kanaal dyk Noord 61, 5642 JA Eindhoven, The Netherlands. Telephone: (31)(40) 817250. FAX: (31)(40) 814 865. Contact: Walter Spierings, Director. President, W. Spierings; Sales Manager, R. Van Oorschot. Company description: Maker of embossed and artistic holograms; holography education.

THE DZ COMPANY. (SEE OUR ADVERTISEMENT ON PAGE 37) Established in 1985. 10 Employees at this address. P.O. Box 5047-R, 181 Mayhew Way, Suite E, Walnut Creek, CA 94596 USA. Telephone: (415) 9354656. FAX: (415) 935 4660. Contact: Robin Christie, Sales Manager. President, Dan Cifelli; VicePresident, Gary Zellerbach; Sales Manager, Robin Christie. Company description: Manufacturers of fast selling holographic products for retail sales and promotions. Many of the items can be imprinted for ad specialty and incentives.

E

EALING ELECTRO-OPTICS INC. New Englander Industrial Park, Holliston, MA 01746 USA. Telephone: (508) 429 8370. Contact: Pauline Lebine. Company description: Manufacturer of mirrors & optics.

EALING ELECTRO-OPTICS. Greycaine Road, Watford, Hertfordshire, England WD2 4PW United Kingdom.

EALING SCIENTIFIC LTD. P.O. Box 238, Pointe Claire-Dorval, Quebec H9R 4N9, Canada. Telephone : (514) 631 1807. Company description: manufactures lasers.

EASTMAN KODAK COMPANY. Sales & Marketing. 1669 Lake Avenue , B-23, Fl. 3, Rochester, NY 14650 USA. Telephone: (716) 722 1066. Description: Manufacturer of holographic equipment & film.

ECOLE NATIONALE SUPERIEURE D'INGENIEURS. c/o Groupe de Laboratoires, C.M .R.S., 5, Av. D'Edimbourg, BLD Marechal-Juin, F-14032 CaenCedex, France. Contact: Jean-Charles Vienot. Description: Scientific & industrial research; holographic non-destructive testing.

E.C. SCHULTZ & COMPANY, Established in 1895.9 Employees at this address. 333 Crossen, Elk Grove Village, IL 60007 USA. Telephone: (312) 640 1190. FAX: (312) 640 1198. Contact: Lynn Schultz, VicePresident. Company description: Our company makes stamping, embossing debossing and applique dies for the graphic industries. Quality craftmanship and 94 years experience joining for innovative, distinctive and exciting effects in todays demanding market.

EDMUND SCIENTIFIC COMPANY. Department H, 101 East Gloucester Pike, Barrington, NJ 08007 USA. Telephone : (609) 5473488. Company description: Mailorder, wholesale, and retail holography products for schools, science fairs, etc.

EIF PRODUCTIONS. EIZYKMAN I FIHMAN. 19 Rue Jean Jacques Rousseau, F- 75001 Paris, France. Telephone: (33)(1) 4236 0631 . Contact: Claudine Eizykman. Company description: Artistic holography; multiplex.

E.I. DUPONT DE NEMOURS & CO., INC. Established in 1802. Over 5000 Employees at this address. Optical Element Venture, Experimental Station, P.O.Box 80352, Wilmington, DE 19880-0352 USA. Telephone: (302) 695 4893. FAX: (302) 695 9631 . Contact: Joyce C. Harkey, Customer Service Representative. Director, A. Wasy D'Cruz; Venture Manager, Lory Galloway; Marketing Manager, Krishna C. Doraiswamy; Senior Product Specialist, Evan D. Laganis. Parent Company: Du Pont. Company description: Du Pont I Optical Element Venture's advanced photopolymer products feature reliable holographic performance, easy dry processing and high environmental stability. Call us to see why our materials are the best choice for your application.

E. I. DUPONT DE NEMOURS & CO., INC., (MAIN HEADQUARTERS) ,1 007 Market Street, Wilmington, DE 19898 USA. Subsidiaries: E.I. Dupont de Nemours & Co,. Optical Element Venture, Wilmington, DE USA.

ELAN BIO-MEDICAL. HOLOGRAPHY LABORATORY, 411 Lewis Road, #417, San Jose, CA 95111 USA. Contact: Michael Gersonde. Company description: Medical holography.

ELECTECH DISTRIBUTION SYSTEMS, PTE. ltd. 605-A, MacPherson Road, #03-03, Citimac Industrial Complex, Singapore 1336. Telephone: (65)2869933. FAX: (65) 2843256. Branch office of Newport Corporation, Fountain Valley, CA USA.

ELECTRO OPTICAL INDUSTRIES, INC. Established in 1964. 80 Employees at this address. 859 Ward Drive, Santa Barbara, CA 93111 USA. Telephone : (805) 964 6701 ext. 280. FAX: (805) 967 8590. Contact: Lloyd Simms, Sales Liaison. President, Arthur J. Cussen. Company description: Manufacturer of infrared test and calibration instrumentation including: collimators, choppers, blackbodies, differential temperature sources, radiometers and custom infrared instrumentation.

ELECTRO OPTICS DEVELOPMENTS LTD. Howards Chase, Pipps Hill Industrial Estate, Basildon, Essex, England, SS14 3BE, United Kingdom. Telephone: (44)(268) 20511. Company description: Equipment & supplies; optics.

ELUSIVE IMAGE. 603 Munger Street # 316, Dallas, TX 75202 USA. Telephone: (214) 720 6060. Contact: Fred Wilbur. Company description: Holography gallery.

ELUSIVE IMAGE. 135 West Palace Avenue, Suite 102, Santa Fe, NM 87501 USA. Telephone: (505) 986 0221 . Contact: Fred Wilbur. Company description: Holography gallery.

ENVIRONMENTAL RESEARCH INSTITUTE OF MICHIGAN. Optical Science Lab., ACD, P.O.Box 8618, Ann Arbor, MI 48107 USA. Telephone: (313) 994 1220. Contact: Juris Upatnieks. Company description: Industrial & academic research. ERBA. 12, rue Libergier, F-511 00 Reims, France. Telephone: (33)(26) 884 452. Contact: Bernard Oilier. Company description: Artistic holography; holography education.

EUROLAUNCH LTD. 2/3 Salisbury Court, Fleet Street, London, EC4Y, England, United Kingdom. Company description: Holographic non-destructive testing.

EVE RITSCHER ASSOCIATES LTD., 73 Allfarthing Lane, Wandsworth, London SW18, England, United Kingdom. Contact: Eve Ritscher. Company description: Artistic holography education.

EXCITEK INC. Established in 1984. 7 Employees at this address. 277 Coit Street, Irvington, NJ 07111 USA. telephone : (201) 372 1669. FAX: (201) 372 8551 . Contact: Brian Turner. President, Brian Turner; Vice-President, Andrew Dietz; Sales Manager, Al Dietz; Customer Service, Donna McGann. Company description: Excitek Inc. buys, sells, repairs and remanufactures Spectra Physics Ion laser systems and plasma tubes with unmatched attention to best quality, service and cost efficiency.

F

FASION MODA. 2803 Third Avenue, Bronx, NY 455 USA. Contact: Stefan Eins. Company descripn: Gallery.

F.AS.T. ELECTRONIC BULLETIN BOARD. P.O. Box 421704, San Francisco, CA 94142-1704 USA. Telephone: (415) 845 8306. FAX: (415) 841 6311 . Contact: Elizabeth Crumley. Company description: Educational materials

FISHER SCIENTIFIC. E.M.D. Division, 4901 West Lemoyne Avneue, Chicago, IL 60657 USA. Telephone: (312) 378 7770. Contact: Sales and Marketing Dept. Company description: Supply science lab equipment, Holography kits, lab manuals, lasers and laser related equipment.

FOCAL IMAGE LTD. P.O.Box 1916, 1 Kelvin Court, London W11 3QR, England, United Kingdom. Telephone: (44)(01) 7273438. Contact: Kaveh Bazargan. Company description: Color holography; medical holography

FORD SCIENTIFIC LABS. 2000 Rotunda Drive, Dearborn, MI 48121 USA. Telephone: (313) 323 1539. Contact: Gordon Brown. Company description: Industrial research; Holographic non-destructive testing.

THE FOREIGN DIMENSION. (SEE OUR ADVERTISEMENT ON PAGE 5). Established in 1989.4 Employees at this address. Suite B, 8'/Fl., Central Mansion, 270-276 Queen's Road Central, Hong Kong, Hong Kong. Telephone: (852)(5) 420 282. FAX: (852)(5) 416 011 Contact: Frederic Schvartzman, General Manager. President, Frederic Schvartzman; Vice-President, Nathalie Aboucar; Sales Manager, F. Schvartzmen; Customer Service, Stephane Denizot. Company description: Specialists in manufacturing all kinds of holographic products (Watches, keyrings, pendants, calculators ...), we are a French managed company offering Hong Kong competitive prices with high quality standards. Contact us for details.

FOUNDATION IDEECENTRUM. P.O. Box 222, 5600 MK, Eindhoven, The Netherlands. Company description: Gallery.

FRED UNTERSEHER & ASSOCIATES HOLOGRAPHY. 3463 State Street, Suite 304, Santa Barbara, CA 93105 USA. Telephone: (805) 568 6997. FAX: (805) 642 4396. Contact: Fred Unterseher.

FREE UNIVERSITY OF BRUSSELS. Department of Applied Physics (ALNA), Faculty of Applied Sciences, Brussels, Belgium. Contact: Stephan Roose. Description: Academic and Scientific research.

FRESNEL TECHNOLOGIES INC. 101 West Morningside Drive, Fort Worth, TX 76110 USA. Telephone: (817) 9267474. FAX: (817) 9267146. Contact: Linda H. Claytor. Company description: Manufactures plastic Fresnel lenses &. lens arrays from its POLY IR® plastics for use into the infrared; also other optical products for use into the ultraviolet from acrylic & other plastics.

FRINGE RESEARCH HOLOGRAPHICS INC.1179A King Street West, Suite 008, Toronto, Ontairo M6K 3C5, Canada. Telephone: (416) 535 2323. Contact: Michael Sowdon, Director. Company description: artistic holography; silver halide holograms; pulse portraits; gallery; workshops; travelling exhibit.

FTI JOFFE, Politechnicheskaya 26, Academy of Sciences of the USSR, 194021 Leningrad, USSR. Contact: G.A. Sobolev. Description: Artistic holography.

FUJI PHOTO OPTICAL CO. , Ltd. No. 324, 1-Chome, Uetake-Machi, Omiya, Japan. Telephone: (81)(04) 86630111 . FAX: (81)(04) 86510521. Contact: Takayuki Saito. Company description: Industrial research, optics.

FUJI ELECTRIC CO. LTD., Mecatronics Division, 1-12-1 Yuraku-cho, Chiyoda-ku, Tokyo 100, Japan. Telephone: (81)(3) 211 7111 . Company description: Manufactures CO_2 lasers and related equipment.

G

GALLERIE ILLUSORIA, Schwarztorstrasse 70, CH-3007 Bern, Switzerland. Contact: Sandro Del-Priete. Company description: Gallery.

GALVANOART. (Subsidiary of Artplay Holographic Studio) H-1191 Budapest, Ady Endre ut. 8, Hungary.

GALVOPTICS LTD. Harvey Road, Basildon, Essex, SS13 1 ES, England, United Kingdom. Telephone: (44)(0268) 728 077. FAX: (44)(0268) 590 445. Contact: R. D. Wale. Company description: Optics; mirrors, lenses.

GARDENER PROMOTION MARKETING. Established in 1980. 6 Employees at this address. 4165 Apalogen Road, Philadelphia, PA 19144 USA. Telephone: (215) 849 4049. Contact: John Gardener. President, John Gardener; Vice-President, Roy Gunther. Parent Company: Gardener Promotion Products Corp. Company description: As the exclusive package goods marketing representative for Holographic Design, we can show you how holography can be used for problem solving or enhancing opportunities compatible with your objectives.

GENERAL HOLOGRAPHICS CORP. (MAIN HEADQUARTERS). 37568 Devoe, Mt. Clemens, MI 48043 USA. Contact: Greg Wright, President.

GENERAL HOLOGRAPHICS CORP. (Branch office) Established in 1986. 1 Employee at this address. 25550 North River Road, Mt. Clemens, MI 48045 USA. Telephone: (313) 468 3430. Contact: Greg Wright, President. Company description: Distributor/wholesaler of holographic products to: industrial; commercial/ retail; and consumer markets.

GENERAL HOLOGRAPHICS, INC. Established in 1978. 6 Employees at this address. P.O. Box 82247, Burnaby B.C., V5C 5P7 Canada. Telephone: (604) 435 6654. FAX: (604) 432 7326. Contact: Paula Simson, Managing Director. President, Bernd Simson; Sales Manager, Paula Simson; Customer Service, Darlene Lafgren. Company description: Manufacturer of dichromate gift items, such as clocks and plates. Distributor of dichromate gift and jewellery items and silver halide wall art. Custom work available. Company logos a specialty.

GENERAL IMAGING CORPORATION. (SEE OUR ADVERTISEMENT ON PAGE 35). Established in 1989. 3 Employees at this address. 1 Industrial Drive South, Lan-Rex Industrial Park, Smithfield, RI 02917 USA. Telephone: (415) 232 2707. FAX: (401) 231 4674. Contact: Rich Zucker|Terry Feeley. Company description: Dupont photopolymer recording material; replication systems; replication facilities for holograms (HOE & HUD); project oriented developmental contracts; supporting technologies.

GEORGE M. WHILEY LIMITED, (MAIN HEADQUARTERS). Established in 1783. 182 Employees at this address. Firth Road, Houston Industrial Estate, Livingston, West Lothian EH54 5DJ, Scottland, United Kingdom. Telephone: (0506) 38611 FAX: (0506) 38262. Contact: B J Sitch, Technical Director. President, G.G. Hall; Vice-President, B.J. Sitch. Parent company: Associated Paper Industries, Silk House, Park Green, Macclesfield, Cheshire SK11 7NU, England, United Kingdom. Branch offices: Aukland, New Zealand Company description: George M. Whiley is a long established manufacturer of stamping foils. We have developed special base materials for Holographic embossing and market these and Holographic foils worldwide.

GERALD MARKS STUDIO. 29 West 26th Street, New York, NY 10010 USA. Telephone: (212) 889 5994. Description: Artistic holography; stereograms; consulting; instruction.

GLOBAL IMAGES, INC. 509 Madison Avenue, Suite 1400, New York, NY 10022 USA. Telephone: (212) 759 8606. FAX: (604) 734 2842. Contact: Walter Clarke. Company description: Manufacturer of holographic embossing machines; equipment for embossing.

GLOBAL IMAGES, INC. 2556 West 2nd Street, Vancouver, British Colombia, V6K 1J8 Canada. FAX: (604) 7342842. Contact: Walter Clarke. Company description: Manufacturer of holographic embossing machines; equipment for embossing.

G.M. VACUUM COATING LAB, INC. 882 Production Place, Newport Beach, CA 92663 USA. Telephone: (714) 6425446. Company description: Plate coating.

GRAY SCALE STUDIOS ITD. Established in 1985. 2 employees at this address. 4500 19th Street, #588, Boulder, CO 80304 USA. Telephone: (303) 442 5889. FAX: (303) 442 5889. Contact: George Sivy, President. Company description: Gray Scale Studios, Ltd. specializes in the design and creation of models and sculptures for Holographic Imaging. Consultant services offered, five years experience, samples of work available upon request.

H

HALO POWER-TRACK -- LIGHTING DIVISION. Mc-Graw-Edison Corporation, 6 West 20th Street, New York, NY 10011 USA. Telephone: (212) 645 4580. Company description: lighting fixtures.

HICKMOTT & AUSTIN HOLOGRAMS. 11 Castelnau, London SW13 9RP, England,United Kingdom. Telephone: (44)(01) 486 5811. Contact: M. Austin. Company description: Artistic holography.

HIGH TECH NETWORK. Skeppsbron 2, S-211 20 Malmo, Sweden. Telephone: (460(040) 350 750. FAX: (46) (040) 237 667. Contact: Christer Agehall. Company description: Artistic holography; security applications.

HOECHST CELANESE CORPORATION. 86 Morris Avenue, Summit, NJ 07901 USA. Telephone: (201) 522 7816. Contact: Gunilla Gilberg. Company descripion: Embossed & artistic holography.

HOL 3 GALERIE FOR HOLOGRAPHIE. Kurfurstendamm 103, 1000 Berlin 31, Federal Republic of Germany. Company description: Gallery & retail shop.

HOLAGE, Established in 1981.1881 Eighth Avenue, San Francisco, CA 94122 USA. Telephone: (415) 564 1840. Contact: Brad D. Cantos. Company de-

scription: Fine art holograms; silver halide holograms.

HOIAR SEEIE KG. Wasserwerksweg 10-14, 0-2960 Aurich 1, Federal Republic of Germany. Telephone: (49) (41) 10005. Company description: Fine art holograms.

HOIAXIS CORPORATION, Established 1984. 499 Farmington Ave, Hartford, CT 06105 USA. Telephone: (203) 232 2030. FAX: (203) 236-3767. Contact: Martin A. Berson. President: Martin Berson, Vice-President: Gary Haber. Company description: Holaxis' specialties include: large format holograms and wide-web embossing of pressure sensitive and hot stamp foils. High quality mastering, photoresist transfers, high speed labeling, and die-cutting are also included.

HOLICON CORPORATION_ (SEE OUR ADVERTISEMENT ON PAGE 9) Established in 1987. 906 University Place, Evanston, Il 60201 USA. Telephone: (312) 491 4310. FAX: (312) 491 7955. Contact: Dr. Hans Bjelkhagen. President, Dr. Max Epstein; Vice-President, Dr. Michel Marhic; Sales Manager, Dr. Michel Marhic; Customer Service, Dr. Hans Bjelkhagen. Subsidiary company: Holographic Industries, Inc. Company description: Holicon Corporation specializes in silver halide holograms, pulse or CW, in particular, portraits. large-format reflection or transmission holograms are made as well as mass production of film holograms.

HOLO 3. Established in 1986. 7 Employees at this address. rue de l'Industrie, 68300 Saint-louis, France. Telephone: (33)(89) 69 82 08. Contact: Mrs. J. Striebig, Deputy Director. President, Prof. P. Smigielski; Vice-President, Mr. A. Weber; Sales Manager, Mr.s J. Striebig. Company description: Non Profit National Organization depending on French Ministry of Research and Technology and transferring optical technologies from the Research Institute of Saint-louis France, towards industrial applications.

HOLO ARP. KAMAKURA INC. 7-10-8 Ginza Chuoku, Tokyo, Japan. Telephone: (81)(03) 574 8307. FAX: (81) (03) 574 8377. Contact: Yumiko Shiozaki. Company description: Artistic holography production; gallery; wholesaler; equipment & supplies for holography; holography education.

HOLOCOM. lange Strasse 51, 0-2117 Kakenstorf, Federal Republic of Germany. Telephone: (49)

(04186) 8510. Contact: Johannes Matthiesen. Company description: Artistic and embossed holography.

HOLOCOM HOLOGRAPHIE. 13, rue Charles V., Faculte des Sciences et des Techniques, F- Paris, France. Telephone: (33)(1) 948) 04 0058. Contact: Alan Baraton. Company description: Artistic holography.

HOLOCOR I.B.F. PRINTING INC. (SEE OUR ADVERTISEMENT ON PAGE 41). Established in 1988. 4 Employees at this address. 95 des Sulpiciens, L'Epiphanie, Quebec JOK 1 JO, Canada. Telephone: (514) 5886801. FAX: (514) 5884898. Contact: JeanRobert Bernier, President. President, Jean-Robert Bernier; Sales Manager, Jean-Robert Bernier; Customer Service, Celine Majeau. Parent company: I.B.F. Printing Inc. Company description: We focus our knowledge in what you want to see: "Holographic micro-engraving". We devote our energy in what you need: "Performance". Holocor® from electroming (shims) to final embossed hologram.

HOLOCRAFTS: DIVISION OF CANADIAN HOLOGRAPHIC DEVELOPMENTS LTD. (SEE OUR ADVERTISEMENT ON PAGE 17) Established in 1979. 20 Employees at this address. Box 1035, Delta, B.C." V4M 3T2 Canada, Telephone: (604) 946-1926. FAX: (604) 946-1648. Contact: Karoline Cullen, Managing Director. President, Gary Cullen; Vice-President, Barry Michelitsch. Company description: Holocrafts specializes in the manufacture of dichromate reflection holograms. We offer prompt delivery of stock and custom production in a variety of shapes, sizes and products. Stereograms are now available!

HOLOCRAFTS OF LONG ISLAND. 227 9th Street, West Babylon, NY 11704-3728 USA. Telephone: (516) 669 0372. Contact: Pat Willard. Company description: Fine art holograms.

HOLODESIGN. 1, Boulevard de la Republique, F-95600 Eaubonne, France. Telephone: (33)91) 39 593 954. Contact: Thierry Garcon. Company description: artistic holography.

HOLODESIGN STUDIES. Rebenstrasse 20, CH-4125 Riehen, Switzerland. Telephone: (41)(61) 672 342. Company description: Marketing consulting.

HOLO-DIMENSIONS INC. 3577 Rue de Bullion, Mon-treal, Quebec, Canada H2X 3A1. Telephone: (514) 845 4419. Company description: Artistic holography.

HOLOFAR LAB (SRL). Piazza Acilia No.3, Int. 3,Rome, Italy 00199. Telephone: (39)(6) 8395 517. Company description: Artistic holography.

HOLOFAX LIMITED. Netherwood Road, Rotherwas Industrial Estate, Hereford HR2 6JZ, England, United Kingdom. Telephone: (44)(432) 278 400. Company description: Silver halide reflection; Photochemistry; Vibration isolation tables.

HOLOFLEX COMPANY. RR 3, Box 381, Urbana, IL 61801 USA. Telephone: (217) 684 2102. Contact: Donald Barnhart. Company description: artistic holography; silver halide holograms.

HOLO GMBH HOLOGRAFIELABOR OSNABROCK. MindernerStr. 205, D-4500 Osnabruck, Federal Republic of Germany. Telephone: (49)(0)(541) 7102 173. FAX: (49)(541) 7102 176. Contact: Vito Orazem. Company description: artistic holography consulting, workshops; teacher training.

HOLOGRAFIA GALLERIA. Jaakonkatu 3, 2nd floor, SF-00100 Helsinki, Finland. Telephone: (358) (06) 941 909. Company description: Gallery, retail shop.

HOLOGRAFIA GALLERIA. (Branch of Starcke KY). c/o Science Center, Oulu, Finland. Telephone: (358) (39) 360 700. FAX: (358)(39) 67 905. Company description: gallery & retail shop

HOLOGRAFICA. 8 Hylda Court, St. Albans Road, NW5, London, England, United Kingdom. Description: Artistic holography.

HOLOGRAFIE - HOFMANN LABOR. Carl-Hermann-Gaiserstrasse 20, 7320 Groppingen, Federal Republic of Germany. Telephone: (49) (07161) 12200. Contact: Martin Hofmann. Company description: Full service artistic holography; buying & selling; holography education; equipment & supplies.

THE HOLOGRAM. P.O. Box 9035, Allentown, PA 18105 USA. Telephone: (215) 434 8236. Contact: Frank DeFreitas, Publisher. Description: Free Newsletter on Holography. Contact for more details.

HOLOGRAMA LABORATOIRE HOLOGRAPHIQUE. 41 rue Mariziano, CH-1227 Geneva, Switzerland. Telephone: (41)(022) 422 144. Contact: Yves Rossignol, company description: Embossed; pulsed portraits.

HOLOGRAMAS DE MEXICO. Established in 1984. 50 Employees at this address. PINO 343, Local 3, Col. Sta. Ma La Ribera, 06400 Mexico, D.F. Mexico. Telephone: (905) 547 9046. FAX: (905) 547 4084. Contact: Dan Lieberman. Company description: Holographic embossing applications.

HOLOGRAM EUROPE SPRL. Avenue Voltaire 137, 1030 Brussels, Belgium.Telephone: (32)(2) 242 7284. Contact: J. B. Boulton. Company description: Retail shop

HOLOGRAM. INDUSTRIES, Established in 1984. 10 employees at this address. 42-44, rue de Trucy, 94120 Fontenay sous bois, France. Telephone: 1 4394 1919. FAX: 1 43940032. Contact: Hugues Souparis, President. Vice-President: Denis Lachaud. Company description: Hologram. Industries produces high quality display and embossed holograms. We have an integrated production line, from lab to embossing. Hologram. Industries initializes graphic holograms and 3D stereograms.

HOLOGRAMM WERKSTATT & GALERIE, GALLEIE FUR HOLOGRAMME, Established in 1984. 2 employees at this address. Via Principale 30,CH - 7649 Castesegna, Switzerland. Contact: Horst Guteunst, Director. Company description: Creative workshop, developments, looking for new and attractive ways for hologram making

HOLOGRAM ONE. 39 Pyrcoft Road, Chertsey, Surrey KT16 9HT, England, United Kingdom. Telephone: (44) (9328 64899. Company description: artistic hologaphy; silver halide transmission holograms; wholesale.

HOLOGRAM ROADSHOW. Longlear House, Warminister, Wiltshire, 12 Queen Square, Bath, Avon BA1 1WU, England, United Kingdom. Telephone: (44) (225) 339 333. Company description: retail shop; travelling exhibit.

HOLOGRAMS AND OTHER STRANGE THINGS. Established in 1987. 2 Employees at this address. 3200 West Oakland Park Boulevard, Lauderdale Lakes, FL 33311 USA. Telephone: (305) 739 9634. Contact: Dennis Drucker, President. Vice-President, Harriet Drucker. Company description: A retail store specializing in holographic products, and other threedimensional and illusory-type items.

THE HOLOGRAM SCHOPPE. P.O. Box 318, 591 Tonawanda, Buffalo, NY 14202 USA. Contact: Maureen McNamara. Company description: retail sales of holography.

HOLOGRAM WORLD. 1212 1/2 Dixon Boulevard, Cocoa, FL 32922 USA. Telephone: (407) 631 3615. Contact: Susan K. Harrison. Company description: Retail shop.

HOLOGRAPHIC APPLICATIONS, Established in 1985. 2 Employees at this address. 21 Woodland Way, Greenbelt, MD 20770 USA. Telephone: (301) 345 4652. FAX: (301) 345 4653. Contact: Suzanne St. Cyr, President. Company description: Technical and Marketing Services for manufactures of holographic products. Educaiton, Design Consultation, Product Development, Vendor Selection, Project Management, and General Contracting for end-users of holography.

HOLOGRAPHIC ART. Established in 1986. 4 Employees at this address. WerderstraBe #73, Bremen 2800, BR Deutschland, Federal Republic of Germany. Telephone: (49)(421) 555 690. FAX: (49)(421) 556 202. Contact: Hartmut Fine. Barbel Rathje, Contact. Company description: Manufacturing & distribution of fine art holographic jewelry. Wholesale of a wide variety of holographic products. Agency offers competent service for any private & commercial need: concept deSign, model making, manufacutring process.

HOLOGRAPHIC CONCEPTS. 14 Cove Road, Forestdale, MA 02644 USA. Telephone: (508) 477 2488. Contact: George Willenborg. Company description: Artistic holography marketing; consulting.

HOLOGRAPHIC CREATIONS. 26- Rue Daniel Stern, Paris 75015, France. Telephone: (33)(1) 45788742. Contact: J-C Raverat de Boisheu. Company description: artistic holography, custom work.

HOLOGRAPHIC DESIGN, INC. (SEE OUR ADVERTISEMENT ON PAGE 11). Established in 1979. 6 Employees at this address. 1084 North Delaware Avenue, Philadelphia, PA 19125 USA. 400 West Erie Street, Chicago, IL 60610 USA. Telephone: (215) 425 9220. FAX: (215) 425 9221. Contact: D. Miller. Branch office: Robert Sherwood Holographic Design, Inc., Chicago, IL USA. Holograpic Products Inc., Richmond, UT USA. Company description: HDI provides holograms for a variety of display, promotional, advertising, packaging, and architectural applications. We offer a complete range of services to take your project from concept to final product.

HOLOGRAPHIC DESIGN SYSTEMS, INC. 1134 West Washington Blvd., Chicago, IL 60607 USA. Telephone: (312) 829 2292. Contact: Robert Billings. Description: Full service artistic holography; marketing consultant.

HOLOGRAPHIC DIMENSIONS, INC., 9235 SW 179 Terrace, Miami, FL 33157 USA. Telephone: (305) 255 4247. Contact: Kevin Brown. Company description: artistic holography

HOLOGRAPHIC IMAGES INC., Established in 1982. 6 Employees at this address. 1301 Dade Boulevard, Miami Beach, FL 33139, USA. Telephone: (305) 531 5465. Contact: Peg Lieberman. President, Larry lieberman; CEO, Frank Millman; Customer Service, Peg Lieberman. Company description: Produces multicolor reflection holograms on film. Specializing in limited edition art holograms - recorded from artwork created by artists from varied media. Custom images for corporations, commissioned editions.

HOLOGRAPHIC INDUSTRIES, INC. (MAIN HEADQUARTERS). Established in 1988. 5 Employees at this address. 3 Warwick Lane, Lincolnshire, IL 60069 USA. Telephone: (312) 945 2670. FAX: (001)(312) 491 7955. Contact: Robert Pricone, President. President, Robert Pricone; Secretary, Max Epstein; Sales Manager, Robert Pricone; Customer Service, Michael Epstein. Parent company: Holicon Corp., Evanston, IL USA. Branch offices: Light Wave Gallery, Chicago, IL USA; Light Wave Gallery, Schaumburg, IL USA. Company description: Holographic Industries designs and operates retail galleries/gift shops in major shopping centers. We produce our own pulse holographic images, and can obtain nearly any holographic product world-wide.

HOLOGRAPHIC PRODUCTS INC. (MAIN HEADQUARTERS). (SEE OUR ADVERTISEMENT ON PAGE 33) Established in 1975. 18 Employees at this address. 755 South 200 West, Richmond, UT 84333 USA. Telephone: (801) 258 2483. FAX: (801) 258 5219. Contact: Dave Rayfield, President. VicePresident, Hollie Rayfield; Customer Service, Marina Heidt. Branch offices: Holographic Design Inc., Philadelphia, PA USA. Company description: At Holographic Products Inc., we manufacture a full line of stock dichromate items. Custom holography is available in sizes up to 14" x 14".

HOLOGRAPHICS AUSTRALIA. Cambria Cottage, 111 New Town Road, Hobart 7008, Australia. Company description: Artistic holography.

HOLOGRAPHIC SERVICE. 10 via Civerchio, 1-20159 Milan, Italy. Telephone: (39)(02) 688 7067. Company description: Marketing consultants

HOLOGRAPHIC SHOP OF MILWAUKEE. 5644 Parking Street, Greendale, WI 53129 USA. Telephone: (414) 421 6767. Contact: George Niedzialkowki. Description: Buying & Selling holograms.

HOLOGRAPHICS INTERNATIONAL. BCM Holographics, London WC1 N 3XX, England, United Kingdom. Telephone: (44)(01) 584 4508. Contact: Sunny Bains. Company descripton: artist's magazine

HOLOGRAPHICS NORTH INC., Established in 1984. 7 Employees at this address. 444 South Union Street, Burlington, VT 05401 USA. Telephone: (802) 658-2275. FAX: (802) 862-6510. Contact: Dr. John Perry, President. VicePresident, Barbara D. Perry; Sales Manager, Josette Noll; Customer Service, Jeff Klute. Company description: Designers/Producers of large format holography up to 44" x 72" (1.1 x1 .8m.) Known worldwide for the highest quality commercial and fine art display work. Design, model building, production, installation and consulting services.

THE HOLOGRAPHIC STUDIO, LTD. Established in 1987. 3 Employees (includes subcontractors) at this address. 2525 York Avenue, Vancouver, British Columbia V6K 1 E4 Canada. Telephone: (604) 734 1614. FAX: (604) 734 2842. Contact: Melissa Crenshaw. Company description: The studio produces quality limited edition multi-color reflection holograms. In addition, we have vast experience in the production of Single color and achromatic refl ection transfers from ruby pulse masters.

HOLOGRAPHIC STUDIOS. 240 East 26th Street, New York, NY 10010 USA. Telephone: (212) 686 9397. FAX: (212) 481 8645. Contact: Jason Sapan. company description: Artistic holography; marketing & lighting consultants; holography workshops.

HOLOGRAPHICS (UK) LTD. 32 Lexington Street, London W1 R 3HR, England, United Kingdom. Telephone: (44)(01) 437 8992. FAX: (44)(01) 494 0386. Contact: Jon Vogel. Company description: Embossed & artistic holograms; pulsed portraits; DCG; large format.

HOLOGRAPHIC VISIONS. 300 South Grand Avenue, Los Angeles, CA 90071 USA. Telephone: (213) 687 7171. Contact: Bill Hilliard. Company description: marketing & lighting consultants; wholesale holography. Holographie Design. Am Kasinogarten 10, D-5600 Wuppertal 1, Federal Republic of Germany. Telephone: (49) 0202 314544. Company description: Fine art holograms

HOLOGRAPHIE KONZEPT. Korberstrasse 3, D-6000 Frankfurt 50, Federal Republic of Germany. Company description: Artistic holography.

HOLOGRAPHIE LABOR I MIKE MIELKE. Georgenslrasse 61, D-8000 Munich, Federal Republic of Germany. Telephone: (49)(89) 271 2989. FAX: (49)(89) 271 1375. Company description: Stereograms; silver halide holograms; DCG; retail shop.

THE HOLOGRAPHY DEVELOPMENT GROUP. The Coach House, 188 Kenilworth Avenue, Toronto, Ontario M4L 396 Canada. Telephone: (416) 691 9381. Fax: (416) 691 0407. Contact: Andrew Laczynski. Company description: Research & development, custom packaging.

HOLOGRAPHY INSTITUTE. P.O. Box 446, Petaluma, CA 94953 USA. Telephone: (707) 778 1497. Contact: P. Pink. President, Jeffrey Murray. Company description: Classes: Holography education for teachers, artists, commercial designers; Workshops for hobbyists-all ages, all levels. Group or individual instruction. Commercial: embossing masters; fine art! special editions; design, consulting, research.

HOLOGRAPHY LTD. Established in 1986. 5 Employees at this address. 21 Hakomemiut Str., Herzlia Pituah, Israel. Contact: David Livneh, Shimon Hameiri. Parent company: The Third Dimension Ltd. Company description: Buying & Selling holograms; educational holography

HOLOGRAPHY NEWS. (MAIN HEADQUARTERS). (SEE OUR ADVERTISEMENT ON PAGE 25) Established in 1987. 3 Employees at this address. 3932 McKinley Street N.W., Washington, D.C. 20015 USA. Telephone: (703) 273 0717. FAX: (703) 273 0745. Contact: Lewis Kontnik. Publisher, Lewis Kontnik; European Editor, Ian Lancaster. Parent company: Reconnaissance, Ltd. Branch office: Surrey, England, United Kingdom. Company description: Holography News-The International Newsletter of the Holography Industry is published ten times a year. It provides regular coverage and features on commercial news, RID activities, patents, corporate developments and conferences.

HOLOGRAPHY NEWS. (SEE OUR ADVERTISEMENT ON PAGE 25) (Branch Office).1 Erica Court, Wych Hill Place, Woking, Surrey AU22 OJB, England, United Kingdom. Telephone: (44)(04) 837 40689. FAX: (44)(04) 83740689. Contact: Ian Lancaster.

HOLOGRAPHY WORKSHOPS--LAKE FOREST COLLEGE (see Lake Forest College Holography Workshops)

HOLOGRAPHY WORKSHOP. Goldsmith College, Millard Bldg., Cormont Road, London SE5 9RG, England, United Kingdom. Telephone: (44)(01) 7333716. Contact: Susan Gamble. Company description: Artistic holography; holography education.

THE HOLOGRAPHY YEARBOOK. Rita Wittig Fachbuchverlag, 10 Chemnitzer Strasse, D-5142 Huckelhoven 1, Federal Republic of Germany. Telephone: (49)(2433) 84412. FAX: (49)(2433) 86356. Contact: Prof. Dr. Siegmar Wittig. Company description: Rita Wittig Publishing provides the broadest range in holography books worldwide. The Holography Yearbook is a comprehensive annual inventory of holography and its applications. Other publications: textbooks and catalogues.

HOLO-IMAGES, INC. 167 Washburn Road, Briarcliff Manor, NY 10510 USA. Telephone: (914) 941 8811 . Company description: Artistic holography.

HOLO-LASER. Established in 1978. 3 Employees at this address. 6, rue de la Mission, Ecole, 25480 Miserey, France. Telephone: (33)(1) 45 315 275. FAX: (33)(1) 48331 702. Contact: Dr. Jean louis, H., Tribillon. Branch offices: Besançon, France; Paris, France. Company description: Embossed holography & equipment; artistic holography; buying & selling; education.

HOLO-LASER. (Branch Office). 4, rue du Refuge, 25000 Besançon, France.

HOLO-LASER. (Branch Office). 12, rue de Vouille, 75015 Paris, France.

HOLOLASER TECH LTD. Established in 1982. 3 Employees at this address. 7 Fraser Avenue, Unit 16, Toronto, Ontario M6K 1Y7 Canada. Telephone: (416) 5380775. FAX: Same. Contact: Glenn Strazds, President. President, Glenn Strazds; Vice-President, Dave Stevens. Parent company: laser Gallery.

HOLOMAGIC INC. 917 17th Avenue SW, Calgary, Alberta, Canada T2T OA4 Telephone: (403) 229 0069. Contact: Ruth Simkin, President. Company description: artistic holography

HOLOMART-- PREMIUM TECHNOLOGY LTD. 9 Brunswick Centre, london WC1 N 1AF, England, United Kingdom. Telephone: (44)(01) 353 4212. FAX: (44) (01) 353 0684. Contact: Tanya. Company description: buying & selling holograms.

HOLOMART PLC, (MAIN HEADQUARTERS). Hamilton House, 1 Temple Avenue, london EC4Y OHA, England, United Kingdom. Telephone : (44)(01) 353 4212. FAX: (44)(01) 353 3325. Contact: Bruce Snyder. Company description: buying & selling holograms

HOLOMEDIA AB/HOLOGRAM MUSEUM. P.O. Box 45012, DroUninggatan 100, 10430 Stockholm, Sweden. Telephone: (46)(08) 105 465. FAX: (46)(08) 107 638. Contact: Mona Forsberg. Company description: Broker for embossed and artistic holography; Buying & selling holograms; Holography education; Gallery

HOLOMEDIA INC., Established in 1977. 7 Employees at this address. 3-15-22, Takaban, Meguro-ku, Tokyo 152 Japan. Telephone: (81) (03) 793 2321. FAX: (81) (03) 793 2322. Contact: Takao Kawahara, Marketing Director. President, Masato Nakajima; Customer service, Hibiki Tsuge. Company description: Holomedia is a reputable company producing Display Dichromate Holograms. High quality, the world's brightest, and in wide sizes (500x500mm).

HOLOMETRIC AB. Bjornasvagen 21, S-113 47 Stockholm, Sweden. Telephone : (46)(08) 790 9780. Contact: Ingegard Dirtoft. Company description: Equipment & supplies; holographic non-destructive testing.

HOLOMEX LTD. Established in 1987. 2 Employees at this address. 4 Borrowdale Avenue, Harrow HA3 7PZ, England, United Kingdom. Telephone : (44)(01) 427 9685. Contact: Michael Anderson, Managing Director. President, Michael Anderson ; Vice-President, Susan Anderson. Company description: The main company product is the Viewcam holographic camera which can make and display silver halide transmission and reflection holograms up to a maximum size of 10in x 10in.

HOLOMORPH VISUALS, INC . 273 de la Gauchetiere W., Montreal, Quebec H27 1 C7 Canada. Telephone: (514) 872 4530. Contact: Kenneth T. Chalk. Company description: Artistic holography

HOLO-OR LTD. Established in 1989. 6 Employees at this address. P.O. Box 1051, Rehovot 76110, Israel. Telephone : (972)(8) 465 089. FAX: (972)(8) 466 378. Contact: Nissim Greisas. President, Grossinger I; Vice-President, Uri levy; Sales Manager, Nissim Greisas. Company description: Holo Or develops and applies holographic technology in applications such as computer generated holograms for CO_2 laser optics, holographic elements for switching and data communication and holographic optical elements for OEM equipment.

HOLOPRINT ROSOWSKI. Postfach 1164, Lindena 23, D-4174 Issum 1, Federal Republic of Germany. Telephone: (49)(02835)1684. Description: Workshops, embossed & artistic holography, buying & selling--wholesale.

HOLOPRODUCTION. Established in 1986. 35 rue Abbatucci, 68330 Huningue, France. Telephone: (33) (89) 69 82 08. Contact: Mrs. J. Striebig, General Manager. Company description: Embossing consultants mass-manufacturing; artistic presentation consultants holography education; medical research; HNDT; la installation; equipment and supplies.

HOLOPUBLIC, KLAUS UNBEHAUN. Established ir 1985. Hirschstrasse 84, D-5600 Wuppertal-2,Federa Republic of Germany. (FRG) . Telephone: 0202 84118. Contact: Klaus Unbehaun. President: Klaus

Unbehaun. Company description: Klaus Unbehaun, owner of "Holopublic", is working as a media journalist (especially commercial holography). He is publishing the newsletter "Holography & 3-D Software", and he is preparing a holography book.

HOLOS ART GALERIE. 4 Place Grenus, 1201 Geneva Switzerland. Telephone: (41)(022) 325 191 . Contact: Pascal Barre. Company description: Gallery, retail sales.

HOLO-SERVICE. Neuensteinerstrasse 19, CH-4153 Basel, Switzerland. Telephone: (41)(061) 502 287. Contact: Edgar Bar. Company description: Artistic holography.

HOLO-SERVICE.FRIES. Eulerstrasse 55, CH-4051 Basel, Switzerland. Telephone: (41)(061) 22647. Contact: Urs Fries. Company description: Artistic holography.

HOLOS GALLERY. (SEE OUR ADVERTISEMENT ON PAGE 21). Established in 1979. 10 Employees at this address. 1792 Haight Street, San Francisco, CA 94117 USA. Telephone : (415) 2214717. FAX: (415) 221 4815. Contact: Gary Zellerbach, President. President, Gary Zellerbach; Sales Manager, Alan Rhody; Customer Service, Kelly Van Sciver. Company description: Holos Gallery is one of the world's oldest and largest distributors of holographic products. We specialize in holographic novelty products, dichromates, film holograms, and excellent new lines of photo-polymer holograms.
Holos-Holos. Peris y Valero, 130 pta. 21, 46006 Valencia, Spain. Telephone: (96) 333 3013. Contact: Vincente Carreton. Company description: artistic holography; marketing consultant.

HOLO/SOURCE CORPORATION. Established in 1985. 5 Employees at this address. 21800 Melrose Avenue, Southfield, MI 48075 USA. Telephone: (313) 355 0412. FAX: (313) 355 0437. Contact: Lee Lacey, President. President, Lee Lacey; Vice-President, Robert Levy; Sales Manager, Bill Seydel; Customer Service, Robert Levy. Company descripn: Holo/Source manufactures fine quality embossed holograms and colorful diffraction grating patterns for catalog and magazine covers, direct mail marketing projects and point-of-purchase displays.

HOLO-SPECTRA. 7742-B Gloria Avenue, Van Nuys, CA 91406 USA. Telephone: (818) 9942577. FAX: (818) 9944709. Contact: R. Arkin. Company description: artistic holography consulting; embossed holography; wholesale buying & selling; silver halide; DCG; lasers, mirrors, lenses; filters, pinholes, isolation tables.

HOLOSYSTEMS INC. P.O. Box 6810, Ithaca, NY 14850 USA. Telephone: (607) 273 1187. Contact: Jonathan Back. Company description: artistic holography.

HOLOTEC CC. P.O. Box 5144, Brackengardens, 1452 Transvaal, South Africa. Telephone: (27)(011) 864 1292. Contact: Mandy Van Der Molen. Company descripiton: Artistic holography.

HOLOTEC PLC. 7 Cameron Road, Seven Kings, Essex, IG1 3DF, England, United Kingdom. Telephone; (44)(01) 5978004. Contact: Janet Ives, Managing Director. Company description: Artistic holography marketing consultants.

HOLOTEK LTD., 300 East River Road, Rochester, NY 14623 USA.Telephone: (716) 4244996. Contact: Charles Kramer. Company description: H.O.E's and scanners.

HOLOTEK, SA Established in 1988. 4 Employees at this address. Carretera de Santander, Granda 47, 33199 Granda-Siero, Asturias, Spain. Telephone : (34)(985) 793526. FAX: (34)(85) 27 1853. Contact: Julio Ruiz Garcia, President. Company description: Holotek works on distribution of embossing and gifts with holograms through big stores and makes custom holograms. Holotek is open to all. For more information please contact us.

HOLOTRON SRL. 46 via Tolstoi, 1-20146 Milan, Italy. Telephone: (39)(02) 479 697. Company description: Marketing consultants.

HOLOVISION. 43 Pall Mall, London SW1 Y 55G, England, United Kingdom. Telephone: (44)91) 839 5622. Company description: Retail shop.

HOLTRONIC. Melchior-Huber Strasse 25, 0-8011 Ottersberg, Post Pliening, Federal Republic of Germany. Telephone: (49)(08121) 81005. Contact: Dieter Basler. Company description: Artistic holography, pulsed portrait, HOEs, embossed holographyl

HOWARD SMITH PRECISION OPTICS. 61 Lancaster Road, New Barnett, Hertfordshire, EN4 BA5, England, United Kingdom. Telephone : (44)(1) 441 7878. Company description: Manufacture mirrors, lenses.

HUGHES AIRCRAFT CO.--LASER PRODUCTS. 6155 El Camino Real, Carlsbad, CA 92008 USA. Telephone: (619) 931 3252. Contact: Marcia Berg. Company description: C02 Lasers for sale.

HYPERSPACE STUDIO. 973 Page Street Studio, San Francisco, CA 94117 USA. Telephone: (415) 431 9581. Contact: J Belk. Company description: Fine art holograms

IAN M. LANCASTER HOLOGRAPHICS CONSULTANCY. Established in 1986. 1 Employee at this address. 1 Erica Court, Wych Hill Park, Woking, Surrey GU22 OJB England, United Kingdom. Telephone: (44)(483) 740689. FAX: (44)(483) 740 689. Company description: Consultant, Curator; European Editor, Holography News; founder, Third Dimension Limited; former Director, Museum of Holography, NY. Specialising in display holography; business development, market studies, marketing concepts, art and display exhibitions.

IBERO GESTAO - GESTAO INTEGRADA E TECNOLOGICA LDA. Established in 1988. 7 Employees at this address. Apartado 1267, 4104 Porto Codex, Portugal. Telephone: (351-2) 301 276. Contact: Filipe Vallada P. Norais, President. Vice-President, Figueroa Gongalves; Sales Manager, L. Abrunhosa, Customer Service: Fatima. Parent Company: InterEuropeia, Portugal. Company description: Artistic holography; buying & selling; marketing & educational consulting ; access to NDT labs.

IBM ALMADEN RESEARCH CENTER. K69/803, 650 Harry Road, San Jose, CA 95120 USA. Telephone: (408) 927 1937. Contact: Glenn Sincerbox. Company description: Manufacturer of HOE's and optical heads; scientific holography research.

IBOU INC. Established in 1984. 2 Employees at this address. CP 214, Cap-de-la-Madeleine, Quebec, G8T 7W2 Canada. Telephone: (819) 295 5229. FAX: (819) 295 5229. Contact: Jean-Pierre Marchand. President, Jean-Pierre Marchand; Sales Managaer, Brigitte Gagnon. Parent company: Graphie (Edition) ; ET (retail & commercial sales). Main Headquarters: Quebec, Canada. Company description: Buying & selling holograms; Consulting; Manufacturefine art & silver halide holograms

IBOU INC. (MAIN HEADQUARTERS) 306 NotreDame, Champlain, Quebec, G8T 7W2, Canada.

ICI AMERICAS. Concord Pike, Wilmington, DE 19897 USA. Telephone: (302) 575 3087. Company description: Optics; HOE's; gratings

ICON HOLOGRAPHIC. 11 Uxbridge Street, London, W8, England, United Kingdom. Company description: artistic holography.

IDHOL. Boite Postale 7, F. 89340 Saint-Agnan, France. Telephone: (33)(16) 8696 1929. Contact: Jacques Bousigue. Company description: Fine art holographics; silver halide holograms; presentation consultant.

ILFORD LIMITED. Established in 1880. Mobberley, Knutsford, Cheshire WA15 7HA, England, United Kingdom. Telephone: (565) 50000. FAX: (44)565 872 734. Contact: Dr. Glenn P. Woodd, Business Development Manager. Parent Company: International Papers Corporation. Company description: Equipment & supplies for holography, film, plates, photochemicals.

ILFORD INC. West 70 Century Road, Paramus, New Jersey 07653 USA. Telephone: (201) 265 6000. Contact: Ek Sachtler, Sales & Inquiries.

ILLINOIS INSTITUTE OF TECHNOLOGY, Mechanical & Aerospace Engineering. Engineering Building #1, Room 2460, Chicago, IL 60616 USA. Telephone (312) 5673249. Contact: Cesar Sciammarella. Company description: Holographic interferometry; industrial holographic research; Non-destructive testing.

ILLINOIS VALLEY MAGNETIC RESONANCE. 4005 Progress Boulevard, Peru, IL 61354 USA. Telephone: (815) 223 8674. Contact: Dr. John L. Mori. Description: Scientific holography research.

IMAC INTERNATIONAL, INC. 1301 Greenwood, Wilmette, IL 60091 USA. Contact: J. Kauffmann. Compa-ny description: Holography marketing consultants.

IMAGES COMPANY. Established in 1982. 14 Employees at this address. P.O. Box 313, Jamaica, NY 11419 USA. Telephone: (718) 706-5003. Contact: Elan Persch, Customer Service. President, John Panico; Vice-President, David Channer; Sales Manager, Ruth Enivoi. Company description: Images Company sells holographic equipment targeted to educational institutions, students and private holographers. Equipment available: development kits, mounting kits for lenses, beamsplitters, mirrors. Spatial filters, display lights, safe lights, filters

IMAGING & DESIGN. Established in 1987. 7 Employees at this address. 1101 Ransom Road, Grand Isand, NY 14072-1459 USA. Telephone: (716) 773 7272. Contact: Keith Allen. Company description: Imaging division distributes film, chemicals, darkroom processing and safety equipment and supplies. Design division direct markets custom or stock emoossed and silver images to ad agencies, converters, corporate end-users

IMPERIAL COLLEGE OF SCIENCE. Optics Section, Blackett Laboratory, London SW7 2BZ, England, United Kingdom. Telephone: (44)(1) 5895111 . Contact: J. Dainty. Company description: Courses in holography; scientific holography research; particle measurement.

ING.-AGENTUR FUR NEUE TECHNOLOGIE IN OPTIK UND PRECISION ENGINEERING. D-7771 Frickingen 2, Federal Republic of Germany. Contact: P. Langenbeck. Company description: Holographic nondestructive testing; industrial research.

INGENIEUR BORO GEIGER. Established 1982. 2 Employees at this address. Dieding 7, D-8017 Ebersberg, Federal Republic of Germany. Telephone: (08092) 6583. FAX: (08092) 31658. Contact: Mr. Thomas Geiger. Company description: Embossed; artistic holography

INSTITUTE OF ART & DESIGN, UNIVERSITY OF TSUKUBA. Established in 1970. 50 Employees at this address. 1-1, Tennodai, Tsukuba, Japan 305. Telephone: (81)(0298) 53 2833. FAX: (81)(0298) 53 6508. Contact: Shunsuke Mitamura, Professor. Description: Artistic holography, Gallery, Holography education

INSTITUTE OF ELECTRONICS BSSR. Established in 1962. 600 Employees at this address. Academy of Sciences-Minsk, 22 Logoiski Trakt, 220841 Minsk-90, USSR. Telephone: (7) Minsk 65 64 74. Contact: Yuri Morgun. President, VA Pilipovich; Vice-President, AA Kovalev. Description: Development & manufacturing of highly coherent monopulse lasers & doublepulse lasers with high spectral radiance based on ruby, YAG, neodymium for applications in holography, holographic interferometry, and holographic systems.

INSTITUTE OF NUCLEAR PHYSICS. Leningradska obl., 188350 Gatchina, USSR. Contact: A.M. Bekker. Description: Scientific research.

INSTITUTE OF OPTHALOMOLOGY. Jud Street, London WC1, England, United Kingdom. Telephone: (44) (01) 3879621. Contact: John Marshall. Company description: Medical holography.

INSTITUTE OF OPTICAL RESEARCH. Royal Institute of Technology, S-100 44 Stockholm, Sweden. Telephone: (46)(08) 790 7283. Contact: Klaus Biedermann. Description: HOE manufacturer; recording materials manufacturer; industrial research; holographic non-destructive testing.

INSTITUTE OF OPTICAL SCIENCE/CENTRAL UNIVERSITY. Chung-Li 32054, Taiwan, R.O.C. Telephone: (886)(3) 425 7681. FAX: (886)(3) 425 8816. Contact: Mr. Tang Yaw Tzong. Description: HOEs, academic research.

INSTITUTE OF PHYSICS. Ukrainian Academy of Sciences, Prospect Nauki 46, 252 650 Kiev 28, USSR. Telephone: (7) 22 2158. Contact: Vladimir Markov. Description; artistic; reflection holography; research in recording materials.

INSTITUTE OF PLASMA PHYSICS AND LASER MICROFUSION, P.O. Box 49,00-908 Wroclaw, Poland. Contact: Zbigniew Sikorsky. Description: academic research

INTEGRAF. (SEE OUR ADVERTISEMENT ON PAGE 13) P.O.Box 586, Lake Forest, IL 60045 USA. Telephone: (708) 234 3756. FAX: (708) 615 0835. Contact: Tung H. Jeong, President. Company description: The main business of Integraf is to distribute holographic film and plates. We also carry prepackaged processing chemicals, and a variety of stock holograms.

INTERCHANGE STUDIOS. 15 Wilkin Street, London NW5 3NG, England, United Kingdom. Telephone: (44)91) 2679421. Company description: workshops

INTEREUROPEIA, (MAIN HEADQUARTERS), Rua Antonio Rodrigues Rocha 248, Vila Nova Gaia, Portugal. Subsidiary company: Ibero Gestao-Gestao Integrada E Tecnologica LDA, Portugal.

INTERFERENCE HOLOGRAM GALLERY. 1179A King Street West, Toronto, Ontario, Canada M6K 3C5. Telephone: (416) 535 2323. Company description: Gallery and retail shop.

ION LASER TECHNOLOGY INC., Established 1983. 35 Employees at this address. 263 Jimmy Dolittle Road, Salt Lake City, UT 84116 USA. Telephone: (801) 5371587. FAX: (801) 537 1590. Contact: Richard G. Collier, VP. President, Lynn Barney; VicePresident, Kevin D. Ostler; VP/Sales Manager, Richard G. Collier; Custormer Service, Don Zane Iii. Company Description: Manufacturer of air-cooled argon lasers.

ISAST/LEONARDO. (SEE OUR ADVERTISEMENT ON PAGE 203) (MAIN HEADQUARTERS) Established in 1981.6 Employees at this address. P.O. Box 75, 1442A Walnut Street, Berkeley, CA 94709 USA. Telephone: (415) 845 8306. FAX: (415) 841 6311. Contact: Kate Sholly. President, Dr. Roger F. Malina; Sales Manager, Tracy Waterman; Customer Service, Kate Sholly. Branch offices: 8 rue Emile Dunois, 92100 Boulogne/Seine, France; 8000 Westpark Drive, McLean, VA 22102 USA. Company description: Publisher of Journal LEONARDO. SpeCial Issues on Holography as an Art Medium; 1989: $45. Holography Theme Pack: $23. Electronic newsletter, database and Directory: Holography Hotline on MCI, WELL.
   ISAST/LEONARDO. (Branch office). 8 rue Emile Dunois, 92100 Boulogne/Seine, France
   ISAST/LEONARDO. (Branch office). 8000 Westpark Drive, McLean, VA 22102 USA.

ISRAMEX CO. LTD., 25, Arlozorov Street, Tel-Aviv, 62-488, Israel. Telephone: (972)(03) 243 333. FAX: (972)(03) 223 202. Branch office of Newport Corporation, Fountain Valley, CA USA.

J

JAEGER GRAPHIC TECHNOLOGY, J.G.T.·· HOLO-FOIL SA, Established 1983. 22 Employees at this address. 20 Avenue des Desirs,B-1140 Brussels, Belgium. Telephone: 00-322-7359551. FAX: 733 1035. Contact: M. Jaeger, President. Vice-President, J. Curci; Sales Manager, R. Doree; Customer Service, H. Majeri. Company description: JGT Brussels is specialized in all kinds of hot stamping technology and runs a separate "HOLO-FOIL" department for holographic & diffraction stampings on all graphic and security materials. Worldwide contacts.

JAMES RIVER PRODUCTS. 5420 Distributor Drive, Richmond, VA 23225 USA. Contact: Drurey Baugn. Telephone: (804) 233 9145. FAX: (804) 231 7891. Company description: Manufacturer of holographic embossing machines.

JAYCO HOLOGRAPHICS. Established in 1986. 15 Employees at this address. 29-43 Sydney Road, Watford, Herts, WD1 7PY England, United Kingdom. Telephone: (44) 923 246 760. FAX: (44) 923 247 769. Contact: Rohit Mistry, President. Company description: Complete production service for embossed holograms. Embossing masters thru to fully finished product. Sixteen years of experience enbles Jayco to offer outstanding quality of product and service at competitive prices.

JODON INC., (MAIN HEADQUARTERS)Established 1963. 15 Employees at this address. 62 Enterprise Drive, Ann Arbor, MI 48103 USA. Telephone: (313) 761 4044. FAX: (313) 761 3322. Contact: Preston Miller, Sales Manager. President: John Gillespie, Vice-President, Mike Gillespie; Sales Manager, Preston Miller. Company description: Manufacture of Helium Neon Lasers, Laser systems, specialty laser tubes, optical and electro-optical instruments and systems. Supplier of Argon and Krypton Lasers, holographics films, plates and chemicals. Engineering services.

THE JOHNS HOPKINS UNIVERSITY. Dept of Physics and Astronomy, Baltimore, MD 21218 USA. Telephone: (301) 338 7385. Contact: Homaira Akbari. Description: Scientific holography research; Particle measurement.

K

KAISER OPTICAL SYSTEMS, INC. P.O.Box 983, 371 Parkland Plaza, Ann Arbor, MI 48106 USA. Telephone: (313) 6658083. Contact: B.J. Chang. Company description: HOEs; H.U.D.s

KAROLINSKA INSTITUTET, School of Dentistry, Box 4064, S-141 04 HUDDINGE, Sweden. Telephone: (46) (08) 7740080. Contact: Hans Ryden. Description: Medical holography.

K.C. BROWN HOLOGRAPHICS. 22 St. Augustine's Road, Camden Town, London NW1 9RN, England, United Kingdom. Telephone: (44)(1) 482 2833. Conact: K.C.Brown. Company description: Pulse porraits; artistic holography.

KENDALL HYDE LTD., Established 1972. 22 Employees at this address. Kingsland Industrial Park,Stroudley Road,Basingstoke, Hants.,RG24 OUG, England, United Kingdom. Telephone: 0256 840830. FAX: 0256840443. Contact: M. D. Kendall, Managing Director. President, M. D. Kendall; VicePrseident, D.J. Hyde, A. Edwards; Customer Service, C. Birch. Company description: Thin film optical coating engineers manufacturing windows, mirrors and beamsplitters up to 3 metres. Coatings for laser applications, conductive coatings and front surface mirrors.

KEYSTONE SCIENTIFIC CO. (SEE OUR ADVERTISEMENT ON PAGE 39) Established in 1985. 4 Employees at this address. P.O. Box 22, Thorndale, PA 19372 USA. Telephone: (215) 380-8092, Toll free: (800) 462 9129. FAX: (215) 384 8093. Contact: Ed Kelly, President. Company description: Manufacturers of automatic film and plate processors, film transports, film and plate holders. Distributors of Agfa, 11- ford and Kodak holographic films, plates and chemicals.

KODAK COMPANY. (See EASTMAN KODAK COMPANY). KOLBE-DRUCK,COLOCO GMBH & CO. KG, (MAIN HEADQUARTERS).Established 1828. 140 Employees at this address. 1m. Industrigelande 50, Postfach 1103, 0-4804 Versmold, Federal Republic of Germany. Telephone: (054232431 (-5) . FAX: 05423 41230. Contact: Sven Deutschmann, Product ManagerHolography. President, Claus-Peter Bohlmann; Sales Manager, Claus-Peter Bohlmann; Customer Service, J6rg Niggebrogge. Subsidiary company: KolbeHolografie- Collection, Coloco Printpartner. Company description: Kolbe-Druck is a printingcompany known for print specialities on plastiC substrates. Complete embossing facilities and application in-house. Kolbe-Holografie-Collection offers a widerange of standard motifs.

KONING EN HARTMAN, Elektrotechniek B.V., Energieweg 1, NL-Delft 2627, Netherlands. Telephone: (31)(015) 609906. FAX: (31)(015) 619 194. Branch office of Newport Corporation, Fountain Valley, CA USA.

KRAFTWERK UNION AG. 0-4330 Mulheim/Ruhr, Federal Republic of Germany. Contact: Gerhard Schoenbeck. Company description: Holographic nondestructive testing; industrial research

KYOTO TECHNICAL UNIVERSITY. Dept. of Photographic Technology, Matsugasaki, Sakyo-ku, Kyoto 606, Japan. Contact: Toshihiro Kubota. Description: Artistic holography: DCG, Color, Reflection holograms.

L

LABOR DR. STEINBICHLER, Established 1980. 12 Employees at this address. Am Bauhof 4, 0-8201 , Neubeuern, Federal Republic of Germany. Telephone: (0049) 8035 1018. FAX: (0049) 8035 1010. Contact: Dr. H. Steinbichler, President. VicePresident, G. Stief; Sales Manager: T. Franz. Company description: Holographic investigations, developments on contract basis; application laboratory for: vibration analysis, non destructive testing , deformation measurements, contour measurements, image analysis; pulsed and CW-lasers, motor test bench, computer based evaluation.

LABOR FOR HOLOGRAFIE, Am Forst 38, Wesel 0-4230, Federal Republic of Germany. Telephone: (49) 281) 52837. Contact: A. Fuchtenbusch. Company description: artistic holography; holography education; fine art holograms.

LAKE FOREST COLLEGE HOLOGRAPHY WORKSHOPS. Sheridan and College Road, Lake Forest, IL 60045 USA. Telephone: (312) 234 3100. Contact: Tung H. Jeong. Description: Each summer during the week after July 4, Lake Forest College offers a 5-day hands-on workshop for partiCipants who have no prior experience in holography. An advanced 5-day workshop follows. Write for information.

LAMBDA ANALYTICAL LABORATORIES. 515 Broad Hollow Road, Melville, NY 11747 USA. Telephone: (914) 654 9117. Company description: Holographic non-destructive testing, optics testing.

LAMBDNTEN OPTICS, Division of Optical Corp. of America.Established 1986. 12 Employees at this address. One Lyberty Way, Westford, MA 01886 USA. Telephone: (508) 692-8140. FAX: (508) 692 9416. Contact: George Olmsted. President: D.A. Johnson, Vice-President: G. Olmsted. Company description: Products: Precision, large aperture (to 36" diam.) aspheric mirrors for holographic production systems.

LAMINEXIHIGH TECH UK LTD. Bromfield Industrial Estates, Mold, Clwyd CH7 1JR, England, United Kingdom. Telephone: (44)(043) 525 9011. Contact: Keith Green. Company description: artistic holography; security holograms.

LASART LTD. Established in 1985. 9 Employees at this address. P.O.Box 703, Norwood, CO 81423 USA. Contact: Steven Siegel, Partner. Partner, August Muth. Company description: Lasart, Ltd. specializes in custom DCG work, from modelmaking, mastering and quality finishing. This includes production and limited edition jewelry, Swiss watches, and medium format composite sculpture.

LASER AFFILIATES. 2047 Blucher Valley Road, Sebastopol, CA 95472 USA. Telephone: (707) 823 7171. FAX: (707) 823 8073. Contact: N. Gorglione. Company description:

tion: Laser Affiliates is an awardwinning non profit organization that designs innovative holographic and laser theatrical productions, installations and exhibitions. Services include curatorial guidance, videotapes and media lectures

LASER APPLICATIONS, INC. DIV OF LASERMETRICS INC. Established 1965. 38 Employees at this address. 12722 Research Parkway, Orlando, FL 32826 USA. Telephone : (407) 380 3200. FAX: (407) 381 9020. Contact person: Joseph Salg, General Manager. President, Robert Goldstein; Vice-President, J.Salg; Sales Manager, B. Bernard; Customer Service, A. Lusigen. Parent Company: Lasermetrics, Inc. New Jersey, USA. Company description: Holographic non-destructive testing; manufacturer ruby/yag lasers; HOE manufactured; Holography equipment.

LASER ARTS. Established in 1985. 1712 Cathedral Street, Plano, TX 75023 USA. Telephone: (214) 423 0158. Contact: M. Talbott. Company description: Holographic consultants and implementers. Commercial utilization of holography, trade shows, unique promotions and museum exhibits (design, build, rent or sell). Venture capitalists consultants. Professionals in business, art, technology and applications.

L.A.S.E.R. CO. 1900 Grove Drive, Haymarket, VA 22069 USA. Telephone: (704) 754 2526. Contact: Jim Bowman. Company description: Fine art holograms; lighting consultant.

LASER ELECTRONICS PTY., LTD., Established 1967. 20 Employees at this address. P.O. Box 359, Southport, Queensland, 4215, Australia. Telephone: 61 75 53 2066. FAX: 61 75 53 3090. Contact: N. Walden, Managing Director. Sales Manager, R.C. Holberton; Customer Service, L. Darcy. Company description: Laser Electronics designs and manufactures an extensive range of lasers and laser systems across seven industry categories including scientific, educational, and research units. Custom systems can also be developed.

LASER FARE LTD. 15 Industrial Lane, Johnston, RI 02919 USA. Telephone : (401) 231 4400. FAX: (401) 231 4674. Contact: Rich Zucker. Company description: Artistic holography; Equipment; HOEs;

LASERFILM ECKHARD KNUTH - MUL TIPLEXHOLOGRAPHIE, Milchstrasse 12, 0-8000 Munich, Federal Republic of Germany. Company description:

Artistic holography.

facture lasers and laser equipment.

LASER FOCUS WORLD. 1 Technology Park Drive, P.O. Box 989,Westford, MA 01886 USA. Telephone: (508) 692 0700. FAX: (508) 692 0525. Company description: Laser trade magazine; annual catalogue.

LASERGRAFICS, Peris y Valero 130, 2a Valencia, Spain. Telephone: (34)(96) 333 3013. Contact: Vincente Carreton. Company description: Artistic holography.

LASER GRAPHICS. Established in 1988. 5 Employees at this address. AG. Dimitriou 150, 546 35 Thessaloniki, Greece. Telephone: (30)(031) 827777. Contact: Yannis Palamas. Xanthippos Vissios. Notis Kaponis. Company description: artistic holography.

LASERGRUPPEN HOLOVISION AB. OsthammarsQatan 69, S - 11528 Stockholm, Sweden. Telephone: (46) (08) 663 9908. Contact: Jonny Gustafsson. Company description: Artistic holography.

LASER HOLOGRAPHICS, INC. 1179 King St. West, unit 111, Toronto, Ontario, Canada M6K 3C5. Telephone: (416) 531 4656. FAX: (416) 530 1594. Contact: Charles Demicher. Company description: Emoossed holography broker; Marketing consultants; wholesale

LASER IMAGE DESIGN. 3031-K Nihi Street, Honolu, Hawaii 96819 USA. Telephone: (808) 848 1990. Contact: Chas Williams. Company description: Brokers for artistic holography.

LASER INNOVATIONS INC. 25 Fisherville Road, Unit 804, Willowdale, Toronto, Ontario M2R 3B7, Canada. Telephone: (416) 861 1747. Company description: art istic holography.

LASER INSTITUTE OF AMERICA. Education Division, 5151 Monroe Street, Toledo, OH 43623 USA. Telephone: (419) 882 8706. Contact: Gerald Glen. Company description: educational holography

LASER INSTRUMENTATION LTD., Unit 1 Bear Court, Daneshill East, Basingstoke, Hampshire RG24 OT, England, United Kingdom. Telephone: (44) (256) 469 572. Company description: Manu-

LASER INTERNATIONAL. 19 Normanton Rise, Holbeck Hill, Scarborough, N Yorks YO11 2XE, England, United Kingdom. Telephone: (44)(0723) 366 096. Contact: Keith Dutton. Company description: Gallery. Laserion Handels GmbH. Postfach 110268, 2800 Bremen 11, Federal Republic of Germany. Company description: Artistic holography; commissions.

LASER IONICS INC., (MAIN HEADOUARTERS). Established 1966. 25 Employees at this address. 701 South Kirkman Road, Orlando, FL 32811 USA. Telephone: (407) 298 1561. FAX: (407) 297 4167. Contact: Drew Nelson, Business Development Manager. President, Richard Demmer; Vice-President, William Newell. Parent Company: Trimedyne, Inc. Company description: Manufacturer of gas ion lasers including Argon, Krypton and mixed gases. Specializing in high power requirements needing stable power in a compact package.

LASERLABBET. Box 521, SE 581 06 Linkoping, Sweden. Telephone: (46) (13) 123 377. Contact: E.A.Jonsson. Company description: Artistic holography.

LASER LABS, INC. 8000 W.110th Street, Suite #115, Overland Park, KS 66210 USA. Telephone: (913) 451 9270. Contact: Steven Craft. Company description: Medical holography.

LASER LIGHT DESIGNS, 2412 Kennedy Way, Antioch, CA 94509 USA. Telephone: (415) 754 3144. Contact: Michael Mallott. Company description: Gallery; Retail shop; Wholesale

LASER LIGHT EXPRESSIONS PTY. LTD. HOLOPTICS. (SEE OUR ADVERTISEMENT ON PAGE 43) Established in 1984. 4 Employees at this address. 3 Gibbons Street, Telopea, New South Wales, Australia 2117. Telephone: (612) 890 1233. FAX: (612) 890 1243. Contact: Rosemary Sturgess, Marketing Manager. President, John A. Tobin. Subsidiary companies: Optical Security Systems; Optical Control and Display & Holoptics. Company description: Ever since 1984, our dedicated team has provided the Australasian region with a range of dependable, innovative holography and diffraction capabilities for display (HOLOPTICS) and security (OPTICAL SECURITY SYSTEMS) based applications.

LASER LIGHT IMAGE. 101 Spring Bank, Hull, HU3 1 BH, England, United Kingdom. Telephone: (44) (0482) 26744. FAX: (44)(0482) 492 286. Contact: Carl Racey. Subsidiary company: Amazing World of Holograms. Company description: Artistic holograms and equipment.

LASER LIGHT LTD. 57 Grand Street, New York, NY 10013 USA. Telephone: (212) 226 7747. Contact: Abe Rezny. Company description: Artistic holography.

LASER LIGHTWORKS, 81A Hatton Square, 16/16A Baldwins Gardens, London EC1 N 7RJ, England, United Kingdom. Telephone: (44)(1) 430 0028. Contact: Peter Thomson. Company description: artistic hologrpahy; holography education.

LASERMEDIA 2046 Armacost Ave., Los Angeles, CA 90025 USA Telephone: (213) 820 3750. FAX: (213) 207 9630. Contact: Mary Slusarski. Company description: Install laser light show exhibitions.

LASERMET LIMITED, Five Oaks, Sway Road, Brokenhurst, Hants S04 27RX, England, United Kingdom. Telephone : Lymington (0590) 23075. Contact: Dr William F. Fagan. Company description : Holographic Non-Destructive testing, Instruments, Research Contracting, Consulting, Metrology, Inspection, Safety.

LASERMETRICS, INC. (MAIN HEADQUARTERS), 196 Coolidge Avenue, Englewood, NJ 07631. Subsidiary company: Laser Applications, Inc. Company description: Industrial research; laser manufacturing.

LASER RESALE INC. 54 Balcom Road, Sudbury, MA 01776 USA Telephone: (508) 443 8484. FAX: (508) 443 7620. Contact: Jack Kilpatrick, System Sales. Company description: Laser Resale provides a marketplace for buying and selling pre-owned lasers, laser systems and associated equipment. Currently available holographic lasers are He:Ne, 15-70 mW, and, argon, 100 mW - 20 W.

LASER SCIENCE LSI. 80 Prospect Street, Cambridge, MA 02139 USA Company description: Laser equipment

THE LASERSMITH, INC. 1000 West Monroe Street, Chicago, IL 60607 USA Telephone: (312) 7335462.

Contact: Steven Smith. Company description: Artistic holography.

LASER TECHNOLOGY, INC. 1055 West Germantown Pike, Norristown, PA 19403 USA Telephone: (215) 631 5043. Contact: Tom Gleason, Sales Manager. President, John Neuman. Company description: Manufacture equipment for laser-based NDT; Holography and Shearography equipment and inspection services.

LASERWORKS. P.O. Box 2408, Orange, CA 92669 USA Telephone: (714) 832 2686. Contact: Selwin Lissack. Description: Holographic artist.

LASING SA, Marques de Pico Velasco, 64, E-28027 Madrid, Spain. Telephone : (34)(01) 268 3643. FAX: (34) (01) 407 3624. Branch office of Newport Corporation, Fountain Valley, CA USA

LASIRIS INC. (SEE OUR ADVERTISEMENT ON PAGE 45) (MAIN OFFICE) Established in 1985. 5 Employees at this address. 3549 Ashby, Ville St. Laurent, Que, Canada H4R 2K3. Telephone: (514) 335 1005. FAX: (514) 335 4576. Contact: Alain Beauregard, President. President, Alain Beauregard; Vice-President, Gaetan Robitaille. Branch office: Quebec. Canada. Company description: Embossed holography; artistic holography; buying & selling; industrial research; HOEs; holographic non-destructive testing.

LASIRIS INC. (SEE OUR ADVERTISEMENT ON PAGE 45) (Branch office). 840 Ste. Therese, Quebec, Quebec. Canada, G1N 157. Telephone: (418) 683 3530. FAX: (418) 682 5594. Contact: Alain Beauregard.

LAWRENCE BERKELEY LABORATORY. University of California. Building 80-101, Berkeley, CA 94720 USA. Telephone: (415) 486 4000. Contact: Malcolm Howells. Description: Industrial & academic holography research.

LAZA HOLOGRAMS. (SEE OUR ADVERTISEMENT ON PAGE 35) (Branch Office) Established in 1983. 4 Employees at this address. 47 Alpine Street, Reading, Berkshire, England RG1 2PY United Kingdom. Telephone: (44) 0734 589 026. FAX: (44) 0734 571 974. Contact: Chris Lambert, Owner. President, Chris Lambert; Customer Service, Carole Lambert. Main Headquarters: 68-72 Katesgrove Lane, Reading, Berkshire, RG1 2ND England. Company description: Specialist mass-producer of high quality film reflection holograms, large or small quantities. Copy service from your master. Full custom service available. Wide range of stock film holograms

LAZA HOLOGRAMS. (SEE OUR ADVERTISEMENT ON PAGE 35) (MAIN HEADQUARTERS) 68-72 Katesgrove Lane, Reading, Berkshire, RG1 2ND England.

LAZART HOLOGRAPHICS. Established in 1985. 2 Employees at this address. 22 Erina Valley Road, Erina, New South Wales 2250, Australia. Telephone: (61)(043) 676 245. FAX: (61)(043) 652306. Contact: Brett Wilson, Director. Company description: Artistic holography; buying & selling holograms.

LAZER WIZARDRY. Established in 1987. 2 Employees at this address. 11022 West Oregon Place, Lakewood, CO 80226 USA. Telephone: (303) 987 9438 Contact: Richard M. Osada, Owner. Company description: Wholesale distribution.

LCPC--LAB CENTRAL DES PONTS ET CHAUSSEES. 58 Boulevard Lefebvre, F-75015 Paris, France. Telephone: (33)(1) 4532 3179. Contact: Jean-Marie Caussignac. Company description: Industrial research; holographic non-destructive testing.

LENINGRAD SUBSIDIARY IN MACHINERY SCIENCE. Academy of Sciences of the USSR, Bolshoi Av. 61, 199178 Leningrad, USSR. Telephone: (7) (247) 9185. Contact: Juri Ostrovsky. Description: Scientific research.

LEONARD KURZ GMBH & CO. Schwabecher Strasse 482, Postfach 1954, D-8510 Firth, Federal Republic of Germany. Telephone: (49)(0911) 71410. Company description: Manufacturer of embossing equipment; broker for hologram embossing.

LES PRODUCTIONS HOLOLAB! 3970, Boulevarde St. Laurent, Montreal, Quebec H2W 1 Y3, Canada. Telephone: (514) 8494325. Contact: Marie-Christiane Mathieu. Company description: artistic holography.

LETTERHEAD PRESS INC. 155 North 120th Street, Dept. HM, Wauwatosa, WI 53226 USA. Telephone: (414) 258 1717. Contact: Mark Mulvaney, President. Company description: Holographic embossing and printing.

LET THERE BE NEON. P.O. Box 337, Canal Street Station, New York, NY 10013 USA. Telephone: (212) 2264883. Contact: Rudy Stern. Company description: Gallery, retail shop.

LICHT-BLICKE-BURO. Bornemannstrasse 10, D-6000 Frankfurt 70, Federal Republic of Germany. Contact: Walter Classen. Company description: Gallery.

LICON IX. (SEE OUR ADVERTISEMENT ON PAGE 32a) Established 1972. 42 Employees at this address. 3281 Scott Boulevard, Santa Clara, CA, 95054 USA. Telephone: (408) 496 0300. FAX (408) 492 1303. Contact: Carmen Jordan, Mng. Marketing Services. President, M.W. Dowley; Sales Manager, Randy Kimball; Customer Service, Greg Springer. Company description: liCON iX, long the recognized leader in Helium Cadmium laser technology, also supplies semiconductor diode laser systems and a recently introduced line of ion lasers.

LIGHT ANGELS. The Corridor, High Street, Bath Spa, Avon BA1 5AJ, England, United Kingdom. Telephone: (44)(0225) 62772. Contact: Mike Watts. Company description: Gallery.

LIGHT CONSTRUCTION, INC. 2154 Dundas Street West, Suite #503, Toronto, Ontario, M6R 1 X3 Canada. Telephone: (416) 533 4692. FAX: (416) 5330572. Company description: Fine art holograms; Large format; independent educational facility.

LIGHT ENGINEERING. 12 New St. Johns, St. Helier, Jersey, Channel Islands, England, United Kingdom. Telephone: (44)(534) 30614. Contact Anthony Hopkins. Company description: Gallery, retail shop.

LIGHT FANTASTIC. Established in 1987. 3 Employees at this address. Tabor Center, Bridge Mart, 1201 16th Street, Denver, CO 80202 USA. Telephone : (303)733 7856. Contact: R. Osada, Owner. Company description: Retail shop and gallery; Buying & selling holograms

LIGHT FANTASTIC PLC. (SEE OUR ADVERTISEMENT ON PAGE 7) (MAIN HEADQUARTERS). Established in 1981. 16 Employees at this address. 4E. F Gelders Hall Road, Shepshed, Leicestershire LE12

9NH, England, United Kingdom. Telephone: (44) (509) 600 220. FAX: (44)(509) 508 795. Contact: Roger C. Knight, Marketing Director. Managing Director, Peter H.L.Woodd; Marketing Director, Roger C. Knight; Marketing & Sales Coordinator, Paula Foulkes-Williams. Branch offices: Light Fantastic, Gallery of Holography, London, England; Light Fantastic, Distribution Centre, Hampshire, England. Company description: Light Fantastic PLC (Est. 1981) is a fully integrated holographic business providing the creative and technical services that produce innovative standard and custom designed holograms of the highest quality.

LIGHT IMPRESSIONS EUROPE PLC. (SEE OUR ADVERTISEMENT ON PAGE 17) 5 Mole Business Park 3, Off Station Road, Leatherhead, Surrey KT22 7BA, England, United Kingdom. Telephone: (44) 0372 386677. FAX: (44) 0372 386 548. Contact: Kenneth Harris. Branch Office: Light Impressions Inc., Santa Cruz, CA USA.

LIGHT IMPRESSIONS, INC. (SEE OUR ADVERTISEMENT ON PAGE 17) Established 1979. 15-20 Employees at this address.149-B Josephine Street, Santa Cruz, CA 95060 USA. Telephone: (408) 458 1991. FAX: (408) 458 3338. Contact: Fred Black, President. President, Fred Black; Sales Manager, Kevin Samson. Branch Office: Light Impressions PLC, Surrey, England. Company description : Light Impressions is an integrated, full-service commercial holography company. We produce custom and stock hologram masters and emboss metalized polyester. Diecutting and hot stamping are also offered.

LIGHT WAVE GALLERY. (Branch of Holographic Industries, Inc.) D-208 Woodfield Mall, Schaumburg, IL 60173 USA. Telephone: (312) 240 5344. Contact: Robert Pricone. Company description: Gallery, retail shop.

LIGHT WAVE GALLERY. (Branch of Holographic Industries, Inc.) North Pier, 435 East IL Street, Chicago, IL 60611 USA. Telephone: (312) 321 1123. Contact: Cindy Helfand. Company description: Gallery, retail shop.

LINDA LAW HOLOGRAPHICS. 8 Crescent Drive, Huntington, NY 11743 USA. Telephone: (516) 351 6056. Company description: Holographic artist; Holography education.

L.I.R.E.R.A. 12, rue Libergier, F-51110 Reims, France.

Telephone: (33) 26 884 452. Contact: Michel Grosmann. Company description: Scientific research; HOEs.

LITTON SYSTEMS CANADA LTD. 25 Cityview Drive, Rexdale, Ontario M9W 5A7 Canada. Contact: Romuaid Pawluczyk. Company description: holographic non-destructive testing; electro optics.

LOS ANGELES SCHOOL OF HOLOGRAPHY. P.O. Box 851, Woodland Hills, CA 91365 USA. Telephone: (818) 703 1111 . FAX: (818) 703 1182. Contact: Jerry Fox. Company description: The Los Angeles School of Holography offers a 3 day class. Students learn all phases of holography, and produce both laser viewable transmission and white light viewable holograms in silver halide format

LOUGH BOROUGH UNIVERSITY OF TECHNOLOGY. Dept. of Physics, Loughborough, Leicesterchire LE11 3TU England, United Kingdom. Telephone: (44)(509) 263 171 . Contact: Nick Phillips. Description: Embossing masters/shims; Scientific, industrial research

LULEA UNIVERSITY OF TECHNOLOGY. Dept. of Mechanical Engineering, S-951 87 Lulea, Sweden. Contact: Nils-Erik Molin. Description: Industrial research; holographic non-destructive testing.

LUMONICS INC., (MAIN HEADQUARTERS) Established 1971. 180 Employees at this address. 105 Schneider Road,Kanata, Ottawa, Ontario K2K 1 Y3, Canada. Telephone: (613) 592 1460. FAX: (613) 592 5703. Contact: Dr. Jim Higgins, President, D. J. James; Vice-President, R.S. Sandwell; Sales Manager, R. Rayman; Customer Service, K. Perkins. Subsidiary Companies: Lumonics Marking Corp.Camarillo, CA USA, Lurnonics Materials Processing Corp. ,Livonia (Detroit) Michigan USA, Lumonics JK Division, Rugby England. Company description: Lumonics is a manufacturer of high power pulsed ruby lasers for portrait holography and engineering holocameras for NDT. Other products include laser marking and materials processing systems.

LUMONICS LTD., Established 1972. 170 Employees at this address. Cosford Lane, Swift Valley, Rugby, Warwickshire, CV21 1QN, United Kingdom. Telephone: (44) 0 788 70321. FAX: (44) 0 788 79824. Contact: J.A. Synowiec, Sales and Marketing Manager, Scientific Products. Parent Company: Lumonics, Inc. Ottowa, Canada. Branch Offices: Lumonics, Brussels. Lurnonics, Munich. Lumonics, Paris. Company description: Lumonics manufactures pulsed lasers for a range of industrial and scientific applications including pulsed ruby lasers for Holography.

LUNO INSTITUTE OF TECHNOLOGY. Department of Physics, Box 118, S-221 00 LUND, Sweden. Telehone: (46)(046) 107656. Contact: Sven-Goran Pettersson. Description: Color H-1; holography education; academic research.

LURE. Institut d'Optique, BP 147, F-91403 Orsay, Cedex, France. Telephone: (33)(1)(69) 416846. Contact: D. Joyeux. Company description: Academic research.

M

J1ACSHANE HOLOGRAPHY/LASER ARTS PROGRAMS. Established in 1985. 2 Employees at this address. 512 West Braeside Drive, Arlington Heights, IL 50004 USA. Telephone: (708) 398 4983. Contact: Jim MacShane, Vice-President, President, Elaine MacShane. Parent company: Laser Arts Educational Programs. Company description: Artistic holography and education; holographic non-destructive testing.

MAGIC LASER. Established in 1985. Quartier de L'horloge, 4 rue Brantome, 75003, Paris France. Telephone: (33)(1) 42743578. FAX: (33)(1) 4774 3357. Contact: Anne-Marie Christakis, Manager. Sales Manager, Thierry Gueguen. Company description: Buying & Selling artistic holography.

MAGIC LASER LABORATORY. 6 rue Marie-Stuart, F-75002, Paris, France.

MAGIC LIGHT HOLOGRAFIE - GALLERIE. Bahnhofsplatz 2, 0-8000 Munich 2, Federal Republic of Germany. Telephone: (49)(089) 595 981. Company description: Artistic holography; Gallery.

AN ENVIRONMENT, INC. P.O. Box 25959, 2041 Saw-

telle Boulevard, Los Angeles, CA 90025 USA Telephone: (213) 477 7922. Contact: Gary Fisher. Company description: Artistic holograms.

MARKEM SYSTEMS LTD. Ladywell Trading Estate, Eccles New Road, Salford, M5 2DA England, United Kingdom. Telephone: (44)(61) 789 8131 . FAX: (44) (61) 707 5315,. Contact: Jane Oliver. President, Jeff Lomax; Vice-President, Ken Williamson. Branch offices: Advance Holographics Laboratories, Loughborough University; Marketm offices at High Wycombe, Kent, Glasgow, Rugby, Halifax, Pendleton. Company description: "One-Stop-Shop"service in embossed hot stamping foil and laminating film, including everything from origination to foil manufacture. Also large format, silver halide service for exhibitions and permanent installation.

MARTINSSON ELEKTRONIK AB. Instrumentvagen 16, Box 9060, S-126 09 HAGERSTEN, Sweden. Telephone: 946)(08) 744 0300. FAX: (46)(08) 7443403. Contact: Per Skande. Company description: Artistic; pulsed portraiture; equipment & supplies.

MARUBUN CORPORATION, 8-1 Nihombashi Odemmacho, Chuo-Ku, Tokyo, 103, Japan. Telephone: (81) (03) 648 8115. FAX: (81)(03) 6489398. Branch office of Newport Corporation, Fountain Valley, CA USA

MARWELL AB. Kyrkbacken 27, S-171 50 Solna, Sweden. Telephone: (46)(8) 838 261 . Company description: Artistic holography; fine art holography.

MASSACHUSETTS INSTITUTE OF TECHNOLOGY. M.I.T. Media Laboratory, Spatial Imaging Group, 20 Ames Street, E15-416, Cambridge, MA 02139 USA Telephone: (617) 253 0632. FAX: (617) 258 6264. Contact: Jane F. White. Description: College holography courses; curriculum development

MATT HANNIFIN CO. P.O. Box 4574, Austin, TX 78765 USA Telephone: (512) 452 7444. Contact: Matt Hannifin. Company description: Holography exhibition installer and spectacle consultant (openings). Previously director of installation-Museum of Holography, NY. Licenced, experienced high-power laser and firework show operator. Manufacturer hand-crafted boomerangs with holograms.

MAZDA MOTOR CORP. 3-1, Shinchi, Fuchu-cho,

Aki-gun, Hiroshima, Japan. Telephone: (81)(082)282 1111 . FAX: (81) (082) 285 9746. Company description: Industrial holography research; Holographic nondestructive testing; Interferometry.

McCAIN MARKETING. Established in 1974. 4 Employees at this address. 10962 North Wauwatosa Road 76W, Mequon, WI 53092, USA. Telephone: (414) 2424023. Contact: Richard McCain, President. President, Richard McCain; Vice-President, Clare McCain; Sales Manager, Richard McCain; Customer Service, Clare McCain. Company description: Act as liason between commercial advertisers (including Fortune 500 companies) and holographers. Educate clients to holography, develop advertising promotions, and educational applications using holography. Recommend professional holographic specialists as needed

MEDIA INTERFACE, LTD. 167 Garfield Place, Brooklyn, NY 11215 USA. Telephone: (718) 788 4012. Contact: Ronald Erikson.Company description: Artistic holograms, holography education consulting.

MEDICAL UNIVERSITY OF SOUTH CAROLINA. Dept. of Anatomy & Cell Biology, 171 Ashley Avenue, Charleston, SC 29425-2203 USA. Telephone: (803) 792 3529. Contact: Ammasi Periasamy. Description: Medical holography.

MELLES GRIOT, (MAIN HEADQUARTERS) Established in 1969. 100 Employees at this address. 1770 Kettering Street, Irvine, CA 92714 USA. Telephone : (714) 261-5600. FAX: (714) 261-7589. Contact: Lisa Tsufura, Technical Manager. Sales Administrator, Paul Kenrick; Vice-President of Marketing & Sales, Kevin Chittim; Sales Manager, Candice Bauccio; Customer Service, Sigi Hennessey. Parent Company: J. Bibby & Sons. Company description: Melles Griot is a major manufacturer of off-the-shelf and custom tables and isolation equipment, laser, lenses, mounting hardware, positioners, polarizers, coated optics, detectors, collimators and spatial filters.

MEREDITH INSTRUMENTS. Established in 1978. 6 Employees at this address. 6403 North 59th Avenue, Glendale, Al 85301 USA. Telephone: (602) 934 9387. Contact: Chad Andersen. President, Dennis Meredith; Sales Manager, Mary Moraine. Company description: Specializing in surplus inventories of HeNe lasers as well as argon and diode lasers, Meredith Instruments is the USA's largest laser discount dealer. Free Catalogue.

MESSERSCHMITT - BOELKOW-BLOHM. IENTRALE, Entwicklung MBB, Postfach 801109, D-8000 Munich 2, Federal Republic of Germany. Company description: Scientific research; color holography.

METAMORFOSI OLOGRAFIA ITALIA SRL. Established in 1983. 10 Employees at this address. Via Lecco 6, 20124 Milano, Italy. Telephone: (39)(2) 204 9943. FAX: (39)(2) 204 1625. Contact: Eva Aprile, Manager. President, Eva Aprile ; Vice-President, Silvia Aprile; Sales Manager, Manuela Polenta; Customer Service, Francesca Cominelli. Company description: Producer of the first and original HOLOTIME, the interchageable hologram watch. This year we are producing small size DCG holograms. Consulting to Italian customs marketing.

METAPLAST ELECTROCHEMICALS CORP., Established 1963. 3 Employees at this address. 67 Whitson Street, Hempstead, New York, 11550 USA. Telephone: (516) 481 4530. FAX: (516) 481 7320. Contact: J. L. Lester, President. Company description: Manufacturer of conductive coatings and spray, plating & electroforming equipment are offered for plating on plastic and other non-conductive surfaces. Electroforming consultants to the aerospace, electronic, phonographic, computer, holography & toy industries.

METROLOGIC INSTRUMENTS, INC. P.O. Box 307, 143 Harding Avenue, Bellmawr, NJ 08031 USA. Telephone: (609) 933 0100. Contact: Marketing Department. Company description : We manufacture mirrors.

MGD MODULATIONS. 1293 Rue de la Visitation, Montreal, Quebec, Canada. Telephone: (514) 598 8860. Company description: fine art holograms.

MICRAUDEL. Established in 1980. 6 Employees at this address. 93, rue Adelshoffen, F-67300 Schiltigheim, France. Telephone: (33) 8881 3293. Contact: Philippe Burger, Directeur Technique. Company description: Electronic and information applications; holographic non-destructive testing.

MINCHIMPROM. 101851 Moscov, USSRContact: N.S. Gafurova. Description: Pulsed portraiture, artistic holography.

MIND'S EYE: HOLOGRAPHIC CONSULTANTS. 17329 lola Street, Granada Hills, CA 91344 USA. Telephone: (818) 360 6023. Contact: Stephen Roth. Company description: Marketing consultant.

MIRAGE HOLOGRAMS LTD. Unit 2 Brook Lane, Business Centre, Brook Lane North, Brentford, Middlesex TW8 OPP England, United Kingdom. Telephone: (44) (1) 568 2454. Contact: Lynne Hesp. Company description: Silver halide hologram maker.

M. I.T. (See Massachusetts Institute of Technology)

MITSUBISHI HEAVY INDUSTRIES LTD., Nagasaki Technical Institute, 1-1 Akunoura-machi, Nagasaki 850-91 Japan. Contact: M. Murata. Description: Holographic non-destructive testing; industrial research.

MITUTOYO MEASURING INSTRUMENTS. 18 Essex Road, Paramus, NJ 07652 USA. Telephone: (201) 368 0525. Contact: Joe Scriff. Company description: Manufacturers of precision measuring instruments including holographic linear tracking systems.

MOELLER WEDEL OPTISCHE WERK. Rosengarten 10, 2000 Wedel, Federal Republic of Germany. Comany description: Artistic holography.

MOSCOW PHYSICAL ENGINEERING INSTITUTE Kashirskoe Shosse 1, Moscow, 115409 USSR. Contact: Alexander Larkin. Description: Artistic holography; scientific holography research.

MUNDAY SPATIAL IMAGING. 39 Pyrcroft Road, Chertsey, Surrey KT16 9HT, England, United Kingdom. Telephone: (44)(0932) 564 899. Contact: Rob Munday. Company description: Artistic holography; buying & selling

MUSéE DE L'HOLOGRAPHIE. 15 a 21 Grand Balcon, Forum des Hailes, BP 180, 75001 Paris, France. Telephone: (33)(1) 4039 9683. Contact: Anne-Marie Christakis, Manager. Sales Manager, Thierry Gueguen. Company description: Gallery; Mail order; Hography-education.

MUSEUM FuR HOLOGRAPHIE & NEUE VISUELLE MEDIEN. Established in 1979. 7 Employees at this address. Pletschmuhelenweg 7, D-5024 Pulheim 1, Federal Republic of Germany. Telephone: (49)(02) 233 385 1053. FAX: (49)(02) 238 52158. Contact: Matthias Lauk, Director. Curator: Hans-Peter Ott. Company description: The first museum of holography in Europe. Permanent showroom including classi-cal holographic artworks. Guided tours. Workshop program. Consultation and organisation of national and international exhibitions. Holography Consulting .

MUSEUM OF HOLOGRAPHY/CHICAGO. 1134 West Washington Boulevard, Chicago, IL 60607 USA. Telephone: (312) 226 1007. Contact: L. Billings. Description: Gallery; Retail shop; Wholesale; Mail order; Holography Education .

MUSEUM OF HOLOGRAPHY. 11 Mercer Street, New York, NY 10013 USA. Telephone: (212) 925 0581. FAX: (212) 840 1663. Contact: Current Director. Description: Gallery

MUSEUM OF THE FINE ARTS RESEARCH & HOLO-GRAPHIC CENTER. 1134 West Washington Boulevard, Chicago, IL 60607 USA. Telephone: (312) 226 1007. Contact: John Hoffmann. Description:Gallery; Hands on Workshops.

N

NANCY GORGLIONE/FINE ART HOLOGRAMS. Established in 1985. 2047 Blucher Valley Road, Sebastopol, CA 95472 USA. Telephone: ((707) 823 7171. FAX: (707) 823 8073. Contact: Nancy Gorglione, Director. Parent company: Cherry Optical Co; Laser Affiliates. Company description: Nancy Gorglione produces unique fine art reflection and transmission hologram composites. Created from multiple masters copied onto glass plates, these large multicolored hologram composites create scenes of lasting sensory appeal.

NASA MARSHALL SPACE FLIGHT CENTER. Space Sciences Laboratory, ES 73, Huntsville, AL 35812 USA. Telephone: (205) 544 7812. Contact: Robert Owen. Company description: Scientific holography research; interferometry.

NATIONAL PHYSICAL LABORATORY. Teddington, Middlesex, England TW11 OLW United Kingdom. Telephone: (44)(01) 977 3222. Contact: D. Robinson. Company description: Scientific and industrial research; holographic non-ciestructive testing.

NEOVISION PRODUCTIONS. P.O. Box 74277, Los Angeles, CA 90004 USA. Telephone: (213) 387 0461. Contact: Bill Hilliard. Company description: Fine art originals.

NEWBOLD WELLS COMPANY. 33 Paul Street, London EC2A 4JU, England, United Kingdom. Telephone: (44) (1) 638 1471. Company description: Artistic holography: Embossed & silver halide holograms.

NEWCASTLE UPON TYNE POLYTECHNIC. Department of Physics, Ellison Building, Newcastle upon Tyne, NE1 8ST. England, United Kingdom. Telephone: (44) (091) 2358453. FAX: (44)(091) 235 8017. Contact: Dr. A.E. MacGregor. Department contacts, Graham Rice; Paul Dunnigan. Description: Comprehensive program of short courses in holography for beginners and advanced students alike in new holographic laboratories. Ongoing program of consultancy and research. Specialise in CW work; equipped with argon-ion lasers

NEW CLEAR IMPORTS LTD. 27 Burrard Street, St. Helier, Jersey, Channel Islands, England, United Kingdom. Telephone: (44)(534) 30614. Contact Anthony Hopkins. Company description: Gallery; retail shop.

NEW DIMENSION HOLOGRAPHICS. 65-72 Pier One, Hickson Road, Sydney, New South Wales 2000, Australia. Telephone: (61)(2) 276063. Contact; Tony Butteriss. Company description: Artistic holography sales; Educational holography consultant.

NEWPORT CORPORATION. (MAIN HEADQUARTERS). Established in 1969. 400 Employees at this address. 18235 Mt. Baldy Circle, P.O.Box 8020, Fountain Valley, CA 92708-8020 USA. Telephone: (714) 9639811. FAX: (714) 963 2015. Contact: Terry Reed, Member Technical Staff. President, Tom Galantowicz; Vice-President, Dean Hodges; Sales Manager, Jim Doty; Customer Service, Frank Aranda. Branch offices: Spectra Physics Pty. Ltd., Victoria, Australia; Aims Optronics SA/NV, Kraainem, Belgium; Antonio A. Santos, Rio de Janeiro, Brazil; Technical Marketing Associates, Ontario, Canada; Superbln Co. Ud., Taipei, Taiwan; BBT Instrumenter ApS, Frederiksberg, Denmark; Photonetics, SA, Marly Le Roi, France.; Advance Photonics, Bombay, India; Isramex Co. Ltd. ,Tel-Aviv, Israel; dB Electronic Instruments S.R.L., Milano, Italy; Marubun Corporation, Tokyo, Japan; Koning en Hartman, Delft, Netherlands; Electech Distribution Systems, Singapore; LaSing SA , Madrid, Spain. Company description: Newport Corpora-

tion is a designer and manufacturer of laser/holographic systems, E/O components, optics, spatial filters, optical & beamsteering instruments, magnetic bases, fiber optic components, vibration isolation systems, and holographic recording materials. Subsidiary companies: Newport Ltd., Herts., United Kingdom; Newport Instruments AG, Schlieren, Switzerland.

NEWPORT GMBH, (EUROPEAN HEADQUARTERS). Bleichstrasse 26, D-6100 Darmstadt, Federal Republic of Germany. Telephone: (49) 061 512 6116. FAX: (49) 061 512 2639. Branch office of Newport Corporation, Fountain Valley, CA. USA.

NEWPORT INSTRUMENTS AG, Giessenstrasse 15, CH-8952 Schlieren, Swtzerland. Telephone: (41) (01) 740 2283. FAX: (41) (01) 7402503. SubSidiary of Newport Corporation, Fountain Valley, CA. USA.

NEWPORT: Kyokuto Boeki Kaisha. 7th Floor, New Otemachi Bldg., 2-1, 2-Chome, Otemachi, Chiyodaku, Tokyo 100-91, Japan. Branch office of Newport Corp., Fountain Valley CA, USA.

NEWPORT LTD., Pembroke House, Thompsons Close, Harpenden, Herts. AL5 4ES, United Kingdom. Telephone: (44) (058) 2769995. FAX: (44) (058) 276 2655. Subsidiary of Newport Corporation, Fountain Valley, CA. USA.

NEWPORT HOLOGRAMS. 3412 Via Oporto, Suite 2, Newport Beach, CA 92663 USA. Telephone: (714) 675 1337. Contact: David Schaffner. Company description: Gallery and retail shop; selling reflection and DCG holograms, jewellry and novelties.

NEW YORK HALL OF SCIENCE, Established in 1964. 100 Employees at this address. 47-01 111th Street, Corona, NY 11368 USA. Telephone: (718) 699 0005. Contact: John Driscoll, Manager, Arts and Technologies Program. Company description: The New York Hall of Science is New York's only handson science and technology museum. Lasers and optics demonstrated daily. Color hologram depicting quantum atom is on display.

NEW YORK HOLOGRAPHIC LABS. P.O. Box 20391, Tompkins Square Station, 176 East 3rd Street, New York, NY 10009 USA. Telephone: (212) 254 9774. FAX: (212) 674 1007. Contact: Daniel Schweitzer. Contact: Samuel Moree. Company description: Private Commissions, Fine Art Holograms, Hands-on Tutorials.

NEW YORK INSTITUTE OF TECHNOLOGY. Center for Optics, 100 Glen Cove Avenue, Glen Cove, NY 11542 USA. Telephone: (516) 686 7863. Contact: Mauric Haliva. Description: Non-Destructive Testing; Industrial research; and Interferometry.

NORGES TEKNISKE HOGSKOIE. Institute for Almen Fysikk, Sem Saelandsv 7, N-7034 TrondheimNTH, Norway. Telephone: (47)(07) 593 422. FAX: (47)(07) 592 886. Company description: Holographic non-destructive testing; scientific holography research.

NORLAND PRODUCTS, INC. 695 Joyce Kilmer Avenue, P.O. Box 145, North Brunswick, NJ 08902 USA. Telephone: (201) 545 7828. Contact: Jean Spalding. Company description: Manufacture and distribute holographic photochemistry.

NORTH AMERICAN HOIOGRAPHICS INC. P.O. Box 451, 103 East Scranton Avenue, lake Bluff, Il 60044 USA. Telephone: (312) 2344244. Contact: Gary lawrence. Company description: Holographic portraits; Holography marketing & applications; Trade shows & exhibits.

NORTHERN ILLINOIS UNIVERSITY. Department of Physics, DeKalb, Il 60115-2854 USA. Telephone: (815) 753 1772. Contact: Thomas Rossing. Description: Scientific holography research; interferometry.

NORTHWESTERN UNIVERSITY, FIBEROPTIC IABORATORY. Dept of Electrical Engineering, Technological Institute, Evanston, Il 60208 USA. Telephone: (312) 491 4427. Contact: Rudy Haidle. Description: Optical processing, Forensic holography.

NORTHWESTERN UNIVERSITY, Dept of Biomedical Engineering, Evanston, Il 60208 USA. Telephone: (312) 491 2946. FAX: (312) 491 4133. Contact: Hans Bjelkhagen. Description: Artistic, Commercial, Interferometry, Scientific, Medical, and Industrial Research, Portraits, Fiberoptics, Pulsed laser, Holographic NonDestructive Testing, Educational Holography.

NOVATOR RESEARCH CENTER. Pl. lesi Ukrainki 1, Kiev-196, 252196 USSR. Telephone: (7)(296)8048. Contact: Georgi Dovgalenko. Description: Medical holography

O

ODHNER HOIOGRAPHICS, Established in 1988. 1 Employee at this address. 833 laurel Avenue, Orlando, Fl32803 USA. Telephone: (407) 8947966. Contact: Jeff Odhner, President. Company description: A manufacturer of custom transmission, reflection, and rainbow holograms on silver halide film; standard sizes are 8"x 1 0" and 4"x 5"; educational and holographic NDT services are also available.

ONTARIO COIIEGE OF ART. 100 McCaul Street, Toronto, Ontario M9W 1 W1, Canada. Contact: Michael Page. Company description: Holography courses

ONTARIO HYDRO. RESEARCH DIVISION, 800 Kipling Avenue, KR 128, Toronto, Ontario M8Z 5S4 Canada. Contact: David Mader. Company description: holography research; holographic non-destructive testing.

ONTARIO SCIENCE CENTRE. 770 Don Mills Road, Don Mills, Ontario, M3C IT3 Canada. Telephone: (416) 429 4100. Company description: Gallery; courses in holography

OP-GRAPHICS (HOLOGRAPHY) ITD. Unit 4: Technorth, 7 Harrogate Road, leeds IS7 3NB, England, United Kingdom. Telephone: (44)(0532) 628 687. FAX: (44) (0943) 467881. Contact: V. love, N. Hardy. Company description: Artistic holography; holography education.

OPTICAL IMAGES. Established in 1974. 3 Employees at this address. 1309 Simpson Way, Suite J, Escondido, CA 92025 USA. Telephone: (619) 746 0976. FAX: (619) 746 0976. Contact: Donald C. Broadbent, Owner. Company description: Optical Images, formerly Broadbent Development lab, is an independent, privately owned holographic facility producing HOE's and display holograms in various recording materials. Donald Broadbent has 24 years experience in holography.

OPTICAL LABORATORY. 3, Rue de Universite, 67000 Strasbourg, France. Description: Artistic holography.

OPTICAL SURFACES LTD. Godstone Road, Kenley, Surrey, CR2 5AA, England, United Kingdom. Telephone: (44)(1) 668 6126. Company description: Artistic holography.

OPTICAL WORKS LTD. 32 The Mall, Ealing, London W5 3TJ, England, United Kingdom. Telephone: (44) (1) 567 5678. Company description: Artistic holography.

OPTICS PLUS INC. 1369 East Edinger Avenue, Santa Ana, CA 92705 USA. Telephone: (714) 972 1948. Contact: John Goetten. Company description: Manufacture optics; precision tool mounts (including lens and mechanical mounts).

OPTILAS B.V. Established in 1976. 23 Employees at this address. P.O. Box 222, 2400 AE Alphen, NO Rijn, The Netherlands. Telephone: (31) 0172031234. FAX: (31) 01720 43414. Contact: A. Kooi, Manging Director. Vice-President, P. Drok. Company description: Sales/ service/engineering of electro-optical and vacuum related products.

OPTIMATION INC., BURR FREE MICROHOLE DIV. (SEE OUR ADVERTISEMENT ON PAGE 26) Established in 1971.4 Employees at this address. P.O. Box 310, Windham, NH 03087 USA. Telephone: (603) 898 1154. FAX: (603) 898-6434: Contact: Dean Jorgensen, President. Company description: Supplies and manufacutres 1) precision pinhole apertures from a few microns in diameter and up. 2} spatial filter instrument (model 205 SF $395.) 3} pin hole xy positioner (model 201 $295.)

OPTISCHE FENOMENON. Nederlandse Stichting Voor, Waarneming & Holografie, Warenarburg 44, NL 2907 CL Capelle aid Ijssel, The Netherlands. Contact: Jan M. Broeders. Telephone: (31)(070) 988 100. FAX: (31)(070) 988 276. Company description: Monthly Newsletter--yearly special issue on lasers and holography; Subsidiary of Dutch Foundation of Perception & Holography.

ORIEL SCIENTIFIC LTD. P.O. Box 31, Leatherhead, Surrey, KT22 7AU, England, United Kingdom. Company description: Artistic holography.

OXFORD HOLOGRAPHICS. Established in 1984. 3 Employees at this address. 71 High Street, Oxford OX1 4BA, England, United Kingdom. Telephone: (44}0865 250 505. FAX: (44) 0865 790353. Contact: Nick Cooper. Company description: Oxford Holographics has both a very well established retail and an expanding wholesale operation, based on keen pricing. We look forward to hearing from you.

OXFORD UNIVERSITY. Department of Engineering Science, Parks Road, Oxford OX1 3PJ England, United Kingdom. Telephone: (44) (0865) 273 805. Contact: D. J. Cooke. Description: Holography education; Industrial research.

P

PACIFIC HOLOGRAPHICS, INC. 1245 Stone Drive, San Marcos, CA 92069 USA. Telephone: (619) 471 9044. Contact: Eric Van Hamersveld. Company description: Main area of business is embossing and photoresist masters; stereogramming; sma" to medium formats.

PENNSYLVANIA STATE UNIVERSITY. Applied Research Laboratory, P.O.Box 30, State College, PA 16804 USA. Contact: C.S. Vikram. Description: Scientific research; particle density testing.

PENNSYLVANIA STATE UNIVERSITY. Dept. of Electrical Engineering, 121 East University Park, Pennsylvania 16802 USA. Contact: Francis Yu. Description: Color holography.

PERCEPTION HOLOGRAPHY. Thornton Marketing Ltd. Aketon Close, Haggs Lane, Follifoot, Harrogate, North Yorks., England HG3 1A2 United Kingdom. Telephone: (44)(93) 782 323. Contact: Mike Burridge. Description: Fine art holography, holography education.

PHASE-R CORPORATION. Box G-90, Old Bay Road, New Durham, NH 03855 USA. Telephone: (603}859 3800. Company description: We manufacture laser equipment. Call for more information.

PHOTOGRAPHER'S FORMULARY. P.O.Box 5105, Missoula, MT 59806-5105 USA. Telephone: (800) 922 5255. Company description: Holographic photochemistry; processing; manufacturer plates & film.

PHOTONETICS SA, 52 Avenue de l'Europe, F-78160 Marly-Le-Roi, France. Telephone: (33)(013) 9163377. FAX: (33)(013) 916 5606. Branch office of Newport Corporation, Fountain Valley, CA USA.

PHOENIX HOLOGRAMS. Established in 1986.3 Employees at this address. Elmar Spreer, Traubenstr. 41 D-2913 - Apen, Federal Republic of Germany. Telephone: (49)044 89 15198. President, Elmar Spreer; Vice-President, Charito Gil; Product and Sales Manager, Tilman Eimers; Customer Service, Jorgen Hintz. Branch office: Tilman Eimers, Ganderkesee, Federal Republic of Germany. Company description: Phoenix Holograms specializes in finding individual solutions for sophisticated problems. We do silver halide & dichromate holograms up to 1 m2 for: architecture, interior decoration/furniture, promotion/corporate design, education, exhibitions.

PHOENIX HOLOGRAMS. (Branch office). Tilman Eimers, Trendelbuscher Weg 1, D-2875 - Ganderkesee 2, Federal Republic of Germany. Telephone: (49) 04221/89298.

PHOTON LEAGUE. Established in 1987. 2 Employees at this address. 110 Sudbury Street-Unit B, Toronto, Ontario M6J 1A7 Canada. Telephone: (416) 531 1224. Contact: Mary Alton, Coordinator. Coordinator, Ken Vincent. Company description: Artist-run, not for profit, holography studio for artists. Photon league is committed to the production of fine art holography, education, dialogue & curation of exhibitions of recent Canadian holographic art.

PHOTONICS DIRECTORY. Optical Publishing Co. Inc. P.O. Box 1146, Berkshire Common, Pittsfield, MA 01201 USA. Telephone: (413) 499 0514. Company description: Publishers of information on Optical components; Optics, Electro-Optics, Lasers, and Imaaging technology.

PHYSICS INSTITUTE. LATVIAN SSR Academy of Sciences, 229021 Riga-Salaspils, USSR. Contact: K.K. Shvarts. Description: Scientific research on recording materials.

PILKINGTON P.E. LTD. Glascoed Road, St. Asaph, Clwyd, England, LL 17 011, United Kingdom. Telephone: (44)(745) 583301. Description: Artistic holography; DCG manufacturer; HOEs.

PLASMA TECHNOLOGY. 1709 East Bayshore Road, #3, Redwood City, CA 94063 USA. Telephone: (415) 579 5458. Company description: Holography equipment; Lasers.

POINT OF VIEW DIMENSIONS, LTD. Established in 1982. 2 Employees at this address. 45-2903 River Drive South, Jersey City, NJ 07310 USA. Telephone: (201) 6268844. Contact: Neal Lubetsky, President. President, Neal Lubetsky; Vice-President, Wendy Barkin. Company description: Point of View Dimensions specializes in the conceptualization, design, and execution of holography (all formats and all sizes) for exhibitions, trade shows, premiums, points of sale, brochures, and annual reports.

POLAROID CORPORATION, Established in 1937. 100 Employees at this address. 730 Main Street-1A, Cambridge, MA 02139 USA. Telephone: (800) 225 1618. FAX: (617) 577 5434. Contact: Phillip E. Mestancik, Marketing Manager. Sales Manager: Robert Brecheisen. Company description: Polaroid manufactures high quality photopolymer holograms for high volume applications, both custom & stock holograms are available. Polaroid's high quality & dependability are products of twenty years of experience in holography.

PORTSON, INC. (LASER IMAGES). Established in 1986. 14 Employees at this address. 9201 Quivira, Overland Park, KS 66215-3905 USA. Telephone: (913) 492 7010. FAX: (913) 492 7099. Contact: Steve Larson. President, Steve Larson; Vice-President, Roger Larson; Sales Manager, Anne Larson. Company description: Manufacturer of stock and custom holograms and holographic products; total in-house production capabilities in dichromate, silver halide, and photo-resist. Stock products include jewelry, watches, calculators, and framed art.

Q

QUe SERA SERA. P.O. Box 29, 9700 AA, Groningen, The Netherlands. Telephone: (31)(050) 140417. FAX: (31)(050) 144142. Contact: H.T. Vogd, Joop DeKens. Company description: Artistic holography; DCG holograms; Wholesale; Workshops; Manufacturer of optics (mirrors/lenses).

R

RADIO SHACK/TANDY CORPORATION. Bilston Road, Wednesbury, West Midlands, WS10 7JN, England. Description: Equipment & supplies.

RAINBOW SYMPHONY INC. Established in 1975. 6 Employees at this address. 22823 Hatteras Street, Woodland Hills, CA 91367 USA. Telephone: (818) 340 7200. FAX: (818) 340 2944. Contact: Mark S. Margolis, President. Company description: Manufacturer of uniquely designed holographic and diffraction products for the gift, novelty, advertising specialty, premium incentive, souvenir and museum markets.

RALCON, Established in 1985. 4 Employees at this address. Box 142, 8501 South 400 West, Paradise, UT 84328 USA. Telephone: (801) 245 4623. FAX: (801) 245 6672. Contact: Richard Rallison, CEO. President, Richard Rallison; Vice-President, Scott Schicker. Company description: Design, development and fabrication of volume holographic optical elements, (HOEs) including gratings, scanners, multifocus devices, heads up and down displays and notch filters formed in dichromated gelatin or photopolymer

RALPH CULLEN HOLOGRAPHICS / UKOS. Established in 1980. 20 Employees at this address. 84 Wimborne Road West, Wimborne, Dorset, BH21 2DP, England, United Kingdom. Telephone: (44)(202) 886 831. FAX: (44)(202) 742 236. Contact: Ralph Cullen, Director. Parent company: U.K. Optical Supplies. Branch office: Ralph Cullen Holographics/UKOS, London, England.

Company description: A ConsultancyDesign Service which in association with UK OPTICAL SUPPLIES (Manufacturing) provide Customized Holographic/ Optical Components. Advice on component selection and laboratory/studios designed to any budget is available.

RANDY JAMES/HOLOGRAPHY. Established in 1976. 1 Employee at this address. P.O. Box 305, Santa Cruz, CA 95061 USA. Telephone: (408) 458 4213. Company description: Commercial and fine art holography since 1974. Extensive backgrou nd in all forms of display holography: design, mastering, and production. Custom quotes, stock price list available.

RAVEN HOLO LTD. Old Saw Mills, Nyewood, Near Petersfield, Hampshire, GU3 15HX, England, United Kingdom. Company description: Fine art holograms

RECONNAISSANCE, LTD. (MAIN HEADQUARTERS). Established in 1984.3 Employees at this address. 3932 McKinley Street N.W., Washington D.C., 20015 USA. Telephone: (703) 273 0717. FAX: (703) 273 0745. Contact: Lewis Kontnik. Publisher, Lewis Kontnik; European Editor, Ian Lancaster. Subsidiary company: Holography News. Branch office: Surrey, England, United Kingdom. Reconnaissance, publisher of Holography News, maintains a comprehensive library on the industry. Reconnaissance organizes seminars, performs Single and multi-client industry analyses and market studies, and assists interested organizations in strategy development.

RECONNAISSANCE, LTD. (Branch Office).1 Erica Court, Wych Hill Place, Woking, Surrey AU22 OJB, England, United Kingdom. Telephone: (44)(04) 837 40689. FAX: (44)(04) 837 40689. Contact: Ian Lancaster.

REAL IMAGE HOLOGRAPHICS (ACK ACK DESIGNS) 31 Clerkenwell Close, London, EC1 England, United Kingdom. Description: Fine art holograms.

REEL IMAGE. P.O. Box 566, Pacifica, CA 94044 USA. Telephone: (415) 355 8897. Contact: Roy Bradshaw, Owner. Company description: Fine art holograms

REGAL PRESS INC .. REGAL HOLOGRAPHICS DIVISION, 129 Guild Street, Norwood, MA 02062 USA. Telephone: (617) 769 3900. Contact: William Duffey. Company description: Holographic embossing, application; Artistic holography.

REVELATION HOLOGRAPHY. 905 East Main Street, Cary, IL 60013 USA. Telephone: (312) 658 3560. Contact: Paul Travosek. Description: Artistic holography.

REYNOLDS METALS CO. Flexible Packing Division, 6603 West Broad Street, Richmond, VA 23230 USA. Telephone: (804) 281 4022. Contact: Rich Patterson. Company description: Embossed holography on packaging materials.

RICHARD BRUCK HOLOGRAPHY. (SEE OUR ADVERTISEMENT ON PAGE 43). Established in 1987. 1 Employee at this address. 3312 West Belle Plaine #2, Chicago, IL 60618 USA. Telephone: (312) 267 9288. Contact: Richard Bruck. Company description: A one-man shop producing quality low-volume runs. Custom and commercial work. Architecture and interior design. Fine art originals. Experienced installation and consultation

RICHMOND HOLOGRAPHIC STUDIOS LTD. (SEE OUR ADVERTISEMENT ON PAGE 51) Established in 1979. 4 Employees at this address. 6 Marlborough Road, Richmond, Surrey, England, TW10 6JR United Kingdom. Telephone: (44)(1) 940 5525. FAX: (44)(1) 948 6214. Contact: Edwina Orr, Director. President, Edwina Orr; Vice-President, David Traynor. Company description: Our principle services include mass produced film holograms up to 500 x 700 mm. suitable for horizontal and vertical displays. Pulsed mastering. Achromatic and multi-colour reflection and transmission copies. Technology transfer consultancy.

RITCHER HOLOGRAMS. Adolf Kolping Strasse 16, D-4050 Monchengladbach, Federal Republic of Germany. Description: Artistic holography.

ROBERT SHERWOOD HOLOGRAPHIC DESIGN, INC. Established in 1985. 4 Employees at this address. 400 West Erie Street, Chicago, IL 60610 USA. Telephone: (312) 944 0784. Contact: K. Kellison, Customer Service. President, Robert Sherwood; Vice-President, Greg Hoskins. Branch office of Holographic Design, Inc., Philadelphia, PA. Company description: At Holographic Design, our designers, holographers, and account service personnel provide you with the highest quality standards this new and exciting technology can offer.

ROBINSON HOLOGRAM LIGHTING SYSTEMS. 5 Hillside Cottages, Owismoor Road, Camberley, Surrey GU15 4SU, England, United Kingdom. Telephone: (44) (0344) 762 739. Contact: Anthony Robinson. Company description: Holography lighting consultant

ROCHESTER INSTITUTE OF TECHNOLOGY. One Lomb Memorial Drive, P.O. Box 9887, Rochester, NY 14623 USA. Telephone: (716) 4752785. Contact: Arnold Lungershausen. Description: Artistic holography education; holography courses.

ROCHESTER PHOTONICS CORPORATION. Established in 1989. 3 Employees at this address. 67 Nettlecreek Street, Fairport, NY 14450 USA. Telephone: (716) 275 5140. Contact: G.M. Morris, President. Vice-President, Dean Faklis. Company description: Designer and manufacturer of diffractive optical elements for use in precision electro-optical systems. Supplier of optical system design services for military and commercial uses.

ROLLS-ROYCE PLC--ADVANCED RESEARCH LABORATORY. P.O.Box 31, Derby DE2 8BJ, England, United Kingdom. Contact: Richard Parker. Company description: Holographic non-destructive testing, industrial research.

ROSEWELL LTD. Blacknest Estate, Bentley, Alton, Hants., England, United Kingdom. Telephone: (44) (44) 4~0 23605. FAX: (44)(44) 420 22517. Contact: Neville McGeorge. Company description: Artistic holography; Embossed holograms.

ROSS BOOKS. Established in 1977. P.O. Box 4340, Berkeley, CA 94704 USA. Telephone: (415) 841 2474. FAX: (415) 841 2695. Contact: Elizabeth T. Yerkes, Editor. President, Franz Ross. Company description: Publisher of holography books and resources; FAX us (number above) or call for our free catalogue (800) 367 0930 (Toll Free within US).

ROTTENKOLBER HOLO-SYSTEM GMBH. Henschelring 15, D-8011 Kirchheim/Munich, Federal Republic of Germany. Telephone: (49)(089) 201 4911. Company description: Holographic non-destructive testing; Industrial research; manufacturer of holo graphic and imagaing systems

ROWLAND INSTITUTE FOR SCIENCE. 100 Cambridge Parkway, Cambridge, MA 02142 USA. Telephone: (617) 4974657. Contact: Jean-Marc Fournier. Description: Scientific holography research.

ROYAL COLLEGE OF ART. Holography Unit, Darwin Building, Kensington Gore, London SW7 2EU, England, United Kingdom. Telephone: (44)(01) 5845020. Contact: Robert Munday. Description: Artistic holography; Portraiture.

ROYAL INSTITUTE OF TECHNOLOGY. Department of Industrial Metrology, S-10044 Stockholm, Sweden. Telephone: (46)(08) 790 7823. FAX: (46)(08) 790 8219. Contact: Lennart Svennson. Description: Interferometry; Industrial and scientific holography research; Holography education; Holographic nondestructive testing.

ROYAL PHOTOGRAPHIC SOCIETY. Salisbury College of Art, Southampton Road, Salisbury, Wiltshire, England, United Kingdom. Telephone: (44)(0722) 23711 ext. 275. Description: Artistic holography; holography education.

ROYAL SUSSEX Hospital. Brighton, England, United Kingdom. Contact: Lawrie Wright. Description: Scientific and Medical Holography research; Interferometry.

RUTHERFORD AND APPLETON LABORATORIES. Chilton, Didcot, Oxon OX11 OOX, England, United Kingdom. Telephone: (44)(0235) 21900. Contact: Robert Sekulin. Company description: Particle measurements; Holographic non-destructive testing.

S

SAAB-SCANIA. S-581 88 LINKOPING, Sweden. Telephone: (46)(013) 129020. Contact: Sven Malmqvist. Company description: Holographic non-destructive testing; scientific holography research.

SANDIA NATIONAL LABORATORIES. Combustion Research Facility. Livermore, CA 94550 USA. Telephone: (415) 422 3138. Contact: Donald Sweeney. Company description: Scientific holography research; Computer-generated holography; Holographic Optical Elements.

THE SCHOOL OF THE ART INSTITUTE OF CHICAGO - HOLOGRAPHY DEPARTMENT. Columbus Drive and Jackson Boulevard, Chicago, IL 60603 USA. Telephone: (312) 443 3786. Contact: Ed Wesley, Visiting Artist. Description: The holography program at The School of The Art Institute is divided into beginning holography and advanced holography. The Department has two Newport tables and two HeNe Lasers (15 mW and 35 mW).

SCHOOL OF HOLOGRAPHY. 263 Montego Key, Novato, CA 94949 USA. Telephone: (415) 824 3769. Contact: Sharon McCormack. Description: Holography education; workshops

SCHOOL OF HOLOGRAPHY/CHICAGO. 1134 West Washington Boulevard, Chicago, IL 60607 USA. Telephone: (312) 226 1007. Contact: L. Billings. Company description: Museum/Curator; Holography Teacher Training; Hands-On Workshops; Curriculum development; Educational materials; Independent educational facility.

SCHULTZ & CO. (see E.C. Schultz & Co.) SCIENCE KIT & BOREAL LABORATORIES. 777 East Park Drive, Tonawanda, NY 14150-6784 USA. Telephone: (716) 874 6020. Company description: Suppliers and mail-order cataloguers of holography educational materials including holography kits, books and more.

SCIENCE & MECHANICS INSTRUMENTS. 605 East 59th Street, Dept. H-2, Brooklyn, NY 11234 USA. Telephone: (718) 531 3381 . Company description: Manufacture and sell light meters and shutters.

SCIENTIFIC COUNCIL ON EXHIBITIONS OF THE USSR, Academy of Sciences, 30, Varilov Street, Moscow USSR. Telephone: (7)(135) 6486. Contact: Larisa Nekrasova. Description: Traveling exhibits, artistic holography.

SCOPE OPTICS LTD. Unit 6 Alston Works, Alston Road, Barnet, Hertfordshire EN5 4EL, England, Unit ad Kingdom. Telephone: (44)(1) 441 2283. Description: Fine art holograms.

SEVCHENKO RESEARCH INSTITUTE ... (see A.N. Sevchenko Research Institute)

SHIPLEY CHEMICAL CO. 1457 McArthur, Whitehall, PA 18052 USA. Telephone: (215) 820 9777. Conact: Stu Price. Company description: Manufacture holographic photochemistry, film, plates and more.

SIEMENS. 186 Wood Avenue South, Iseling, NJ 8830 USA. Telephone: (201) 321 3400. Conlact: Marketing department. Company description: Manufacturer of lasers and components.

SIEMENS LTD. Siemens House, Windmill Road, Sunbury-on-Thames, Middlesex, TW16 7HS, England, United Kingdom. Company description: Manufacturer of lasers and components.

SILLCOCKS PLASTICS INTERNATIONAL. (MAIN HEADQUARTERS). Established in 1930. 310 Snyder Avenue, Berkeley Heights, NJ 07922 USA. Telephone: (800) 922 0958. FAX: (201) 665 1856. Contact: Shari Spiro, Holographic Consultant. President, John Herslow; Vice-President, Vic Berkowitz; Sales Manager, Jack Fenimore; Holographic Customer Service, Shari Spiro. Subsidiary company: SPI Identification, Inc.; Daylux, Inc. Company description: Sillcocks applies embossed foil to sheeted pvc vinyl creating holographic and diffraction foil credit. promotional, and business cards, buttons, rulers, key tags, calenders, dazers, danglers, tent-signs, with custom printing, laminating, diecutting.

S.1. VAVILOV, State Optics Institute, 199064 Leningrad, USSR. Contact: A. Prostev. Description: artistic holography; recording materials research.

SMITH & MCKAY PRINTING CO. INC., Established in 1919. 25 Employees at this address. 96 North Almaden Boulevard, San Jose, CA 95110-2490 USA. Telephone: (408) 292 8901 FAX: (408) 292 0417. Contact: Dave McKay, President. President, Dave McKay; Sales Manager, John Rogers; Customer Service, Stephanie Louie. Parent Company: Holographic Impressions. Company description: Hot stamp foil holograms onto paper products. Dimensional printing and fine lithography. Assist coordination of printing projects featuring embossed holography. Publish

quarterly educational newsletter. Hold holography seminars for graphic designers.

SOCIETY FOR PHOTO-OPTICAL INSTRUMENTATION ENGINEERS (S.P.I.E.). P.O. Box 10, Bellingham, WA 98227 USA. Telephone: (206) 676 3290. Contact: Sales & Marketing Department. Company description: Seminars, conferences on Holography; publishes news in OE Reports.

SOI - SOCIETY OLOGRAFICA ITALIA. Via degli Eugenii 23, 00178 Roma, Italy. Telephone: (39)(6) 799 0452. Company description: Organize & produce travelling exhibits.

SORO ELECTRO-OPTICS, SA 26 rue Berthollet, 94110 Arcueil, France-. Company description: Design & manufacture optics.

SOVISKUSSTVO vlo MEZHDUNARODNAYA KNIGA. Established in 1968. 20 Employees at this address. Art holography department, 141120 Tryazino, Moscow region, USSR. Telephone: (7) 238 4600. FAX: (7) 230 2117. Contact: German B. Avksentjev. Director. Dep't head, Valery A. Vanin. Parent company: Platan. Branch offices: contact for info. Company description: Our firm offers: DCG pendents; lamps with holograms; art reflection holograms (size 102x127 mm, and 180x240 mm.).

SPACE AGE DESIGNS INC., Established in 1984. P.O. Box 72, Carversville, PA 18913 USA. Telephone: (215) 2978490. Contact: Valli Rothaus, President. Description: Embossing; Artistic holography; Marketing consultants.

SPECAC Limited. 6A River House, Lagoon Road, St. Mary Cray, Orpington, Kent, BR5 3QX, England, United Kingdom. Telephone: (44)(1) 689 7317. Description: Artistic holography.

SPECTRAL IMAGES. 15 Wanderdown Way, Ovingdean, Brighton BN2 7BX, England, United Kingdom. Telephone: (44)(0273) 34844. Company description: Medical research.

SPECTRA-PHYSICS INC., (MAIN HEADQUARTERS). 3333 North First Street, San Jose, CA USA. Parent company: CIBA-Geigy. Branch offices: New Jersey, USA; Mountain View, CA USA; United King dom; France; Belgium; Netherlands; Federal Republic of Germany; Itely; Japan; Australia; Hong Kong ; Switzerland.

SPECTRA-PHYSICS INC., LASER PRODUCTS DIVISION. Established in 1961. Over 500 Employees at this address. 1250 West Middlefield Road, Box 7013, Mountain View, CA 94039-7013 USA. Telephone: (415) 961 2550. FAX: (415) 969 4084. Contact: Randy Heyler, Marketing Manager, Gas Lasers. General Manager, Henry Massenburg. Parent company: CIBA-Geigy.

Branch offices: New Jersey, USA; Mountain View, CA USA; United Kingdom ; France; Belgium; Netherlands; Federal Republic of Germany; Itely; Japan; Australia; Hong Kong; Switzerland. Company description: Spectra-Physics is the world's largest supplier of CW and pulsed gas and solid state laser systems, including a comprehensive optical accessories line and a worldwide customer service network.

SPECTRA-PHYSICS AG. (Branch Office). Schweizergasse 39, 4054 basel, Switzerland. Telephone: (41) (061) 541 154. FAX: (41)(061) 541 129.

SPECTRA-PHYSICS B.V. (Branch Office). Prof. Dr. Dorgelolaan 20, P.O. box 2264, 5600 Cg Eindhoven, The Netherlands. Telephone: (31)(040) 451 855. FAX: (31)(040) 439 922.

SPECTRA-PHYSICS B.V.BA (Branch Office). North Trade Building, Noorderlaan 133, 2030 Antwerp, Belgium. Telephone: (32)(03) 542 6203. Main headquarters in San Jose, CA USA.

SPECTRA-PHYSICS GmbH, (Branch Office) Siemensstrasse 20, 0-6100 Darmstadt, Federal Republic of Germany. Telephone: (49)(061) 517 080. FAX: (49) (061) 51 708 232. Main headquarters in San Jose, CA USA.

SPECTRA-PHYSICS INC., LASER PRODUCTS DIVISION. (Branch Office). 366 South Randolphville Road, Piscataway, NJ 08854-4175. Telephone: (800) 631 5693. FAX: (201) 981 0029. Main headquarters in San Jose, CA USA.

SPECTRA-PHYSICS K.K. (Branch Office). 15-8 nanpeidai-cho, Shibuya-ku, Tokyo 150, Japan. Telephone: (81)(03) 7705411. FAX: (81)(03) 770 4197. Main headquarters in San Jose, CA USA.

SPECTRA PHYSICS LIMITED. (Branch Office). Established in 1965. 50 Employees at this address. Boundary Way, Hemel Hempstead, Herts. HP2 7SH, England, United Kingdom. Telephone: 0442 232322. FAX: 0442 68538. Contact: Mrs. Molly Helm, Sales Secretary. President, Mr. Jon Tomkins; Sales Manager, Anthony Brown. Main headquarters in San Jose, CA USA.

SPECTRA PHYSICS PTY. LTD. (Branch Office). 826 Mountain Highway, Bayswater,Victoria, Australia. Telephone: (61)(03) 7295155. FAX: (61)(03) 720 4256.

Branch office of Newport Corporation, Fountain Valley, CA USA; Spectra-Physics Inc., San Jose, CA USA.

SPECTRA-PHYSICS S.A. (Branch Office). Avenue de Scandinavie, ZA de Courtaboeuf, BP 28, 91941 LES ULiS Cedex, France. Telephone: (33)(1) 6907 9956. FAX: (33)(1) 69076093. Main headquarters in San Jose, CA USA.

SPECTRA-PHYSICS S.R.L. (Branch Office). via Derna 1, 1-20132 Milano, Italy. Telephone: (39)(02) 285 3665. FAX: (39)(02) 282 7445. Main headquarters in San Jose, CA USA.

SPECTROGON. Established in 1976. 550 County Avenue, Secaucus, NJ 07904 USA. Telephone: (201) 867 4888. FAX: (201) 867 2191. Contact: Sam Ponzo. President, Sam Ponzo; Sales Manager, Sam Ponzo. Parent Company: Spectrogon AB, Box 2076, S-18302 TABY, SWEDEN. Company description: Manufactures & designs interference filters for the IR, visible & UV spectral regions; narrow and bandpass, long and shortwavepass, isolation-line, & ND filters; atmospheric windows; diffraction gratings; AR & metallic coatings; laser optics.

SPECTROGON AB, (MAIN HEADQUARTERS) Box 2076, S-18302 Taby, Sweden. Subsidiary company: Spectrogon, Secaucus NJ, USA

SPECTROLAB LTD. Established in 1962. 200 Employees at this address. P.O. Box 25, Newbury, Berkshire, RH6 8BQ, England United Kingdom. Telephone: 0635 248080 FAX: 0635 248745. Contact: J. Ford. Company description: Manufacture holographic systems including laser tables, optics, optical and laser positioning systems.

SPECTRON DEVELOPMENT LABORATORIES. 3303 Harbor Boulevard, Suite G-3, Costa Mesa, CA 92626 USA. Telephone: (714) 549 8477. Contact: Colleen Fitzpatrick. Company description: Scientific holography research; Interferometry; Thermoplastic recording material research; Holographic nondestructive testing.

S.P.I.E. (See SOCiety for Photo-Optical Instrumentation Engineers).

SPINDELER & HOYER GmbH. Postfach 3353, Koenigsallee 23, D-3400 Goettingen, Federal Republic of Germany. Telephone: Contact: Rainer Lessing. Company description: Manufactures H-1 s.

SPOT. Agentur fur Holographie, An St. Katharinen 2, 0-5000 Koln 1, Federal Republic of Germany. Description: Holography marketing consultant,

SPYCATCHER LIMITED. 2 Croft Cottages, Wattisfield, Diss Norfolk 1 P22 1 NS, England, United Kingdom. Telephone: (44)(0359) 5122. FAX: (44)(0359) 890 121. Company description: Artistic holography; embossed holography; wholesale.

STANFORD UNIVERSITY--Oepartment of Mechanical Engineering, Building 570 Room 571 C, Stanford, CA 94301 USA. Telephone: (415) 7233243. Contact: Joseph Goodman. Description: Optical diagnostics.

STARCKE, KY. (Main Headquarters) Established in 1983. 4 Employees at this address. P.O. Box 22, SF-32811, Peipohja, Finland. Telephone: (358) 393 60700. FAX: (358) 396 7905. Contact: Mr. Ari-Veli Starcke. Vice-President, Mrs. Milla-Ritta Starcke; Customer Service, Ms. Mikaela Renvall. Subsidiary: Oy Foil It Ltd. Branch offices: Starcke KY Branch, Helsinki, Finland; Holografia Galleria, Oulu, Finland. Company description: Starcke KY is the leading company selling holograms in Scandinavia

STARCKE, KY, (Branch office). Unionikatu 45 111 Floor, SF-00170 Helsinki, Finland. Telephone: (358) 0135 5044. FAX: (358) 0135 5306. Contact: Mr. AriVeli Starcke.

STARLIGHT HOLOGRAPHIC INC. (SEE OUR ADVERTISEMENT ON PAGE 39) (MAIN HEADQUARTERS), Established in 1987. 5 Employees at this address. 73 Stable Way, Kanata, Ontario K2M 1 A8, Canada. Telephone: (613) 235 0440. FAX: (613) 237 9208. Contact: Stephen Leafloor, President. VicePresident, Susan Skrypnyk. Parent company: Starmagic Holographic Gallery, Ontario Canada. Company description: Starlight Holographies Inc. is one of Canada's leading representatives of the international holographic community, in both stock and custom holography. Its subsidiaries, "Starmagic Holographic Gallery" are currently opening across Canada.

STARMAGIC HOLOGRAPHIC GALLERY. (SEE OUR ADVERTISEMENT ON PAGE 39) 47 Clarence Street, Ottawa, Ontario K1 N 9K1, Canada. Telephone: (613) 235 0440. Branch office of Starlight Holographic Inc., Ontario, Canada.

STATE RESEARCH AND PROJECT INSTITUTE OF CHEMICO-PHOTOGRAPHIC INDUSTRY, 125167 Moskow, USSR. Description: Materials research.

STEINBICHLER OPTOTECHNIK GMBH, Established in 1987. 9 Employees at this address. Am Bauhof 4, D-8201 Neubeuern, FRG Federal Republic of Germany. Telephone: 8035 1017 FAX: 8035 1010. Contact: Dr. H. Steinbichler. President, Dr. H. Steinbichler; Vice-President, J. Engelsberger; Sales Manager, T. Anzenberger; Customer Service, W. Sixt. Company description: Holographic vibration analyzers, image analysis systems, instant holographic camera processing thermoplastic film, optic contour measuring systems, speckle measuring systems, holographic non-destructive test equipment.

STEUER KG GmbH & Co. Postfach 100327, Ernst May Strasse 7, D-7022-Leinfelden-Echterdingen, Federal Republic of Germany. Telephone: (49)(0711) 753 143. Company description: Manufacturer of holographic embossing machines.

STEVE PROVENCE HOLOGRAPHY. 15220 Fern Avenue, Boulder Creek, CA 95006 USA. Telephone: (408) 338 2051 . Contact: Steve Provence. Subsidiary company: Active Image, Boulder Creek, CA USA.

STOLTZ AG. Tafernstrasse 15, CH-5405 Baden Dattwil, Switzerland. Telephone: (41)(056) 840 151 . Contact: Beat Ineichen. Company description: Holographic non-destructive testing.

STUDIO FOR HOLOGRAPHIE. Established in 1985.1 Employee at this address. MoosstraBe 27, 0-8031 Eichenau, Federal Republic of Germany (FRG). Telephone: 08141 70831 . Contact: Dr. Carlo Schmelzer. Company description: Products: mastering and copy services (rainbox/reflection), production of mass-run embossed holograms, open stock images, art-pieces.

STUDIO WEIL-ALVARON. Established in 1933. Ostra Tullgatan 8, S-211 28 Malmo, Sweden. Telephone: (46) (40) 129 956. Contact: Lektor H. Herman Weil. Description: Hans Weil's inventions were made in the period 1933-1937, while Gabor invented holography in 1948, the laser was invented 1962 and the first laser- illuminated hologram was exposed as late as 1964.

SUPERBIN CO. LTD, 5F-3, 792, Tun Hua South

Road, P.O. Box 59555, Taipei, Taiwan 106. Telephone: (886)(02) 733 3920. FAX: (886) (02) 732 5442. Branch office of Newport Corporation, Fountain Valley, CA USA.

SUPERIOR TECHNOLOGY IMPLEMENTATION. Established in 1988. 4 Employees at this address. Hjortekaersvej 99B, 2800 Lyngby, Denmark. Telephone: (45)(45) 933 358. FAX: (45)(45) 930 353. Contact: Knud Banck-Hammerum, GM. President, Knud Banck-Hammerum, GM; Vice-President, Per Ibsen; Sales Manager, Knud Banck-Hammerum, GM; Customer Service, Jan Stensborg. Company description: Embossed holography; Artistic holography; Education; HOE manufacturer; Equipment and supplies.

SWEDE HOLOPRINT. Duvhoksgatan 6A, Malmo 21460, Sweden. Telephone: (46) (040) 898 21. Contact:
Bjorn Wahlberg. Company description: Artistic holograpy; Artistic marketing consultant.

SWISS FEDERAL INSTITUTE OF TECHNOLOGY. Laboratory of Photoelasticity, Ramistrasse 101, CH-8092 Zurich, Switzerland. Telephone: (41)(0380 246 000. Contact: Walter Schumann. Description: Holographic non-destructive testing; Scientific and industrial research.

SYDNEY COLLEGE OF THE ARTS. Dept. of Holography, 58 Allen Street, Glebe, Sydney, Australia. Description: Holography courses.

SYNCHRONICITY HOLOGRAMS. Established in 1984. 1 Employee at this address. Box 4235, Lincolnville, ME 04849 USA. Telephone: (207) 763 3182. Contact: Arlene Jurewicz. Company description: Synchronicity Holograms provides outreach educational presentations on the artistic, scientific and perceptual aspects of holography. Presentation and workshops can be geared to all age levels and abilities of audience.

SYSTEMS GROUP OF TRW INC. One Space Park, Redondo Beach, CA 90274 USA. Telephone: (213) 8124321. Contact: Customer Service Dept. Company description: Scientific holography research; Interferomtery; Particle measurement; Industrial research.

T

T.A.1. INCORPORATED. 12021 South Memorial Parkway, Huntsville, AL 35803 USA. Telephone: (205) 881 4999. FAX: (205) 880 8041. Contact: Loy Shreve, Director. Company description: Manufacture and testing of optic and laser inspection equipment.

TECHNICAL MARKETING ASSOCIATES, 6695 Millcreek Drive, Unit #1, Mississauga, Ontario, Canada L5N 5M5. Telephone: (416) 826 7752. FAX: (416) 826 8225. Branch office of Newport Corporation, Fountain Valley, CA USA.

TECHNICAL UNIVERSITY OF BUDAPEST. Institute of Precision Mechanics and OptiCS, Applied Biophysics Laboratory, H-1621 Budapest, Hungary. Contact: Pal Greguss. Description: Medical holography research.

TECHNICAL UNIVERSITY OF EINDHOVEN. Faculty of Architecture, Calibre Institute, P.O. Box 513, Eindhoven, NL-5600MB, The Netherlands. Contact: Geert T. A. Smelzer. Description: Academic research, computer generated holograms

TECHNICAL UNIVERSITY OF WROCLAW. Institute of Physics, Wybrvzeze Wyspianskiego 27, PL-50-370 Wroclaw, Poland. Contact: Henryk Kasprzak. Description: Academic, Scientific research.

TECHNOEXAN/SEMICON, INC. 2300 Drew Road, Mississauga, Ontario L5S 1 H4 Canada. Telephone: (416) 6727271. FAX: (416) 677 7299. Contact: Director. Company description: The Soviet company Technoexan, a joint venture of the Physical Technical Institute Joffe (FTI) in Leningrad and Semicon GmbHAustria, makes Denisyuk holograms available for western customers.

TECHNOEXAN/SEMICON. Morellenfeldgasse 41, A-8010 Graz, Austria. Telephone: (43)(0316) 38 25 41 .

FAX: (43) (0316) 38 24 03. Contact: Christian Haydvogel, Director. Company description: The Soviet Company Technoexan, a joint venture of the Physical Technical Institute Joffe (FTI) in Leningrad and Semicon GmbH-Austria, makes Denisyuk holograms available for western customers.

TECHNORTH. Unit 4 - 7 Harrogate Road, Leeds LS7 3NB England, United Kingdom. Telephone: (44)(532) 328 687. Company description: Artistic holography.

TEXTILE GRAPHICS, INC. (SEE OUR ADVERTISEMENT ON PAGE 25). (Branch office) 201 North Fruitport Road, Spring Lake, MI 49456 USA. Telephone: (616) 8425626. Contact: Jan Bussard.

TEXTILE GRAPHICS, INC. (MAIN HEADQUARTERS). (SEE OUR ADVERTISEMENT ON PAGE 25). Established in 1987. 10884 South Street, P.O. Box 68, Nunica, MI 49448 USA. Telephone: (616) 837 8048. Contact: Jan Bussard, President. Branch Office: Spring Lake, MI USA. Company description: Artistic integration of holographic art into silk screen printing on textiles for the creation of a total graphic, with a patented bonding process using heat and pressure to withstand 100 washings/dryings.

THIRD DIMENSION ARTS INC. Established in 1983. 8 Employees at this address. 1241 Andersen Drive, Suites C & 0, San Rafael, CA 94901 USA. Telephone: (415) 485 1730. FAX: (415) 485 0435. Contact: Tim LaDuca. President, Dara Haskell; Sales Manager, Tim LaDuca; Customer Service, Mary Kelly. Parent company: Holocrafts of Canada. Company description: Third Dimension Arts Inc. manufacturers of dichromate jewelry, gifts, and 3-D arts trademark hologram watches. Suppliers to: the gift, jewelry, museum, and entertainment industry; (licencing) markets. Custom designs welcome!

THIRD DIMENSION LTD, 1855 Charter Lane, Suite C, Lancaster, PA 17601 USA. Telephone: (717) 393 6400. FAX: (717) 3930182. Contact: Jacque Phillips. Main Headquarters: Third Dimension Ltd, London, England.

THIRD DIMENSION LTD. Established in 1982. 12 Employees at this address. 4 Wellington Park Estate, Waterloo Road, London NW2 7JW, England, United Kingdom. Telephone: (01) 208 0788. FAX: (01) 450 6528. Contact: Solomon Balas. President, Solomon Balas; Vice-President, Christopher Eyles. Branch Of-

fices : Third Dimension Ltd., Lancaster PA, USA. Company description: Third Dimension Ltd, is the world leading manufacturer of silver halide film display holograms, offers over 80 stock images, Custom Holograms, portraiture reductions for Galleries, Retail and Marketing Aids.

THREE DIMENSIONAL IMAGERY, LTD. 3031-K Nihi Street, Honolulu, HI 96819 USA. Telephone: (808) 373 1810. Company description: Artistic holography.

THREE-D LIGHT GALLERY. 107 The Commons, Ithaca, NY 14850 USA. Telephone: (607) 273 1187. Contact: Eve Walter. Company description: Custom orders for artistic holography; holography gallery.

TJING LING INDUSTRIAL RESEARCH. 130 Keelung Road, Section III, Taipei, Taiwan. Telephone: (86)(20 704 1856. Description: Fine art originals.

TNO INSTITUTE OF APPLIED PHYSICS. Department of Optics, P.O. Box 155, NL-2600 AD Delft, The Netherlands. Contact: Ruud L. van Renesse. Description: Academic and industrial holography research.

TOKAI UNIVERSITY. Department of Electro Photo Optics, 1117 Kitakaname Hiratsuka City, Kanagawa 259-12, Japan. Contact: Hidetoshi Katsuma. Description: Artistic holography; scientific research.

TOKYO UNIVERSITY, Medical Division, 2nd Surgical Research, Toyko, Japan. Description: Medical holography

TOPPAN PRINTING CO. LTD., Central Research Institute. 5, 1-chome Taito, Taito-ku, Tokyo 110, Japan. Contact: K. Ohnuma.

TOPPAN PRINTING CO. LTD. (Branch office). 1 Embarcadero Center, Suite 21 06-MP, San Francisco, CA 94111 USA. Telephone: (415) 9827733. FAX: (415) 956 2551 . Contact: Mr Ozawa, Sales & Marketing. Company description: Full holographic Services, design, execution, print mass-manufacturing, embossing, application onto products.

TOUCHWOOD HOLOGRAPHICS, Established in 1985. 1 Employee at this address. 50 Sugworth Lane, Radley, Abingdon, Oxon, OX14 2HY, England, United

Kingdom. Telephone: 0865 735874. Contact: Mr. George W. Clare. Chairman, Mr. G.W. Clare. Paren Company: touchwood Sports Ltd. Main headquarters: 426 Abingdon Road, Oxford, OX1 4XN England. Branch Offices: 4 High Street, Abingdon; 107 St. Aldates, Oxford. Company description: Small experimental laboratory investigating holographic applications for advertising and promotions in the normal sports trade. One-off experimental commissions can be undertaken. Sorry, no long runs or repeat work.

TOWNE LABORATORIES, INC. (SEE OUR ADVERTISEMENT ON PAGE 37) Established in 1956. 65 Employees at this address. P.O. Box 460-HM, One U.S. Highway 206, Somerville, NJ 08876-0460 USA. Telephone: (201) 722 9500 FAX: (201) 722-8394. COntact: Charles Ondrejik, Sales Manager. President, J.J. Obzansky; M.A. Obzansky; Sales Manager, C. Ondrejik. Company description: Towne Laboratories is a producer of fine quality precision holographic photoplates with or without a sub-layer of IRON-OXIDE and precision spun striation free photoresist in sizes to 18"x18".

TRANSFER PRINT FOILS INC. P.O.Box 518, 9 Cotters Lane, East Brunswick, NJ 08816 USA Telephone: (201) 238 1800. Contact: Charlie Yetka. Manufacure holographic foils: embossing of foils.

TREND. Miramarska 85, 41000 Zagreb, Yugoslavia. Telephone: (38)(041) 511 426. Contact: Dalibor Vukicevic. Company description: Gallery.

TRIDIMENSIONALE HOLOGRAMS. Alberto, Alcocer, 38-2D, 28016 Madrid, Spain. Telephone: (34) (481) 290 745. Contact: Daniel Weiss. Company Description: Artistic holography; Pulsed portraiture.

TRI-ESS SCIENCES:STUDENT SCIENCE SERVICE. 1020 West Chestnut Street, Burbank, CA 91506-1623 USA. Telephone: (818) 247 6910. Company description: Educational materials and laser equipment supplier.

TRILONE HOLOGRAPHIE CORP. Established in 1986. 7 Employees at this address. 4200 Boulevard St. Laurent, Montreal, Suites 305-307, Quebec H2W 2R2 Canada. Telephone: (514) 845-6992. FAX: (514) (849) 8706. Contact: Armand Kessous, Director of Marketing. President, Hendry H. Danan; Vice-President, Gerard Allon; Sales Manager, Armand Kessous;

Customer Service: Henriette Allon. Company description: Top mass producer of large size holograms on film. Produc line of holoposters™ available for interior decoration, point-fo-purchase advertising & indoor billboard advertising. Trademarks are: holoposter, holodisplay, holomedia. Brochure available.

U

ULTRAFINE, 16 Foster Road, Chiswick, London W4 4NY, England, United Kingdom. Telephone: (44)(01) 995 2303. Company description: Holographic non-destructive testing; scientific and industrial research.

U.K. GOLD PURCHASE DBA HOLOGRAMS UNLIMITED. (SEE OUR ADVERTISEMENT ON PAGE 29). 7907 NW 53 Stree, Miami, FL 33166 USA. Subsidary company: U.K. Gold Purchasers, Inc. dba Holograms Unlimited, Corpus Christi TX, USA

U.K. GOLD PURCHASERS, INC. DBA HOLOGRAMS UNLIMITED. (SEE OUR ADVERTISEMENT ON PAGE 29). Established in 1979. 2 Employees at this address. 5858 S.P.I.D. Sunrise Mall, Corpus Christi, TX 78412 USA. Telephone: (512) 993 5211. Contact: Marvin Uram, President. President, Marvin Uram: Vice-President: Amy Uram; Sales Manager: Marvin Uram; Customer Service: Amy Uram, Parent company: U.K. Gold Purchasers Inc., Miami FL USA. Company description: If its holographic, we have it, or want it! One stop wholesale distributor for varied products the public can afford and will buy. All items at factory to you prices.

UK OPTICAL SUPPPLIES. (SEE OUR ADVERTISEMENT ON PAGE 21) Established in 1980. 20 Employees at this address. 84 Winmborne Road West Wimborne, Dorset BH21 2DP, Enlgand, United KIngdom. Tlephone: (44)(0202) 886 831. FAS: (44) (0202) 742 236. Contact: Ralph Cullen< Director. President, R. Cullen. Company description: Supplying probably the World's largest selection of HOlographic Optical components which are: Best Quality; Best Value; Designed by Experienced Holographers plus compponent selection and laboratory/sutdio set-up adice freely available.

UNIPHASE VETREIBS-GMBH. Established in 1984. 5 Employees at this address. LeiBstr. 8, 8152 Feldkirchen-Westerham, Federal Republic of Germany. Telephone: (49)08063/9036. FAX: (49) 08063/7663. Contact: Werner Bleckwendt, Managing Director. Company description: Manufacturer of lasers

UNITED TECHNOLOGIES RESEARCH CENTER. 1100 Employees at this address. Silver Lane, East Hartford, CT 061 08 USA. Telephone: (203) 727 7060. FAX: (203) 727 7852. Contact: Dr. Karl A. Stetson. Parent company: United Technologies Corporation. Company description: UTRC electronic holography systems display real-time fringe patterns on TV comparable to photographic holography. They also provide data output for quantitative analysis. Complete systems or retrofits are available.

UNIVERSIDADE DO PORTO. Laboratorio de Fisica, Praca Gomes Teixeira, P-4000 Porto, Portugal. Contact: Oliverio Soares. Description: Holographic nondestructive testing; Academic holography research.

UNIVERSITA 01 ROMA-LA SAPIENZA. Dipartimento di Fisica, Piazzale Aldo Moro, 2,1-00185 Roma, Italy. Contact: Paolo De Santis. Description: Scientific research.

UNIVERSITAT ERLANGEN - NURNBERG. Physikalisches Institute, Erwin-Rommel Strasse 1, 0-8520 Erlangen, Federal Republic of Germany. Telephone: (49) (09131) 857 408. Contact: Adolf Lohmann. Description: Scientific holography research ; HOE; computer generated holography.

UNIVERSITe DE FRANCHE-COMTE. Laboratoire d'Optique, LA 214, UFR Sciences et des Techniques, F-25030 Besancon Cedex, France. Description: Scientific holography research.

UNIVERSITE DE NEUCHATEL. Institut de Microtechnique, 2, rue A L Breguet, CH-2000 Neuchatel, Switzerland. Telephone: (41)(038) 246 000. Contact: Rene Dandliker. Description: Industrial research.

UNIVERSITe DE PARIS-SUD. Institute d'Optique, F-91405 Orsay, France. Telephone: (33)(9) 416 750. Contact: Serge Lowenthal. Description: Scientific holography research.

UNIVERSITE LAVAL. Dept. Physique-COPL, Pavilion Vachon, Ste-Foy, Quebec, G1 K 7P4 Canada. Contact: Roger A. Lessard. Description: Holography education; workshops.

UNIVERSITe LOUIS PASTEUR. Department de Spetrophysics. 7 rue de L'Universite, 67000 Strasbourg, France. Description: Scientific holography research.

UNIVERSITY ESSEN. Fachbereich 7/Physik, Universitatsstrasse 2, 0-4300, Essen 1, Federal Republic of Germany. Telephone: (49)(0201) 183 3019. Contact: Detlef Leseberg. Description: Scientific holography research; HOE; computer generated holography.

UNIVERSITY GENT, Workshop Holography. Established in 1968. 2 Employees at this address. 41 St. Pieters Nieuwstraat, B 9000, Gent, Belgium. Telephone: (32)(0) 91 233821 , ext. 2465. FAX: (32)(0) 91 237326. Contact: Pierre M. Boone, PhD. Parent company: Center for Applied Research in Art and Technology, Gent, Belgium. Description: We are a small optical group in a 70-person Strength of Materials Laboratory, mainly doing holographic interferometry, but also commissioned work for display, museums, etc.

UNIVERSITY OF ABERDEEN, Dept. of Engineering, Kings College, Aberdeen AB9 2UE, Scotland, United Kingdom. Description: Holographic non-destructive testing, industrial research.

THE UNIVERSITY OF ALABAMA IN HUNTSVILLE, Center for Applied Optics, Huntsville, AL 35899 USA. Telephone: (205) 895 6029. Contact: John Caulfield. Description: Scientific holography research.

UNIVERSITY OF ALiCANTE. Department of Applied Physics, Centro de Holografia, Facultad de Ciencias, Alicante Apdo 99, Spain. Telephone: (34)(566) 1200 ext 1147. Contact: A. Fimia. Description: Artistic holography; HOEs; workshops.

THE UNIVERSITY OF ARIZONA, Optical Science Center, Tucson, AZ 85721 USA. Telephone: (602) 621 2836. Contact: Robert Shannon. Description: Industrial and scientific Holography research; Holographic interferometry; Holographic non-destructive testing.

UNIVERSITY OF BOLOGNA. via Fiumazzo 347, 1-

48010 Belricetto (RA), Italy. Contact: Pier Luigi Capucci. Description: Artistic holography research & education.

UNIVERSITY OF CALIFORNIA. SAN DIEGO. Dept. Electrical and Computer Engineering, La Jolla, CA 92093 USA. Contact: Sing Lee. Description: Holography research

UNIVERSITY OF CALIFORNIA. Advanced Imaging Center, College of Engineering, Santa Barbara, CA 93106 USA. Telephone: Contact: John Landry. Description: Holography research

UNIVERSITY OF DAYTON, Research Institute. 300 College Park, Dayton, OH 45469 USA. Telephone: (513) 2293221. Contact: Lloyd Huff. Description: Scientific research, Industrial research; courses.

THE UNIVERSITY OF MICHIGAN. College of Engineering, Chrysler Center, Ann Arbor, MI 48109-2092 USA. Telephone: (313) 763 5464. Contact: Charles Vest. Description: Holographic interferometry; Particle measurement; Holographic scientific & industrial research; Holographic non-destructive testing.

UNIVERSITY OF MICHIGAN, Department of Electrical and Computer Engineering; EECS Building, Ann Arbor, MI 48109-2122 USA. Telephone: (313) 764 9545. Contact: Emmet Leith. Description: Scientific holography research; Design H.O.E.s; Courses on holography.

UNIVERSITY OF MUNICH. Institute of Medical Optics, Thresienstrasse 37, D-8000 Munich 2, Federal Republic of Germany. Contact: R. Rohler. Description: Medical holography; scientific holography research.

UNIVERSITY OF MUNSTER. Ear, Nose and Throat Clinic, Kardinal von Galen Ring 10, D-4400 Munster, Federal Republic of Germany. Telephone: (49)(0251) 836861. FAX: (49)(0251) 836960. Contact: Gert von Bally. Description: Medical holography; Interferometry.

UNIVERSITY OF OXFORD--HOLOGRAPHY GROUP, Department of Engineering Science, Parks Road, Oxford OX1 3PJ, England, United Kingdom. Contact: Paul Hubel. Description: Scientific holography research; workshops.

UNIVERSITY OF ROCHESTER. Institute of Optics, Rochester, NY 14627 USA. Telephone: (607) 275 4151. Contact: Brian Thompson. Description: Scientific and industrial holography research; interferometry; particle testing & measurement.

UNIVERSITY OF SOUTHERN CALIFORNIA. Department of Physics, University Park, Los Angeles, CA 90089-0484 USA. Telephone: (213) 743 6391. Contact: Jack Feinberg. Description: Scientific holography research; Interferometry.

UNIVERSITY OF STRATCHCLYDE. Mechanical Engineering Group, Glasgow, Scotland, United Kingdom. Contact: P. Waddell. Description: Scientific research; industrial research.

UNIVERSITY OF STUTTGART. Institute of Applied Optics, Pfaffenwaldring 9, D-7000 Stuttgart 80, Federal Republic of Germany. Telephone: (49)(0711) 685 6075. Contact: Hans Tiziani. Description: Scientific holography research; interferometry.

UNIVERSITY OF TOKYO. Faculty of Engineering, Hongo 7-3-1, Bunkyo-ku, Toyko, Japan. Contact: T. Uyemura. Description: Scientific and Medical holography research; Interferometry.

UNIVERSITY OF WISCONSIN, College of Engineering. 432 North Lake Street, Madison, WI 53706 USA. Telephone: (608) 262 1299. FAX: (608) 263 3160. Contact: Francis P. Drake, Program Director. Description: Courses on Laser System Design. Covers: Laser operation, techniques for using & modifying laser output; types of lasers combined with material on scanning, modulation, & detection of laser radiation; designing practical laser systems.

UNIVERSITY OF ZAGREB. Institute of Physics, Bijenicka 46, 41000 Zagreb, Yugoslavia. Telephone: (38) (041) 271 211. Description: Industrial, Holographic non-destructive testing; medical holography.

V

VINCENNES UNIVERSITY. 1002 North First Street, Vincennes, IN 47591 USA. Telephone: (812) 885- 5294. Contact: Richard Duesterberg. Description: Offering holography workshops for high school teachers, & college level courses in holography. We have 4 research- grade optical tables, as well as argon and krypton lasers. Call for more details

VINTEN ELECTRO OPTICS LTD. Unit 28 Ashfield Way, Whetstone, Leicester, LE8 3NU, England, United Kingdom. Telephone: (44)(533) 867 110. Description: Manufacture/Design mirrors and optics.

VNIKTK KUL TURA. ul. Intusiastov 34, 1051118 Moskow, USSR. Contact: O.B. Serov. Description: Marketing consultant.

VOLVO-FL YGMOTOR. S-461 81 Trollhattan, Sweden. Telephone: (46) (0520) 94471. Contact: Robert Frankmark. Company description: Holographic nondestructive testing.

VOLKSWAGEN AG. Forschung und Entwicklung, Messtechnik 0-3180, Wolfsburg 1, Federal Republic of Germany. Telephone: (49)(05) 361 221. Contact: Armin Felske. Company description: Industrial research, Interferometry; Holographic non-destructive testing.

W

WASEDA UNIVERSITY. Dept. of Applied PhYSics, School of Science and Engineering, 3-4-1, Ohkubo, Shinjuku-Ku, Tokyo 160, Japan. Telephone: (81)(03) 209 3211. Description: Medical holography research.

WAVEFRONTS. Established in 1974. 1 Employee at this address. 2428 Judah Street, San Francisco, CA 94122 USA. Telephone: (415) 664 0694. Contact: Louis Brill, Director. Company description: Involved in developing & expanding market & sales efforts for holographic retaillwholesale product lines. Assist in preparation of promotions and collateral sales materials, identify potential sales markets & implementation of sales.

WAVE GUIDES INC. 521 East 12th Street #11 , New York, NY 10009 USA. Telephone : (212) 9873713. Contact: Hale Aust , Owner. Company description: Artistic holography

WAVE MECHANICS. 1535 North Ashland Avenue, Second Floor, Chicago, IL 60622 USA. Telephone; (312) 3844860. Contact: Deni Drinkwater-Welch. Description: Artistic holographer; silver halide transmission and reflection; consultant.

WENTWORTH LABORATORIES LTD. Sunderland Road, Sandy, Bedfordshire, SG19 1 RB, England, United Kingdom. Telephone: (44) (0767) 81221 . FAX: (44) (0767) 291 951 . Contact: M.F. Horgan. Company description: Manufacturer of holography Lab equipment, including labware and isolation tables.

WENYON & GAMBLE. 8 Berry Street, London EC1V OAU, England, United Kingdom. Telephone : (44)(1) 251 1797. Company description: Fine art holography; Curriculum development.

WHILEY, George M. Ltd. (see George M. Whiley Limited)

WHITE LIGHT WORKS, INC. (SEE OUR ADVERTISEMENT ON PAGE 41). P.O. Box 851 ,Woodland Hills, CA 91365 USA. Telephone: (818) 703 1111 . FAX: (818) 703 1182. Contact: Jerry Fox. Company description: White Light Works is a full service holographic production company specializing in low cost embossed holograms and Multiplex "People Stopper" holographic deisplays for trade shows and POP applications.

WHITE TIGER HOLOGRAMS. Johannes Verhulststraat 45, 1071 MS Amsterdam, The Netherlands. Telephone: (31)(20) 797 182. FAX: (31)(20) 790896. Contact: Neil Walker. Company description: Artistic holography: Stereograms, Embossed holograms, bro ker for DCGs.

WHOLE HOGRAPHY. Established in 1987.1 Employee at this address. 4142 Bellefontaine Street, Houston, TX 77025-1105 USA. Telephone: (713) 667 3325. Contact: Michael E. Crawford. Company description: Broker for silver halide holograms

THE WHOLE PICTURE, A GALLERY OF HOLOGRAPHY. Established in 1987. 4 Employees at this address. 634 Parkway, Gatlinburg, TN 37738 USA. Telephone: (615) 436 3650. Contact: Jim Kelly, Owner. Company description: Gallery

WISE INSTRUMENTS. Unit 9, Hollins Business Centre, Marsh Street, Stafford, ST16 3BG, England, United Kingdom. Telephone: (44)(0785) 223 535. Contact: Peter Wise.Company description: Manufactures optics

WOBER DESIGN HOlOlAB AUSTRIA .. (SEE OUR ADVERTISEMENT ON PAGE 15) Established in 1985. Kahlenbergstral3e 6, A-3042 WOrmla, Austria. Telephone: (43) 02275 8210. FAX: (43) 22 758210. Contact: Irmfried Wober. Parent company: Amblehurst Ltd., England. Company description: Our Holography Laboratory, founded in 1985, is the first in Austria. We are the biggest hologram producers in town. We organize exhibitions in Austria and Germany. and sell embossed holograms.

WONDERS OF HOLOGRAPHY GAllERY. (SEE OUR ADVERTISEMENT ON PAGE 64a) Established in 1986. 25 Employees at this address. P.O. Box 1244, Jeddah 21431, Saudi Arabia. Telephone: (966) (2) 652 0052. FAX: (966)(2) 651 1325. President, MA Baghdadi; Vice-President, A.M. Baghdadi; Sales Manager, MA Salam; Customer Service, AA Khan. Company description: We are the first gallery in the Arab World, and the first live laser show company .. Please contact us - we are distributors for several laser and holographic companies all over the world.

WORCESTER POLYTECHNIC INSTITUTE. Mechanical Engineering Department, 100 Institute Road, Worcester, MA 01609-2280 USA. Telephone: (508) 831 5536. FAX: (508) 831 5483. Contact: Ryszard Pryputniewicz. Description: Scientific, Medical & Industrial holography research; Interferometry; Holographic non-destructive testing.

WORLD ART PROJECT. 10247 40th Street West, Webster, MN 55088 USA. Telephone: (507) 744 2913. Company description: Artistic holographyartistic environments- mixed media; architectural holography; international collaborations.

WOTAN LAMPS LTD. 1 Gresham Way, Durnsford Road, London, SW19, England, United Kingdom. Telephone: (44)(1) 947 1261. Company description: Equipment for holography; H-1 producers.

WRIGHT PETTERSON AIR FORCE BASE. Structure Division AFSC, Dayton, OH 45433 USA. Telephone: (513) 255 6104. Contact: Frank Adams. Description: Scientific and industrial holography research; Holographic non-destructive testing; Interferometry.

WYKO Corporation. 1955 East Sixth Street, Tucson, AZ 85719 USA. Telephone: (602) 325 5000. Contact: James Wyant. Company description: Scientific holography research; Interferometry and analysis.

X

X-IAL. Established in 1987. 6 Employees at this address. "Les Algorithmes", Parc d'innovation, F-67400 Illkirch, France.Telephone: (33) 88 67 44 90. FAX: (33) 88 67 40 66. Contcat: Dr. Christian D. Liegeois, Director. Parent company: Pirelli. Company description: Stereograms for embossing; HOEs designed and manufactured

Y

YORK UNIVERSITY. Department of Physics, 4700 Keele St., North York, Ontario, Canada M3J 1 P3. Contact: Dr. S.B. Joshi. Description: Holography courses.

3

3D GALLERY. Established in 1987. 4 Employees at this address. 207 Queen's Quay West, Toronto, Ontario, M6J 1A7 Canada. Telephone: (416) 359 0417. Contact: Brian Postnikoff, President. President, Brian Postnikoff; Vice-President, Avi McCaffrey. Company description: 3D Gallery is Canada's premier exponent of the 3D cult. From holographic pop-art, to the laser fish hook, 3D Gallery is the 3D solution to the 2D blues.

3D MEDIA. Kyrkvagen 24, S-910 36 SAVAR, Sweden. Telephone: (46)(090) 98141 . Contact: Hans Hallstrom. Company description: Artistic holography consulting.

3M--OPTICS TECHNOLOGY CENTER. 1331 Commerce Street, Petaluma, CA USA. 94952 USA. Telephone: (707) 765 3240. Contact: Valerie. Company description: Manufacturer of Holographic Optical Elements and other optics.

NAMES AND ADDRESSES

INDIVIDUALS

A

ABENDROTH, Detlev. President, AKS HolographieGalerie GmbH. Potsdamer StraBe 10, 4300 Essen 1, Federal Republic of Germany.

ABOUCAR, Nathalie. Vice-President, The Foreign Dimension, Suite B, 8./Fl., Central Mansion, 270-276 Queen's Road Central, Hong Kong

ABRAMS, Claudette. 22 Bayview Avenue, Wards Island, Toronto, Ontario, M5J 1Z1 Canada. Description: Artistic holographer.

ABRUNHOSA, L. Sales Manager, Ibero Gestao Integrada e Tecnologica LDA. Apartado 1267,4104 Porto Codex, Portugal.

ADAMS, Frank. Contact for Wright Petterson Air Force Base. Structure Division AFSC, Dayton, OH 45433 USA.

AGEHALL, Christer. Contact for High Tech Network, Skeppsbron 2, S-211 20 Malmo, Sweden.

AITES, Edward. Contact for Aites Lightworks. 2148 North 86th Street, Seattle, WA 98103 USA.

AKBARI, Homaira. Contact for The Johns Hopkins University. Dept of Physics and Astronomy, Baltimore, MD 21218 USA.

ALEXANDER. 1323 14th Street, Apt. L, Santa Monica, CA 90404 USA. Telephone: (213) 393 9846. FAX: (213) 451 5291. Description: Holographic artist.

ALLEN, Jeffrey. Established in 1970. 1 Employee at this address. P.O. Box 2003, Sausalito, CA 94965 USA. Telephone: (415) 459 8232. Description: All types, full service consulting. Twenty years experience, including direct involvement with the first dichromate and embossed holograms. Designed, produced,manufactured,& marketed: fine art multiplexes, environments, jewelry, gifts, etc.

ALLEN, Keith. Contact for Imaging & Design. 1101 Ransom Road, Grand Island, NY 14072-1459 USA.

ALLON, Gerard & Henriette. Vice-President, Customer Service, Trilone Holographie Corp. 4200 Boulevard St. Laurent, Montreal, Suites 305-307, Quebec H2W 2R2 Canada.

ALTON, Mary. Coordinator, Photon League. 110 Sudbury Street-Unit B, Toronto, Ontario M6J 1A7 Canada.

ANDERSEN, Chad. Contact for Meredith Instruments. 6403 North 59th Avenue, Glendale, AZ. 85301 USA.

ANDERSON, Michael & Susan. President, Vice President, Holomex. 4 Borrowdale Avenue, Harrow HA3 7PZ, England, United Kingdom.

ANDERSON, Steve. Proprietor, ArtKitek.122 Myrtle

Avenue, Cotati, CA 94931 USA.

ANDREWS, John. Contact for Advanced Holographics, LTD. 243 Lower Mortlake Rd.,Unit 11, Richmond, Surrey TW9 2LL, England, United Kingdom.

ANDREWS, Matthew, MA (RCA). 7A Brunswick Park, Camberwell, London, SE5 7RH, England, United Kingdom, Telephone: 703 1254. Description: Freelance Holographer

ANGELSKY, Oleg V. Contact for Chernovtsy State University, 2 Kotsyubinsky Str., 274012 Chernovtsy, USSR.

ANTHONY, S. Lee. Vice-President, Cambridge Stereographics Group. P.O. Box 159, Kendall Square Station, Cambridge, MA 02142-0002 USA.

ANZENBERGER, T. Sales Manager, Steinbichler Optotechnik GmbH. Am Bauhof 4, 0-8201 Neubeuern, FRG Federal Republic of Germany.

APRILE, Eva, Sylvia. President, Vice-President, Metamorphosi Olografia Italia SRL. Via Lecco 6, 20124 Milano, Italy.

ARANDA, Frank. Customer Service, Newport Corporation, 18235 Mt. Baldy Circle, P.O.Box 8020, Fountain Valley, CA 92708-8020 USA.

ARKIN, R. Contact for Holo-Spectra. 7742-B Gloria Avenue,Van Nuys, CA 91406 USA.

ARMSTRONG, Frank. Sales Manager, Aerotech Inc., Electro Optical Division, 101 Zeta Drive, Pittsburgh, PA 15238 USA.

ATTWOOD, David. Contact for Lawrence Berkeley Laboratory. University of California. Building 80-101 , Berkeley, CA 94720 USA.

AUST, Hale. Owner, Wave Guides Inc. 521 East 12th Street, New York, NY 10009 USA.

AUSTIN, M. Contact for Hickmott & Austin Holograms, 11 Castelnau, London SW13 9RP, England, United Kingdom.

AVKSENTJEV, German B. Director, Soviskusstva v/o exhdunarodnaya Kniga. Art holography department, 41120 Tryazino, Moscow region, USSR

B

BACK, Johnathan. Contact for Holosystems Inc. P.O. Box 6810, Ithaca, NY 14850 USA.

BAGDAN, Szilvia. Customer Service, Artplay Holographic Studio. H-1191 Budapest, Ady Endre ut. 8, Hungary.

BAGHDADI, A.M. Vice-President, Wonders of Holography Gallery, P.O. Box 1244, Jeddah 21431 , Saudi Arabia.

BAGLEY, Sheila. Customer Service, A.H. Prismatic, LTD. New England House, New England Street, Brighton, BN1 4GH England, United Kingdom.

BAINS, Sunny. Contact for Holographics International, BCM Holographics, London WC1 N 3XX, England, United Kingdom.

BALAS, Solomon. President, Third Dimension Ltd. 4 Wellington Park Estate, Waterloo Road, London NW2 7JW, England, United Kingdom

BALOGH, Tibor. President, Artplay Holographic Studio. H-1191 Budapest, Ady Endre ut. 8, Hungary.

BANCK-HAMMERUM, Knud. GM, Superior Technology Implementation. Hjortekaersvej 99B, 2800 Lyngby, Jenmark.

BANFANTI, F. Sales Manager, CISE SpA Technologie Innovative. via Reggio Emilia, 39, 20090 Segrate, Milano, Italy. Mailing address: P.O. Box 12081, 1-20134 Milano, Italy.

BAR, Edgar. Contact for Holo-Service, Neuensteinerstrasse 19, CH-4153 Basel, Switzerland.

BARATON, Alan. Contact for Holocom Holographie. 13, rue Charles V., Faculte des Sciences et des Techniques, F- Paris, France.

BARKER, Graham. Contact for Architectural Glass & Holography, 1 South Street, Great Waltham, Chelmsford, Essex CM3 1 DF, England, United Kingdom.

BARKIN, Wendy. Vice-President, Point of View Dimensions, Ltd. 45-2903 River Drive South, Jersey City, NJ 07310 USA.

BARNEY, Lynn. President, Ion Laser Technology Inc. 263 Jimmy Dolittle Road, Salt Lake City, UT 841 16 USA.

BARNHART, Donald. Contact for Holoflex Company, RR 3, Box 381, Urbana, IL 61801 USA.

BARRE, Pascal. Contact for Holos Art Galerie. 4 Place Grenus, 1201 Geneva, Switzerland.

BASLER, Dieter. Contact for Holtronic. MelchiorHuber Strasse 25, D-8011 Ottersberg, Post Pliening, Federal Republic of Germany.

BAUCCIO, Candice. Sales Manager, Melles Griot, 1770 Kettering Street, Irvine, CA 92714 USA

BAUGN, Drurey. Contact for James River Products. 5420 Distributor Drive, Richmond, VA 23225 USA.

BAZARGAN, Keveh. Contact for Focal Image Ltd. P.O.Box 1916, 1 Kelvin Court, London W11 3QR, England, United Kingdom.

BEAUREGARD, Alain. President, Lasiris Inc. 3549 Ashby, Ville St. Laurent, Que, Canada H4R 2K3.

BEECK, Manfred-Andreas. Contact for Volkswagen AG. Forschung Messtechnik, Postfach D-3180, Wolfsburg 1, Federal Republic of Germany.

BEKKER, A.M. Contact for Institute of Nuclear Physics, Leningradska obl., 188350 Gatchina, USSR.

BELK, Joseph. Owner, Hyperspace Studio. 973 Page Street Studio, San Francisco, CA 94117 USA.

BENTON, Stephen. Contact for Massachusetts Institute of Technology. M.I.T. Media Laboratory, Spatial Imaging Group, 20 Ames Street, E15-416, Cambridge, MA 02139 USA.

BENYON, Margaret. Established in 1968. Holography Studio, 40 Springdale Avenue, Broadstone, Dorset, BH18 9EU, England United Kingdom. Telephone: (44)(0202) 698 067. Description: Independent holographic artist.

BERG, Marcia. Contact for Hughes Aircraft Co. Laser Products. 6155 El Camino Real, Carlsbad, CA 92008 USA.

BERKHOUT, Rudie. 223 West 21st Street, Apt. B, New York, NY 10011. Telephone: (212) 255 7569. Description: Artist.

BERKOWITZ, Vic. Vice-President, Sillcocks Plastics International. 310 Snyder Avenue, Berkeley Heights, NJ 07922 USA.

BERNARD, B. Sales Manager, Laser Applications, Inc. 12722 Research Parkway, Orlando, FL 32826 USA.

BERNIER, Jean-Robert. President, Holocor LB.F. Printing Inc. 95 des Sulpiciens, L'Epiphanie, Quebec JOK 1JO, Canada.

BERSON, Martin. President, Holaxis Corporation. 499 Farmington Ave, Hartford, CT 06105 USA.

BIEDERMANN, Klaus. Contact, Institute of Optical Research, Royal Institute of Technology, S-100 44 Stockholm, Sweden

BILLINGS, R. or L. Contact for Museum of Holography/ Chicago; School of Holography/ Chicago. 1134 West Washington Boulevard, Chicago, IL 60607 USA.

BIRCH, C. Customer Service, Kendall Hyde Ltd. Kingsland Industrial Park, Stroudley Road, Basingstoke, Hants., RG24 OUG, England, United Kingdom.

BJELKHAGEN, Dr. Hans. Contact for Holicon Corporation. 906 University Place, Evanston, IL 60201 USA; Northwestern University, Biomedical Engineering, Evanston, IL 60208 USA.

BLECKWENDT, Werner. Contact for Uniphase Vetreibs GmbH. LeiBstr. 8, 8152 FeldkirchenWesterham, Federal Republic of Germany.

BLYTH, Jeff. Contact for Brighton Imagecraft. 7 Bath Street, Brighton, East Sussex, BN1 3TB, England, United Kingdom.

BOHLMANN, Claus-Peter. President of Kolbe-Druck, Coloco GmbH & Co. KG. 1m. Industrigel11nde 50, Postfach 1103, D-4804 Versmold, Federal Republic of Germany.

BOONE, Pierre. Contact for University Gent, Workshop Holography. 41 St. Pieters Nieuwstraat , B 9000, Gent, Belgium.

BOTOS, Stephen A. President, Aerotech Inc., Electro Optical Division, 101 Zeta Drive, Pittsburgh, PA 15238 USA.

BOULTON, J.B.Contact for Hologram Europe sprl. Avenue Voltaire 137, 1030 Brussels, Belgium

BOUSIGUE, Jacques. Contact for IDHOL. Boite Postale 7, F. 89340 Saint-Agnan, France

BOWMAN, Jim. Owner, LAS.E.R. Co. 1900 Grove Drive, Haymarket, VA 22069 USA.

BOXALL, Ossie. President, Applied Holographics PLC, Braxted Park, Great Braxted, Malden, Essex CM8 3XB, England, United Kingdom.

BOYD, Patrick. Established in 1986. 1 Employee at this address. 18 Whiteley Road, London SE19 1JT, England United Kingdom. Telephone: (44)(01) 670 4160. FAX: (44)(78) 263 2178. Parent company: Space Time Holographic Applications. Description: Creative holographer specialising in large format pulsed holography of fashion imagery (commissions for Zandra Rhodes in London & Gallerias Preciados in Madrid)

BRADSHAW, Roy. Owner, Reel Image, P.O. Box 566, Pacifica, CA 94044 USA.

BRAGINTON, Mary. Contact, Coulter Optical Company P.O. Box K Dept. MP, 54121 Pinecrest Road, Idyllwild, CA 92349 USA.

BRANDIGER, F.J. Vice-President, CVI Laser Corpoation, CVI West. 470 Lindbergh Avenue, Livermore, CA 94550 USA.

BRANDT JEN , Henry A., Hank A. President, Vice: President, Brandtjen & Kluge, Inc, 539 Blanding Woods Road, St. Croix Falls, WI 54024 USA.

BRECHEISEN, Robert. Sales Manager, Polaroid Corporation, 730 Main Street-1A, Cambridge, MA 02139 USA.

BRILL, Louis. Director, Wavefronts. 2428 Judah Street, San Francisco, CA 94122 USA.

BROADBENT, Donald. Owner, Optical Images. 1309 Simpson Way, Suite J, Escondido, CA 92025 USA.

BRODEL, D. Contact for Ascot Laser Picture Studio, 27 Upper Village Road, Sunninghill, Ascot, Berkshire SL5 7 AJ England, United Kingdom.

BROEDERS, Jan M. Owner, Optische Fenomenon. Nederlandse Stichting Voor, Waarneming & Holografie, Warenarburg 44, NL 2907 CL Capelle aid Ijssel, The Netherlands.

BROWN, Gordon. Contact for Ford Scientific Labs. 2000 Rotunda Drive, Dearborn, MI48121 USA.

BROWN, K.C. Contact for K.C. Brown Holographics. 22 St. Augustine's Road, Camden Town, London NW1 9RN, England, United Kingdom.

BROWN, Kerry J. Contact for Artigliography Co. 7130 Mohawk West Drive, Indianapolis, IN 46236 USA.

BROWN, Kevin. Contact for Holographic Dimensions, Inc., 9235 SW 179 Terrace, Miami, FL 33157 USA.

BRUCK, Richard. Owner, Richard Bruck Holography. 3312 West Belle Plaine #2, Chicago, IL 60618 USA.

BRYNGDAL, Olof. Contact for University of Essen. Pherdemarkt 4, D-4300 Essen, Federal Republic of Germany.

BUELL, D. Marketing Department, Applied Holographics Corp., 1721 Fiske Place, Oxnard, CA 93033 USA.

BUNTS, Frank/Flatiron Studio. 15 West 24th Street 7th Floor, New York, NY 10010 USA. Telephone: (212) 645 5173. Company description: Fine Art using interference patterns to produce a sense of depth and movement. - Work in the collections of The Museum of Holography, N.Y.; The Philadelphia Museum of Art; The Library of Congress.

BURCH, J.M. Contact for College of Manufacturing, Cranfield Institute of Technology, Cranfield, Bedford MK43 OAL, England,United Kingdom.

BURGER, Philippe. Directeur Technique, Micraudel. 93, rue Adelshoffen, F-67300 Schiltigheim, France.

BURGMER, Brigitte. Contact, Deutsche Geselischaft for Holografie e.V. LerchenstraBe 142 a, D-4500 Osnabrück, Federal Republic of Germany.

BURNS, Joseph. Contact for Burns Holographics, P.O. Box 377, Locust Valley, NY USA. Telephone: (914) 356 5203. Description: Holographic artist

BURRIDGE, Mike. Contact for Perception Holography. Thornton Marketing Ltd. Aketon Close, Haggs Lane, Follifoot, Harrogate, North Yorks., England HG3 1 A2 United Kingdom.

BUSSARD, Jan. Contact for Textile Graphics, Inc. 201 North Fruitport Road, Spring Lake, MI 49456 USA.

BUSSAUT, Laurent. Research Director, Art, Science and Technology Institute--Holography Society. 2018 R Street, N.W.,Washington D.C. 20009 USA.

BUTTER ISS, Tony. Contact for New Dimension Holographics. 65-72 Pier One, Hickson Road, Sydney, New South Wales 2000, Australia.

BYRNE, Kenneth G. 177 East 4th Street, Brooklyn, NY 11218 USA. Telephone: (718) 633 4195. Description: Holographic artist.

C

CANTOS, Brad D. Contact for Holage. 1881 Eighth Avenue, San Francisco, CA 94122 USA.

CAPUCCI, Pier Luigi. Established in 1984. 1 Employee at this address. University of Bologna, via Fiumazzo 347, 1-48010 Belricetto (RA), Italy. Telephone: (39) 0545 77 296. Contact: Dr. Pier Luigi Capucci, Visual Communications. Description: My services are: Art and communication exhibitions curator, consultant and critic. Expert in display/ embossed commercial images for subject, structure, disposition in the final setting. Media languages expert for the best image making/employing.

CARLSSON, Torgny E. Vickervagen 4, 14569 Norsborg, Sweden. Telephone: (46)(753) 85659. Description: Holographic artist.

CARLTON, David. 8934 Tarragon Court, Manassas, VA 22110 USA. Telephone: (703)361 9443. Description: Artist.

CAROLA, Maggie. Customer Service, Crown Roll Leaf, Inc. 91 Illinois Ave., Paterson, NJ 07503 USA.

CARRETON, Vincente. Contact for Lasergrafics, Peris y Valero 130, 2a Valencia, Spain.

CASDIN-SILVER, Harriet. Contact, Casdin-Silver Holography. 99 Pond Avenue, Suite D403, Brookline, MA 02146 USA.

CASE, Steven. Contact for University of Minnesota, 4-174 Electrical Engineering/Computer Science Building, 200 Union Street, S.E., Minneapolis, MN 55455 USA.

CASSEN, J.Jay. President, Diversified Graphics, Ltd. 5433 Eagle Industrial Court, Hazelwood, MO 63042 USA.

CAULFIELD, John. Contact for University of Alabama in Huntsville, Center for Applied Optics, Huntsville, AL 35899 USA.

CAUSSIGNAC, Jean-Marie. Contact for LCPC--Lab Central des Ponts et Chaussees. 58 Boulevard Lefebvre, F-75015 Paris, France.

CHALK, Kenneth. Contact for Holomorph Visuals, Inc. 273 de la Gauchetiere W., Montreal, Quebec H27 1C7 Canada.

CHANG, B.J. Contact for Kaiser Optical Systems, Inc. P.O.Box 983, 371 Parkland Plaza, Ann Arbor, MI 48106 USA.

CHANNER, David. Vice-President, Images Company. P.O. Box 313, Jamaica, NY 11419 USA.

CHEN, David. Vice-President, Acme Holography, 12 Sunset Road, West Somerville, MA, Boston Area. USA.

CHERRY, Greg. Contact, Cherry Optical Company, 2047 Blucher Valley Road, Sebastopol, CA 95472 USA.

CHITTIM, Kevin. Vice-President, Melles Griot,1770 Kettering Street, Irvine, CA 92714 USA

CHOUDRY, Amar. Contact for University of Alabama in Huntsville, Center for Applied Optics, Huntsville, AL 35899 USA.

CHRISTAKIS, Anne-Marie. Manager, Magic Laser. Quartier de L'horloge, 4 rue Brantome, 75003, Paris France; Manager, Musee de L'holographie.

CHRISTE-WIL TON, Lars. Contact for High Tech Network, Skeppsbron 2, S-211 20 Malmo, Sweden.

CHRISTIE, Robin. Sales Manager, The DZ Company. P.O. Box 5047-R, 181 Mayhew Way, Suite E, Walnut Creek, CA 94596 USA.

CIFELLI, Dan. President, The DZ Company. P.O. Box 5047-R, 181 Mayhew Way, Suite E, Walnut Creek, CA 94596 USA.

CLARE, G.W. Chairman, Touchwood Holographics, 50 Sugworth Lane, Radley, Abingdon, Oxon, OX14 2HY, England, United Kingdom.

CLARKE, Walter. President, Global Images, Inc. 509 Madison Avenue, Suite 1400, New York, NY 10022 USA.

CLASSEN, Walter. Contact for Licht-Blicke-Buro. Bornemannstrasse 10, D-6000 Frankfurt 70,Federal Republic of Germany.

CLAUS, Patrick. Customer Service, Micraudel. 93, rue Adelshoffen, F-67300 Schiltigheim, France.

CLAY, Burton. Contact for Holographix Inc. 87 Second Avenue, Burlington, MA 01803 USA.

CLAYTOR, Linda H. Contact for Fresnel Technologies Inc. 101 West Morningside Drive, Fort Worth, TX 76110 USA.

COBLIJN, Alexander. Owner, Que Sera Sera. P.O. Box 29, 9700 AA, Groningen, The Netherlands.

COBURN, Joe. President, Coburn Corporation. 1650 Corporate Road West, Lakewood, NJ 08701 USA.

COHN, Dr. Gerald. Contact for Abbott Laboratories. Department 93F, (Building AP-9), Routes 43 and 137, Abbott Park, IL 60064 USA.

COLLIER, Richard G. Contact for Ion Laser Technology Inc. 263 Jimmy Dolittle Road, Salt Lake City, UT 84116 USA.

COLLINS, Jonathan. Vice-President, Atelier Holographique de Paris, 13, Passage Courtois, F-75011 Paris, France. Telephone: (33)(1) 43796918.

COMINELLI, Francesca. Customer Service, Metamorphosi Olografia Italia SRL. Via Lecco 6, 20124 Milano, Italy.

CONLAN, J. Contact for Ashland Electric ProductsVisual Electronics Corp., 80-39th Street, Brooklyn, NY 11232 USA.

CONNER, Arlie. 1514 South East Salmon, Portland, OR 97214 USA. Telephone: (503) 239 0545. Description: Research and development, large display format, artistic and commissions. Expert in liquid crystal devices as well.

CONNORS, Betsy. Contact for Acme Holography.12 Sunset Road, Somerville, MA 02144 USA.

COOKE, D.J. Contact for Oxford University. Department of Engineering Science, Parks Road, Oxford OX1 3PJ England, United Kingdom.

COOMBES, Angela. Contact for Dovecote Studio, Witham Friary, Frome, Somerset , England, United Kingdom.

COOPER, Nick. Contact for Oxford Holographics. 71 High Street, Oxford OX1 4BA, England, United Kingdom.

CORNELIUS, Mark. Customer Service, Advanced Holographics Corp. 2469 East Fort Union Blvd. Suite 108, Salt Lake City, UT 84121 USA.

COSSETTE, Marie-Andree. 1145 Avenue des Laurentrentides Apt.2, Quebec, Quebec G15 3C2, Canada.

Telephone: (418) 6872985. Description: Artist

COWLES, Susan Ann. 123 Grove Green Road, E11 4ED, England, United Kingdom. Description: Artist; fine art holograms.

CRAFT, Steven. Contact for Laser Labs, Inc. 8000 W.110th Street, Suite #115, Overland Park, KS 66210 USA.

CRAWFORD, Michael E. Contact for Whole Hography, 4142 Bellefontaine Street, Houston, TX 77025-1105 USA.

CRENSHAW, Melissa. Contact for The Holographic Studio, Ltd. 2525 York Avenue, Vancouver, British Columbia V6K 1 E4 Canada.

CROSBY, Paul. Sales & Marketing Dir. ,Coherent. 3210 Porter Drive, Palo Alto, CA 94306 USA.

CROSS, Lloyd. P.O. Box 145, Point Arena, CA 95468 USA. Description: Holography Consultant

CRUMLEY, Elizabeth. Contact for FAS.T. Electronic Bulletin Board, P.O. Box 421704, San Francisco, CA 94142-1704 USA.

CUCHE, Denis. Contact for Ciba-Geigy AG. Research Center P&A, CH-1701 Fribourg, Switzerland.

CUELI, Manuel. Vice-President, Crown Roll Leaf, Inc. 91 Illinois Ave., Paterson, NJ 07503 USA.

CULLEN, Ralph. Director, UK Optical Supplies; Ralph Cullen Holographics /UKOS. 84 Wimborne Road West, Wimborne, Dorset BH21 2DP, England, United Kingdom.

CULLEN, Gary & Karoline. President, Managing Director, Holographic Developments LTD. Box 1035, Delta, B.C., V4M 3T2 Canada.

CURCI, J. Vice-President, Jaeger Graphic Technology,

J.G.T--Holofoil SA 20 Avenue des Desirs,B-1140 Bru. ssels, Belgium

CUSSEN, Arthur J. President, Electro Optical Industries, Inc. 859 Ward Drive, Santa Barbara, CA 93111 USA.

CVETKOVICH, Thomas J. Owner, Chromagem. 573 South Schenley, Youngstown, OH 44509 USA.

D

DA-HSIUNG, Hsu. Contact: Beijing institute of posts and Telecommunications. Department of Applied Physics, Holography Laboratory, Beijing 10080, People's Republic of China.

DAINTY, J. Contact for Imperial College of Science. Optics Section, Blackett Laboratory, London SW7 2BZ, England, United Kingdom.

D'AMATO, Salvatore, President. American Bank Note Holographics, Inc. 500 Executive Boulevard, Elmsford, New York 10523 USA.

DANAN, Henry H. President, Trilone Holographie Corp. 4200 Boulevard St. Laurent, Montreal, Suites 305-307, Quebec H2W 2R2 Canada.

DANDLIKER, Rene. Contact for Universite de Neuchatel, Institut de Microtechnique, 2, rue A L Breguet, CH-2000 Neuchatel, Switzerland.

DARCY, L. Customer Service, Laser Electronics Pty., Ltd. P.O. Box 359, Southport, Queensland, 4215, Australia.

DAVIS, Frank. President, Angstrom Industries, Inc. 3202 Argonne, Houston, TX 77098 USA.

DAYUS, Ian. Sales Manager, A.H. Prismatic, LTD. New England House, New England Street, Brighton, BN1 4GH England, United Kingdom.

D'CRUZ, A Wasy. Director, E.1. Dupont De Nemours 3.. Co. Inc Optical Element Venture, Experimental Station, P.O. Box 80352, Wilmington, DE 19880-0352 SA.

DEEM, Rebecca. Established in 1975. 3463 State Street, Suite 304, Santa Barbara, CA 93105 USA. Telephone: (805) 568 6997 FAX: (805) 642 4396. Description: Fine art works incorporating holographysingle sculptural and installation pieces combining relection holograms with mixed media-limited edition Nail mounted pieces-curatorial, catalog and lecture services.

DEFREITAS, Frank. Publisher, The Hologram. P.O. Box 9035, Allentown, PA 18105 USA Telephone: (215) 434 8236. Description: Limited edition holograms; workshops ; publishes 'The Holo-Gram' newsletter (free) ; global networking of enthuSiasts, collectors and practitioners; bibliographies and title searches (current, rare, out-of-print books/patents manuscripts).

DEKENS, Joop. Contact for Que Sera Sera, P.O. Box 29, 9700 AA, Groningen, The Netherlands.

DELEPIERE, Marc. President, Micraudel. 93, rue Adelshoffen, F-67300 Schiltigheim, France.

DEL-PRIETE, Sandro. Contact, Gallerie Illusoria, Schwarztorstrasse 70, CH-3007 Bern, Switzerland

DEMICHER, Charles. Contact for Laser Holographics, Inc. 1179 King St. West, Unit 111, Toronto, Ontario, M6K 3C5 Canada.

DEMMER, Richard. President, Laser Ionics Inc. 701 South Kirkman Road, Orlando, FL 32811 USA

DENISYUK Yu. Contact for FTI Joffe, Politechnicheskaya 26, Academy of Sciences of the USSR, 194021 Leningrad, USSR.

DENIZOT, Stephane. Customer Service. The Foreign Dimension, Suite B, 8'/FL, Central Mansion, 270-276 Queen's Road Central, Hong Kong.

DE SANTIS, Paulo. Contact for Universita di Roma--La

Sapienza,. Dipartimento di Fisica, Piazzale Aldo Moro, 2, 1-00185 Roma, Italy.

DESCHAUMES, H. Sales Manager, Agfa Gevaert N.V. Holography Film Dept, Septestraat 27, B2510, Mortsel, Belgium

DEUTSCHMANN, Sven. Product ManagerHolography, Kolbe-Druck, Coloco GmbH & Co. KG. 1m. Industrigelande 50, Postfach 1103, D-4804 Versmold, Federal Republic of Germany.

DEWAR, David.39 Rue D'Oradour, Luxembourg 2266, Luxembourg. Description: Artist

DE WINNE, R. Product Manager, Agfa Gevaert N.V. Holography Film Dept, Septestraat 27, B2510, Mortsel, Belgium.

DE WOLF, E. President, Agfa Gevaert N.V.Holography Film Dept, Septestraat 27, B2510, Mortsel, Belgium.

DIDIER, Leduc. 152 rue de la Roquette, 75011 Paris, France. Telephone: 43-79-52-48. Description: Artist Independent

DIETRICH, Edward. 2036 West Haddon, Chicago, IL 60622 USA Telephone: (312) 292 0770. Description: Consultant.

DIETZ, Andrew, Al. Vice-President, Sales Manager, Excitek, Inc. 277 Coit Street, Irvington, NJ 07111 USA

DINSMORE, Sydney. Artist / Educator. The Holograpic Studio. 2525 York Avenue #1, Vancouver, B.C. V6K 1 E4, Canada. Telephone: (604) 734 1614. FAX: (614) 734 2842. Description: Specialize in multicolour reflection transfers from pulse masters. Extensive experience as curator and project co-ordinator for fine art holography.

DIRTOFT, Ingegard. Contact for Holometric AB, Bjo nasvagen 21, S-113 47 Stockholm, Sweden.

DISHMAN, Rick. Contact for American Holographic.

P.O.Box 1310, 521 Great Road, Littleton, MA 01460 USA.

DOBRANYI, Zsuzsa. Vice-President, Artplay Holographic Studio. H-1191 Budapest, Ady Endre ut 8., Hungary.

DOMINGUEZ, Carmenza. Schweizerstrasse 77, 6000 Frankfurt 70, Federal Republic of Germany. Description: Artistic holographer.

DOOLAN, S. Customer Service, Amblehurst Ltd., 52 Invincible Road, Farnborough, Hants GU14 7QU, England, United Kingdom.

DORAISWAMY, Krishna C. Marketing Manager, E.I. Dupont De Nemours & Co. Inc Optical Element Venture, Experimental Station, P.O.Box 80352, Wilmington, DE 19880-0352 USA.

DOREe, R. Sales Manager, Jaeger Graphic Technology, J.G.T--Holofoil SA 20 Avenue des Desirs,B- 1140 Brussels, Belgium

DOTY, Jim. Sales Manager, Newport Corporation, 18235 Mt. Baldy Circle, P.O. Box 8020, Fountain Valley, CA 92708-8020 USA.

DOVGALENKO, Georgi. Contact for Novator Research Center, PI. Lesi Ukrainki 1, Kiev-196, 252196 USSR. '

DOWLEY, M.W. President, LiCONix, 3281 Scott Boulevard, Santa Clara, CA, 95054 USA.

DRAKE, Francis P. Program Director, University of Wisconsin, College of Engineering. 432 North Lake Street, Madison, WI 53706 USA.

DRINKWATER-WELCH, Deni. Contact for Wave Mechanics. 1535 North Ashland Avenue, Second Floor, Chicago, IL 60622 USA.

DRISCOLL, John. Contact for New York Hall Of Science. 47-01 111 th Street, Corona, NY 11368 USA

DROK, P. Vice-President, Optilas B.V. P.O. Box 222, 2400 AE Alphen, AID Rijn, The Netherlands

DRUCKER, Dennis & Harriet. President, Vice-President,
Holograms and Other Strange Things. 3200 West Oakland Park Boulevard, Lauderdale Lakes, FL 33311 USA.

DUBOV, Philip. 5401 Bee Cave Road" Austin, TX 78746 USA. Telephone: (512) 3275961. Description: Artist; fine art holography.

DUDDERAR, Thomas. Contact for AT&T Bell LabsMaterials Group. 600 Mountain Avenue, Murray Hill, NJ 07974-2070 USA.

DUESTERBERG, Richard. Contact for Vincennes University. 1002 North First Street, Vincennes, IN 47591 USA.

DUFFEY, William. Contact for Regal Press Inc., Regal Holographics Division, 129 Guild Street, Norwood, MA 02062 USA.

DUNNIGAN, Paul. Contact for Newcastle Upon Tyne Polytechnic, Department of Physics, Ellison Building, Newcastle upon Tyne, NE1 8ST. England, United Kingdom.

DUSTON, Deborah A. President, Duston Holographic Services. 90 Sherbrooke Avenue, Ottawa, Ontario, K1Y 1 R9 Canada.

DUTTON, Keith. Contact for Laser International, 19 Normanton Rise, Holbeck Hill, Scarborough, N Yorks Y011 2XE, England, United Kingdom.

DYENS, Georges M. 1293 Rue de la Visitation, Montreal, Quebec H2L 3B6 Canada. Telephone: (514) 598 8860. Description: Artistic holographer.

E

EDWARDS, A. Vice-President, Kendall Hyde Ltd. Kingsland Industrial Park, Stroudley Road, Basingstoke, Hants., RG24 OUG, England, United Kingdom.

EIMERS, Tilman. Sales Manager, Phoenix Holograms. Trendelbuscher Weg #1, 2875 Ganderkesse 2, Federal Republic of Germany.

EINS, Stefan. Contact for Fasion Moda. 2803 Third Avenue, Bronx, NY 10455 USA.

EIZYKMAN, Claudine. Contact, EIF Productions. Eizykman/ Fihman. 19 Rue Jean Jacques Rousseau, F-75001 Paris, France.

EMMONS, Larrimore B. Contact for Armstrong World Industries. P.O. Box 3511. Lancaster, PA 17604 USA.

ENGELSBERGER, J. Vice-President, Steinbichler Optotechnik GmbH. Am Bauhof 4, D-8201 Neubeuern, FRG Federal Republic of Germany.

ENIVOI, Ruth. Sales Manager, Images Company. P.O. Box 313, Jamaica, NY 11419 USA.

ENNOS, Tony. Contact for Ultrafin. 16 Foster Road, Chiswick, London W4 4NY, England, United Kingdom.

EPSTEIN, Max. President, Holicon Corporation, Inc. 3 Warwick Lane, Lincolnshire, IL 60069 USA; Northwestern University, Dept. of Electrical and Biomedical Engineering, Technological Institute, Evanston, IL 60208 USA.

EPSTEIN, Michael. Customer Service, Holographic Industries, Inc. 3 Warwick Lane, Lincolnshire, IL 60069 USA.

ERF, Robert. Contact for United Technologies Research Center, Optics and Acoustics 86, Silver Lane, East Hartford, CT 06108 USA.

ERIKSON, Ron. Contact for Media Interface, Ltd. 167 Garfield Place, Brooklyn, NY 11215 USA

EVANS, John. Vice-President, British Aerospace PLC. Sowerby Research Centre, FPC: 267, P.O. Box 5, Filton, Bristol BS12 70W, England, United Kingdom.

EYLES, Christopher. Contact, Third Dimension Ltd. 4 Wellington Park Estate, Waterloo Road, London NW2 7JW, England, United Kingdom

F

FAGAN, William. Contact for Lasermet Limited, Five Oaks, Sway Road, Brokenhurst, Hants S04 27RX, England, United Kingdom.

FAKLIS, Dean. Vice-President, Rochester Photonics Corporation. 67 Nettlecreek Street, Fairport, NY 14450 USA.

FARGION, Daniele. Contact for Universita di Roma-La Sapienza,. Dipartimento di Fisica, Piazzale Aldo Moro, 2, 1-00185 Roma, Italy.

FEELEY, Terry. Contact for General Imaging Corporation. 1 Industrial Drive South, Lan-Rex Industrial Park, Smithfield, RI 02917 USA.

FEFERMAN, Bennett J. 3610 Misty Oak Drive, Apt. 140, Melbourne, FL 32901 USA. Telephone: (407) 984 8894. Description: Artistic holographer.

FEINBERG, Jack. Contact for University of Southern California. Department of Physics, University Park, Los Angeles, CA 90089-0484 USA

FELSKE, Armin. Contact for Volkswagenwerk AG, Forschung und Entwicklung, Messtechnik D-3180, Wolfsburg 1, Federal Republic of Germany.

FENIMORE, Jack. Sales Manager, Sillcocks Plastics International. 310 Snyder Avenue, Berkeley Heights, NJ 07922 USA.

FEROE, James, 1420 45th Street, Emeryville, CA 94608 USA. Telephone: (415) 6589787. Description: Artistic holographer.

FIMIA, A. Contact for University of Alicante. Department of Applied Physics, Centro de Holografia, Facultad de Ciencias, Alicante Apdo 99, Spain.

FINé, Hartmut. Contact for Holographic Art, Werderstraße #73, 2800 Bremen 1, BR Deutschland/Federal Republic of Germany.

FISCHER, Julian. Klugstrasse 49, 8000 Munchen 19, Federal Republic of Germany. Telephone: (49)(83) 1572682. Description: Artist.

FISHER, Gary. Contact for Man Environment, Inc. P.O. Box 25959, 2041 Sawtelle Boulevard, Los Angeles, CA 90025 USA

FITZPATRICK, Colleen. Contact for Spectron Development Laboratories, 3303 Harbor Boulevard, Suite G-3, Costa Mesa, CA 92626 USA.

FOLTZ, Susannah D. 1824 Silver S.E. Albuquerque, NM 87106 USA. Telephone: (505) 277 2616. Description: Artist.

FORD, J. Contact for Spectrolab Ltd. P.O. Box 25, Newbury, Berkshire, RH6 8BO, England United Kingdom.

FORNARI, Arthur David. 813 Eighth Avenue, Brooklyn, NY 11215 USA. Telephone: (718) 965 3956. Description: Artistic holographer; silver halide transmission & reflection holograms.

FORSBERG, Mona. Contact for HoloMedia AB/Hologram Museum. P.O. Box 45012, Drottninggatan 100, 10430 Stockholm, Sweden.

FOULKES-WILLIAMS, Paula. Marketing & Sales Coordinator, Light Fantastic PLC. 4E1F Gelders Hall Road, Shepshed, Leicestershire LE12 9NH, England, United Kingdom.

FOURNIER, Jean-Marc. Contact for Rowland Institute for Science. 100 Cambridge Parkway, Cambridge, MA 02142 USA.

FOX, Jerry. Contact for White Light Works, Inc.; L.A. School of Holography, P.O. Box 851 ,Woodland Hills, CA 91365 USA

FRANKMARK, Robert. Contact for Volvo-Flygmotor, S-461 81 Trollhattan, Sweden.

FRANZ, T. Sales Manager for Labor Dr. Steinbichler, Am Bauhof 4, D-8201, Neubeuern, Federal Republic of Germany.

FREUND, Art. Owner, Art Freund Holography. 124 Brookwood Drive, Santa Cruz, CA 95065 USA.

FRIES, Urs. Contact for Holo-Service.Fries, Eulerstrasse 55, CH-4051 Basel, Switzerland.

FRIGON, Jacques. 2375 Fullum, Montreal, Ouebec, Canada H2K 3P3. Telephone: (514) 521 4270. Description: Artist.

FRUIT, Vikki. Vice-President, Angstrom Industries, Inc. 3202 Argonne, Houston, TX 77098 USA.

FUCHTENBUSCH, A. Contact for Labor for Holografie, Am Forst 38, Wesel D-4230, Federal Republic of Germany.

G

GAFUROVA, N.S. Contact for Minchimprom, 101851 Moscov, USSR.

GAGNON, Brigitte.Sales Manager, IBOU Inc. CP 214, Cap-de-la-Madeleine, Quebec, G8T 7W2 Canada.

GALANTOWICZ, Tom. President, Newport Corporation, 18235 Mt. Baldy Circle, P.O.Box 8020, Fountain Valley, CA 92708-8020 USA.

GALLERNEAULT, Christine. Contact for Alcan International Ltd.Kingston Research & Development, P.O.Box 8400, Kingston, Ontario K7L 4Z4, Canada.

GALLOWAY, Lory. Venture Manager, E.1. Dupont De Nemours & Co. Inc Optical Element Venture, Experimental Station, P.O.Box 80352, Wilmington, DE 19880-0352 USA.

GAMBLE, Susan. Contact for Holography Workshop. Goldsmith College, Millard Bldg., Cormont Road, London SE5 9RG, England, United Kingdom.

GARCIA, Julio Ruiz. President, Holotek, SA Carretera de Santander, Granda 47, 33199 Granda-Siero, Asturias, Spain.

GARCON, Thierry. Contact for Holodesign. 1, Boulevard de la Republique, F-95600 Eaubonne, France.

GARDENER, John. President, Gardener Promotion Marketing. 4165 Apalogen Road, Philadelphia, PA 19144 USA.

GASKILL Jack, Contact for The University of Arizona, Optical Sciences Center, Tucson, Arizona 85721 USA.

GAUCHET, Pascal. President, Atelier Holographique de Paris, 13, Passage Courtois, F-75011 Paris, France. Telephone: (33)(1) 43 79 6918.

GAUTHIER, Henry. President, Coherent. 3210 Porter Drive, Palo Alto, CA 94306 USA.

GEIGER, Thomas. Contact for Ingenieur BOro Geiger. Dieding 7, D-8017 Ebersberg, Federal Republic of Germany.

GERSON DE, Michael. Contact for Elan Bio-Medical. Holography Laboratory, 411 Lewis Road, #417, San Jose, CA 95111 USA.

GILBERG, Gunilla. Contact for Hoechst Celanese Corporation. 86 Morris Avenue, Summit, NJ 07901 USA.

GILLES, Jean. 14 Rue des Quatre Vents, 25000 Besançon, France. Telephone: (33) (81) 503 139. Description: Artistic holographer.

GILLESPIE, John, Mike. President, Vice-President, Jodon Inc. 62 Enterprise Drive, Ann Arbor, MI 48103 USA.

GIPP, R. Sales Manager, Applied Holographics Corp., 1721 Fiske Place, Oxnard, CA 93033 USA.

GITIOES, George. 54 Brighton Street, Bundeena, NSW 2230 Australia. Telephone: (61)(2) 523 197. Description: H-1 Maker.

GLEASON, Tom. Sales Manager, Laser Technology, Inc. 1055 West Germantown Pike, Norristown, PA 19403 USA

GLEN, Gerald. Contact for Laser Institute of America, Education Division, 5151 Monroe Street, Toledo, OH 43623 USA.

GOCHA, Dexter. Sales & Marketing Manager, Apollo Lasers, Inc. P.O. Box 2730, Chatsworth, CA 91311 USA.

GOETTEN, John. Contact for Optics Plus Inc. 1369 East Edinger Avenue, Santa Ana, CA 92705 USA.

GOLDBERG, Bruce. 32 Farragut Avenue, San Francisco, CA 94112 USA. Telephone: (415) 584 1197. Description: Consulting and education services.

GOLDSTEIN, Robert. President, Laser Applications, Inc. 12722 Research Parkway, Orlando, FL 32826 USA.

GONçALVES, Figueroa. Vice-President, Ibero Gestao Integrada e Tecnologica LDA. Apartado 1267, 4104 Porto Codex, Portugal.

GOODMAN, Joseph. Contact, Stanford University-Department of Mechanical Engineering. Building 570 Room 571 C, Stanford, CA 94301 USA.

GORGLIONE, N. Contact for Cherry Optical Co.; Laser Affiliates; Nancy Gorglione / Fine Art Holograms; 2047 Blucher Valley Road, Sebastopol, CA 95472 USA

GREEN, Keith. Contact for Laminex/High Tech UK Ltd., Bromfield Industrial Estates, Mold, Clwyd CH7 1JR, England, United Kingdom.

GREENAN, Jay. President, Advanced Holographics Corp. 2469 East Fort Union Blvd. Suite 108, Salt Lake City, UT 84121 USA.

GREGUSS, Pal. Institute of Precision Mechanics and Optics, Applied Biophysics Laboratory,Technical University Budapest, H-1621 Budapest, Hungary.

GREISAS, Nissim. Sales Manager, Holo-Or Ltd. P.O. Box 1051, Rehovot 76110, Israel.

GROSMANN, Michel. Contact for L.I.R. E.R.A.12, rue Libergier, F-5111 0 Reims, France ..

GROSSINGER, I. President, Holo-Or Ltd. P.O. Box 1051, Rehovot 76110, Israel ..

GUEGUEN, Thierry. Sales Manager, Magic Laser.

Quartier de L'horloge, 4 rue Brantome, 75003, Paris France ..

GUIGNARD, M. Contact for Aerospatiale-Division Helicopteres, 2 avenue Marchel-Cachin, F-93126 La Courneuve Cedex, France ..

GUISE, Rodney. Contact for Hologram World, 64 The Promenade, Blackpool, Lancashire, England, United Kingdom.

GULBRANSEN, Keith. Customer Service, Advanced Holographics Corp. 2469 East Fort Union Blvd. Suite 108, Salt Lake City, UT 84121 USA ..

GULLAMON, Antoni Pinol. Anselmo Clave 74, Olesa de Montserrat, 08640 Barcelona, Spain. Telephone: (34)(3) 778 2299. Description: Holographic artist.

GUNTHER, Roy. Vice-President, Gardener Promotion Marketing. 4165 Apalogen Road, Philadelphia, PA 19144 USA.

GUSTAFSSON, Jonny. Contact for Lasergruppen Holovision AB. Osthammarsgatan 69, S - 115 28 Stockholm, Sweden.

GUTEKUNST, Horst. Director, Hologramm Werkstatt & Galerie, Gallerie Fur Hologramme. Via Principale 30, CH - 7649 Castesegna, Switzerland

H

HABER, Gary. Vice-President, Holaxis Corporation. 499 Farmington Ave, Hartford, CT 06105 USA.

HAHN, Dr. Yu. H. President, CVI Laser Corporation, CVI West. 470 Lindbergh Avenue, Livermore, CA 94550 USA.

HAIDLE, Rudy. Contact for Northwestern University,

Fiberoptic Laboratory, Dept of Electrical Engineering, Technological Institute, Evanston, IL 60208 USA.

HAJIME, Yamashita. Contact for Mazda Motor Corp. 3-1, Shinchi, Fuchu-cho, Aki-gun, Hiroshima, Japan.

HALIVA, Mauric. Contact for New York Institute of Technology. Center for Optics, 100 Glen Cove Avenue, Glen Cove, NY 11542 USA.

HALL, G.G. President, George M. Whiley Limited. Firth Road, Houston Industrial Estate, Livingston, West Lothian EH54 5DJ, Scottiand, United Kingdom.

HALLSTROM, Hans. Contact for 3D Media, Kyrkvagen 24, S-910 36 SAVAR, Sweden.

HAMEIRI, Shimon. Contact, Holography Ltd. 21 Hako-memiut Str., Herzlia Pituah, Israel.

HANNIFIN, Matt. Owner, Matt Hannifin Co. P.O. Box 4574, Austin, TX 78765 USA.

HANSEN, Matthew E. 741 East Gorham Street, Madison, WI 53703 USA. Telephone: (608) 255 3580. Description: Holographic artist.

HARDY, Nick. Contact for Op-Graphics (holography) Ltd. Unit 4: Technorth, 7 Harrogate Road, Leeds LS7 3NB, England, United Kingdom.

HARKEY, Joyce C. Customer Service, E.1. Dupont De Nemours & Co. Inc Optical Element Venture, Experimental Station, P.O.Box 80352, Wilmington, DE 19880-0352 USA.

HARIDAS, P. Contact for Massachusetts Institute of Technology, Laboratory for Nuclear Science, Cambridge, MA 02139 USA.

HARIHARAN, Parameswaran. Contact for CSIRO, Division of Chemical Physics, P.O.Box 218, Lindfield 2070, Australia.

HARRIS, Kenneth. Contact for Light Impressions Europe Pic., 5 Mole Business Park 3, Off Station Road, Leatherhead, Surrey KT22 7BA, England, United Kingdom.

HARRIS, Nick. 711 East 13th Street, Houston, TX 77008 USA. Telephone: (713) 861 2865. Description: Artistic holographer; workshops; consulting.

HARRISON, Susan. Contact for Hologram World, 1212112 Dixon Boulevard, Cocoa, FL 32922 USA.

HASHIMOTO, Daijo. Contact for Toyko University, Medical Division, 2nd Surgical Research, Toyko, Japan.

HASKELL, Dara. President, Third Dimension Arts, Inc. 1241 Andersen D(ive, Suites C & D, San Rafael, CA 94901 USA.

HAYDVOGEL, Christian. Director, Technoexan/ Semicon. Morellenfeldgasse 41, A-801 0 Graz, Austria.

HEATON, Lorna. 3998 Rue de Bullion, Montreal, Quebec, Canada H2W 2E4. Telephone: (514) 845 5403. Description: Artist

HEIDT, Marina. Customer Service, Holographic Products, Inc. 755 South 200 West, Richmond, UT 84333 USA.

HEIN, Mrs. Elke. Sales Manager, die dritte dimension. Frankfurter StraBe 132 -134, D-6078 Neu Isenburg, Federal Republic of Germany.

HELFAND, Cindy. Contact for Light Wave, North Pier, 435 East Illinois Street, Chicago, IL 60611 USA.

HENNESSEY, Sigi. Customer Service, Melles Griot, 1770 Kettering Street, Irvine, CA 92714 USA

HERFORTH, Pamela. Sales Manager, Crown Roll Leaf, Inc. 91 Illinois Ave., Paterson, NJ 07503 USA.

HERSLOW, John. President, Sillcocks Plastics International. 310 Snyder Avenue, Berkeley Heights, NJ 07922 USA.

HESLOP, Lynne. Contact for Mirage Holograms Ltd. Unit 2 Brook Lane, Business Centre, Brook Lane North, Brentford, Middlesex TW8 OPP England, United Kingdom.

HEWIN, Brian. Contact for Spycatcher Limited. 2 Croft Cottages, Wattisfield, Diss Norfolk 1 P22 1 NS, England, United Kingdom.

HEYLER, Randy. Marketing Manager, Gas Lasers, Spectra-Physics Inc., Laser Products Division. 1250 West Middlefield Road, Box 7013, Mountain View, CA 94039-7013 USA

HICKMAN, Hope. Contact for Bob Mader Photography. P.O.Box 796728, Richardson, TX 75080 USA.

HICKMOTT, A. Contact for Hickmott & Austin Holograms, 11 Castelnau, London SW13 9RP, England, United Kingdom.

HIGGINS, Dr. Jim. Contact for Lumonics Inc. 105 Schneider Road, Kanata, Ottawa, Ontario K2K 1 Y3, Canada.

HIGGINS, Sam. Contact for Spectral Images, 15 Wanderdown Way, Ovingdean, Brighton BN2 7BX, England, United Kingdom.

HILDEBRAND Percy. Contact for Spectron Development Laboratories, 3303 Harbor Boulevard, Suite G-3, Costa Mesa, CA 92626 USA.

HILLIARD, Bill. Contact for Holographic Visions. 300 South Grand Avenue, Los Angeles, CA 90071 USA; contact for Neovision Productions, P.O. Box 74277, Los Angeles, CA 90004 USA

HINTZ, Jürgen. Contact for Phoenix Holograms, Traubenstr. 41, D-2913 - Apen, Federal Republic of Germany.

HOBART, James. CEO, Coherent. 3210 Porter Drive, Palo Alto, CA 94306 USA.

HODGES, Dean. Vice-President, Newport Corporation, 18235 Mt. Baldy Circle, P.O.Box 8020, Fountain Valley, CA 92708-8020 USA.

HOFFMANN, John. Contact for Museum of the Fine Arts Research & Holographic Center. 1134 West Washington Boulevard, Chicago, IL 60607 USA.

HOFMANN, Martin. Contact for Holografie - Hofmann Labor. Carl-Hermann-Gaiserstrasse 20, 7320 Groppingen, Federal Republic of Germany.

HOGMOEN, Kare. Contact for Det Norske Veritas, Research Division, P.O. Box 300, N-1322 Hovik, Oslo Norway.

HOLBERTON, R.C.Sales Manager, Laser Electronics Pty., Ltd. P.O. Box 359, Southport, Queensland, 4215, Australia.

HOLMES, Burton. Contact for Burton Holmes International, 1004 Larrabee Street, West Hollywood, CA 90069 USA.

HOPKINS, Anthony. Contact for Light Engineering; New Clear Imports. 12 New St. Johns Road, St. Helier, Jersey, Channel Islands, England, United Kingdom.

HORGAN, M.F. Contact for Wentworth Laboratories, Ltd. Sunderland Road, Sandy, Bedfordshire, SG19 1 RB, England, United Kingdom.

HOSKINS, Greg. Vice-President, Robert Sherwood Holographic DeSign, Inc. 400 West Erie Street, Chicago, IL 60610 USA.

HOUDE-WALTER, Susan. Contact, University of Rochester, Institute of Optics, Rochester, NY 14627 USA.

HOWELLS, Malcolm. Contact for Lawrence Berkeley Laboratory. University of California. Building 80-101, Berkeley, CA 94720 USA.

HOWLETT, Glenn. Rr Box 2873, Warren, VT 05674 USA. Telephone: (802) 496 2576. FAX: (802) 496 6488. Description: Holographic artist; reflection holograms.

HUBEL, Paul. Contact for University of Oxford--

Holography Group, Department of Engineering Science, Parks Road, Oxford OX1 3PJ, England, United Kingdom

HUDSON, Philip M.G. Director, Amblehurst Ltd., 52 Invincible Road, Farnborough, Hants GU14 7QU, England, United Kingdom.

HUFF, Lloyd. Contact for University of Dayton, Research Institute. 300 College Park, Dayton, OH 45469 USA.

HYDE, D.J. Vice-President, Kendall Hyde Ltd. Kingsland Industrial Park, Stroudley Road, Basingstoke, Hants., RG24 OUG, England, United Kingdom.

I

IAN, R. Contact for Advanced Environmental Research. Route 1, Box 1830, Woolwich, ME 04579 USA.

IBSEN, Per. Vice-President, Superior Technology Implementation. Hjortekaersvej 99B, 2800 Lyngby, Denmark.

INEICHEN, Beat. Contact for Stoltz AG, Tafernstrasse 15, CH-5405 Baden Dattwil, Switzerland.

ISHII, Setsuko. 1-23-26-404 Kohinata, Bunkyo-Ku, Tokyo, Japan. Telephone: (81)(03) 945 9017. Description: Holographic artist; making DCG, Transmission, Embossed holograms.

IVES, Janet. Contact for Holotec Pic. 7 Cameron Road, Seven Kings, Essex, IG1 3DF, England, United Kingdom.

J

JACKSON-SMITH, Rosemary. P.O.Box 2850, Key Largo, Florida 33037 USA. Telephone: (51 6) 324 3000. Description: Artist.

JACOBSON, Dr. Alex. Contact: CVI Laser Corporation, CVI West. 470 Lindbergh Avenue, Livermore, CA 94550 USA.

JAEGER, M. President, Jaeger Graphic Technology, J.G.T--Holofoil S.A. 20 Avenue des Desirs,B-1 140 Brussels, Belgium

JAIN, Dr. Anil. Contact for APA Optics, Inc. 2950 Northeast 84th Lane, Blaine, MN 55432 USA.

JAMES, D.J. President, Lumonics Inc. 105 Schneider Road,Kanata, Ottawa, Ontario K2K 1Y3, Canada.

JAMES, Randy. Owner, Randy James/Holography. P.O. Box 305, Santa Cruz, CA 95061 USA.

JAOUDE, Majdeline A. Contact for Associates of Science and Technology (AST) Inc., 2450 Lancaster Road, Suite 36, Ottawa K1 B 5N3, Canada.

JEONG, Tung H. President, Integraf; Contact fo r Lake Forest Holography Workshops. P.O. Box 586, Lake Forest, IL 60045 USA.

JOHNSON, D.A. President, Lambda/Ten Optics, Division of Optical Corp. of America. One Lyberty Way, Westford, MA 01886 USA.

JOHNSON, E.A. Contact for Laserlabbet, Box 521, SE 581 06 Linkoping, Sweden.

JORDAN, Carmen. Mng. Marketing Services, LiCO

Nix, 3281 Scott Boulevard, Santa Clara, CA, 95054 USA.

JORGENSEN, Dean. President, Optimation INC., Burr Free Microhole Div.

JOSHI, Dr. S.B. Contact for York University, Department of Physics, 4700 Keele St., North York, Ontario, Canada M3J 1 P3.

JOYEUX, D. Contact for LURE. Institut d'Optique, BP 147, F-91403 Orsay, Cedex, France.

JUNG, Dieter. Viuonvillestrasse 11, 0 10000 Berlin 41, Federal Republic of Germany. Telephone: (49) (30) 7718431. Description: Holographic artist.

JUPTNER, Werner. Contact for Bremer Institute for Angewandte. Strahltechnik, BIAS, Ermlandstraf3e 59, 0-2820 Bremen 71, Federal Republic of Germany.

JUREWICZ, Arlene. Contact for Synchronicity Holograms, Box 4235, Lincolnville, ME 04849 USA.

K

KAC, Eduardo. Rua Hilario de Gouveia, 110/1002, Rio de Janeiro RJ Cep 22040, Brazil. (55)(021) 237 6012. Description: The basis of my work is verbal syntax. I make holopoems; poems conceived in space and time, that change their features according to the beholder's viewpoint.

KAPONIS, Notis. Contact for Laser Graphics. AG. Dimitriou 150,54635 Thessaloniki, Greece.

KASPRZAK, Henryk. Contact, Technical University of Wroclaw. Institute of Physics, Wybrvzeze Wyspianskiego 27, PL-50-370 Wroclaw, Poland

KATSUMA, Hidetoshi. Contact for Tokai University, Department of Electro Photo Optics, 1117 Kitakaname Hi-

ratsuka City, Kanagawa 259-12, Japan.

KAUFFMANN, J. Contact for Imac International, Inc. 1301 Greenwood, Wilmette, IL 60091 USA.

KAUFFMAN, R.E. Sales Manager, Amblehurst Ltd., 52 Invincible Road, Farnborough, Hants GU14 7QU, England, United Kingdom.

KAUFMAN, Andreas. Graf-von-Galen Strasse 5, 0- 4800 Bielefeld 1, Federal Republic of Germany. Telephone: (49)(521) 102 269. Description: Holographic artist.

KAWAHARA, Takao. Marketing Director, Holomedia Inc. 3-15-22, Takaban, Meguro-ku, Tokyo 152 Japan.

KELLISON, Kirt. Contact for Robert Sherwood Holographic Design, Inc. 400 West Erie Street, Chicago, IL 60610 USA.

KELLY, Ed. President, Keystone Scientific Co. P.O. Box 22, Thorndale, PA 19372 USA.

KELLY, Jim. Contact for The Whole Picture, A Gallery of Holography. 634 Parkway, Gatlinburg, TN 37738 USA.

KELLY, Mary. Customer Service, Third Dimension Arts, Inc. 1241 Andersen Drive, Suites C & 0, San Rafael, CA 94901 USA.

KENDALL, M.D. President, Kendall Hyde Ltd. Kingsland Industrial Park, Stroudley Road, Basingstoke, Hants., RG24 OUG, England, United Kingdom.

KENRICK, Paul. Sales Administrator, Melles Griot, 1770 Kettering Street, Irvine, CA 92714 USA

KESSOUS, Armand. Marketing Director, Trilone Holographie Corp. 4200 Boulevard St. Laurent, Montreal, Suites 305-307, Quebec H2W 2R2 Canada.

KHAN, A.A. Customer Service, Wonders of Holography Gallery. P.O. Box 1244, Jeddah 21431, Saudi Arabia.

KHANNA, Shyam. Contact for Columbia University, Deptartment of Otolaryngology, 630 West 168th Street, New York, NY 10032 USA.

KILERI, Marilyn. Contact for Three Dimensional Imagery, Ltd. 3031-K Nihi Street, Honolulu, HI 96819 USA.

KILPATRICK, Jack. System Sales, Laser Resale Inc. 54 Balcom Road, Sudbury, MA 01776 USA.

KIMBALL, Randy. Sales Manager, LiCONix, 3281 Scott Boulevard, Santa Clara, CA, 95054 USA.

KINCEL, David. Vice-President, Aerotech Inc., Electro Optical Division, 101 Zeta Drive, Pittsburgh, PA 15238 USA.

KING, E. Contact for Advanced Environmental Research. Route 1, Box 1830, Woolwich, ME 04579 USA.

KLEMPNER, Jonathan. Contact for Light Harmonics Inc. 93 Lake Shore Drive, Oakland, NJ 07436 USA.

KLINE, Richard. Contact for Central Michigan University, Art Department. Mt Pleasant, MI 48859 USA

KLUTE, Jeff. Customer Service, Holographics North Inc. 444 South Union Street, Burlington, VT 05401 USA.

KNIGHT, Roger C. Marketing Director, Light Fantastic PLC. 4E/F Gelders Hall Road, Shepshed, Leicestershire LE12 9NH, England, United Kingdom.

KOCHER, Clive. Contact for Royal Photographic Society, Salisbury College of Art, Southampton Road, Salisbury, Wiltshire, England, United Kingdom.

KODERA, Tokio. Contact for DAI Nippon Printing Co. Ltd.Cental Research Institute 12, 1-Chome IchigayaKagacho, Shinjuku-ku, Tokyo 162, Japan.

KONTNIK, Lewis. Publisher, Holography News.3932 McKinley Street N.W., Washington, D.C. 20015 USA.

KOOI, A. Manging Director, Optilas B.V. P.O. Box 222, 2400 AE Alphen, ND Rijn, The Netherlands

KORIL, Jerry. Contact for Creative Label, 2450 Estes Drive, Dept. M, Elk Grove Village, IL 60007 USA.

KRAMER, Charles. Contact for Holotek Ltd., 300 East River Road, Rochester, NY 14623 USA.

KREMER, Peter. Vice-President, AKS HolographieGalerie GmbH. Potsdamer StraBe 10, 4300 Essen 1, Federal Republic of Germany

KRIEG, Christine. Marketing Assistant. Coherent. 3210 Porter Drive, Palo Alto, CA 94306 USA.

KUBOTA, Toshihiro. Contact for Kyoto Technical University. Dept. of Photographic Technology, Matsugasaki, Sakyo-ku, Kyoto 606, Japan.

KURZEN, Aaron. P.O. Box 3233, Stony Creek, CT 06405 USA. Telephone: (203) 488 4711. Description: Fine art holograms

L

LACEY, Lee. Contact for Holo/Source Corporation. 21800 Melrose Avenue, Suite 7, Southfield, MI 48075 USA.

LACHAUD, Denis. Contact for Hologram. Industries, 42-44, rue de Trucy, F-94120 Fontenay sous bois, France.

LACOSTE, Russell. Vice-President, Sales and Marketing. American Bank Note Holographics, Inc. 500 Executive Boulevard, Elmsford, New York 10523 USA.

LACZYNSKI, Andrew. Contact for The Holograp Development Group. The Coach House, 188 Kenilworth Avenue, Toronto, Ontario M4L 396 Canada.

LA DUCA, Tim. Sales Manager, Third Dimension Arts, Inc. 1241 Andersen Drive, Suites C & D, San Rafael, CA 94901 USA.

LAFGREN, Darlene. Customer Service, General Holographics, Inc. P.O. Box 82247, Burnaby B.C., V5C 5P7 Canada.

LAFRENIERE, Anik. 4060 Boulevard St. Laurent, Montreal, Quebec, H2W 1 Y9 Canada. Telephone: (514) 2869619. Description: artist.

LAGAN IS, Evan D. Senior Product Specialist, E.I. Dupont De Nemours & Co. Inc Optical Element Venture, Experimental Station, P.O.Box 80352, Wilmington, DE 19880-0352 USA.

LAMBERT, Chris & Carole.Contacts for Laza Holograms. 47 Alpine Street, Reading, Berkshire RG1 2PY, England, United Kingdom.

LANCASTER, Ian. Contact for Holography News; Ian Lancaster Holographics Consultancy, 1 Erica Court, Wych Hill Park, Woking, Surrey GU22 OJB England, United Kingdom.

LANDRY, John. Contact for University of California. Advanced Imaging Center, College of Engineering, Santa Barbara, CA 93106 USA.

LANG, Tim. Contact for JK Lasers Ltd. Cosford Lane, Swift Valley, Rugby, Warwickshire CV21 1 QN, England, United Kingdom.

LANGENBECK, P. Contact for Ing.-Agentur for neue Technologie in Optick und Precision Engineering. D-7771 Frickingen 2, Federal Republic of Germany.

LARKIN, Alexander. Contact for Moscow Physical Engineering Institute. Kashirskoe Shosse 1, Moscow, 115409 USSR.

LARSON, Steve, Roger, Anne. Contacts for Portson, Inc. (Laser Images). 9201 Quivira, Overland Park, KS 66215-3905 USA.

LAUK, Matthias. Contact for Museum fOr Holographie & neue visuelle Medien. PletschmOhelenweg 7, D- 5024 Pulheim 1, Federal Republic of Germany.

LAW, Linda. Owner, Linda Law Holographics. 8 Crescent Drive, Huntington, NY 11743 USA.

LAWRENCE, Gary. Contact for North American Holographics Inc. P.O. Box 451, 103 East Scranton Avenue, Lake Bluff, IL 60044 USA.

LEAFLOOR, Stephen. President, Starlight Holographic Inc. 73 Stable Way, Kanata, Ontario K2M 1 A8, Canada.

LEBINE, Pauline. Contact, Ealing Electro-Optics Inc. New Englander Industrial Park, HOlliston, MA 01746 USA

LEE, Sing. Contact for University of California. San Diego. Dept. Electrical and Computer Engineering, La Jolla, CA 92093 USA.

LEFLOCH, H.C. Contact for Aerospatiale-Ets D'Aquitaine, Saint-Medard-en-Jalles, F-33165 Bordeaux, France.

LEIS, H.G. Contact for Daimler Benz AG. D-7000 Stuttgart 60. Federal Republic of Germany.

LEITH, Emmett. Contact for University of Michigan, Department of Electrical and Computer Engineering; EECS Building, Ann Arbor, MI 481 09-2122 USA.

LEKKI, Walter J. President, Corion Corp. 73 Jeffrey Ave., HOlliston, MA 01746 USA

LESAR, Christopher J. 100 Hickory Lane, Lancaster, OH 43130 USA. Telephone : (614) 654 0862. Description: Holographic artist.

LESEBERG, DetJef. Contact for University Essen, Fachbereich 7/Physik, Universitatsstrasse 2, D-4300, Essen 1, Federal Republic of Germany.

LESSARD, Roger A. Professor, Universite Laval.

Dept. Physique-COPL, Pavilion Vachon, Ste-Foy, Quebec, G1 K 7P4 Canada.

LESSING, Rainer. Contact: Spindeler & Hoyer GmbH. Postfach 3353, Koenigsallee 23, D-3400 Goettingen, Federal Republic of Germany.

LESTER, J.L. President, Metaplast Electrochemicals Corp. 67 Whitson Street, Hempstead, NY 11550 USA.

LEV, Steven. Vice-President, Chromagem. 1871 Selma, Youngstown, OH 44503 USA.

LEVINE, Chris. 32 Lexington Street, London W1 R 3HR, England, United Kingdom. Telephone: (44)(1) 4378992. Description: Artist

LEVINE, Jeffrey. Contact for Another Dimension! AD 2000,948 State Street, New Haven, CT 06511 USA ..

LEVY, Robert. Vice-President, Holo/Source Corporation. 21800 Melrose Avenue, Southfield, MI 48075 USA

LEVY, Uri. Vice-President, Holo-Or Ltd. P.O. Box 1051, Rehovot 76110, Israel..

LIEBERMAN, Dan. Contact for Hologramas De Mexico. PINO 343, Local 3, Col. Sta. Ma La Ribera, 06400 Mexico, D.F. Mexico.

LIEBERMAN, Larry & Peg. Contact for Holographic Images, Inc., 1301 Dade Boulevard, Miami Beach, FL 33139 USA.

LIEGEOIS, Dr. Christian D. Contact for X-IAL and Arbeitskreis Holografie B.V. (FRG); "Les Algorithmes", Parc d'innovation, F-67400 Illkirch, France.

LIGHT, Gail. 1521 Revere Circle, Schaumburg, IL 60193 USA. Telephone: (312) 3rr9545. Description: Artist; Fine art originals.

LIJN, Liliane. 99 Camden Mews, London EC2A 2AA, England, United Kingdom. Telephone: (44)(1) 485 8524. Description: artist.

Names & Addresses 181
LINCOLN, Margaret. Contact for Light Angels, The Corridor, High Street, Bath Spa, Avon BA1 5AJ, England, United Kingdom.

L1SSACK, Selwin. Contact for Laserworks. P.O. Box 2408, Orange, CA 92669 USA.

LIVNEH, David. Contact, Holography Ltd. 21 Hakomemiut Str., Herzlia Pituah, Israel.

LLOYD, S. 655 Sixth Street, Trafford, PA 15085 USA. Description: Consultant.

LOHMANN, Adolf. Contact for Universitat ErlangenNurnberg, Physikalisches Institute, Erwin-Rommel Strasse 1, D-8520 Erlangen, Federal Republic of Germany.

LOKBERG, Ole. Contact for Norges Tekniske Hogskole, Institute for Almen Fysikk, Sem Saelandsv 7, N-7034 Trondheim-NTH, Norway.

LOMAX, Jeff. President, Markem Systems Ltd .. Ladywell Trading Estate, Eccles New Road, Salford, M5 2DA England, United Kingdom.

LONG, Kathryn. Contact for Light Impressions, Inc. 149-B Josephine Street, Santa Cruz, CA 95060 USA.

LOUIE, Stephanie. Customer Service, Smith & McKay, 96 North Almaden Boulevard, San Jose, CA 95110-2490 USA.

LOVE, V. Contact for Op-Graphics (holography) Ltd. Unit 4: Technorth, 7 Harrogate Road, Leeds LS7 3NB, England, United Kingdom.

LOWENTHAL, Serge. Contact for Universite de Paris-Sud, Institute d'Optique, F-91405 Orsay, France.

LUBETSKY, Neal. Owner, Point of View Dimensions, LTD. 45-2903 River Drive South, Jersey City, NJ 07310 USA.

LUNGERSHAUSEN, Arnold. Contact for Rochester Institute of Technology. One Lomb Memorial Drive,

P.O. Box 9887, Rochester, NY 14623 USA.

LUSIGEN, R. Customer Service, Laser Applications, Inc. 12722 Research Parkway, Orlando, FL 32826 USA.

LYSOGORSKI, Charles. 271 Keyes Road, Honeoye Falls, NY 14472 USA. Telephone: (716) 533 1258. Description: Technical and artistic holographer; Fabrication and consulting.

M

MACGREGOR, A.E. Contact for Newcastle Upon Tyne Polytechnic, Department of Physics, Ellison Building, Newcastle upon Tyne, NE1 8ST. England, United Kingdom.

MACSHANE, Jim & Elaine. Contact for MacShane Holography/Laser Arts Programs. 512 West Braeside Drive, Arlington Heights, IL 60004 USA.

MADDUX, Gene. Contact for Wright Petterson Air Force Base, Structure Division AFSC, Dayton, OH 45433 USA.

MADER, Bob. Owner, Bob Mader Photography. P.O.Box 796728, Richardson, TX 75080 USA.

MAGGI, Ruggero. Contact for Amazon, C. So Sempione 67, 20149-Milano, Italy.

MAJEAU, Celine. Customer Service, Holocor I.B.F. Printing Inc. 95 des Sulpiciens, L'Epiphanie, Quebec J0K 1 J0, Canada.

MAJERI, H. Customer Service, Jaeger Graphic Technology, J.G.T.--Holofoil S.A. 20 Avenue des Desirs,B-1140 Brussels, Belgium

MALINA, Dr. Roger F. President, ISAST/Leonardo, P.O.

Box 75, 1442A Walnut Street, Berkeley, CA 94709 USA.

MALMQVIST, Sven. Contact for Saab-Scania, S-581 88 L1NKOPING, Sweden.

MALOTI, Michael. Contact for Laser Light Designs, 2412 Kennedy Way, Antioch, CA 94509 USA.

MANDERSCHEID, Thierry. Contact for Citroen Industrie. 35, rue Grange Dame Rose, F-92360 Meudon-la-Foret, France.

MANGUM, Gary. Vice-President, Advanced Holographics Corp. 2469 East Fort Union Blvd. Suite 108, Salt Lake City, UT 84121 USA.

MARCHAND, Jean-Pierre. Contact for IBOU Inc. CP 214, Cap-de-la-Madeleine, Quebec, G8T 7W2 Canada.

MARGOLIS, Mark S. President, Rainbow Symphony Inc. 22823 Hatteras Street, Woodland Hills, CA 91367 USA.

MARHIC, Michael. Vice-President, Holicon Corporation. 906 University Place, Evanston, IL 60201 USA; Northwestern University, Dept. of Electrical Engineering, Technological Institute, Evanston, IL 60208 USA.

MARKOV, Vladimir. Contact, Institute of Physics. Ukrainian Academy of Sciences, Prospect Nauki 46, 252650 Kiev 28, USSR

MARSHALL, John. Contact for Institute of Opthalomology, Jud Street, London WC1 , England, United Kingdom.

MASCIS, Michael. Sales Manager, Corion Corp. 73 Jeffrey Ave., Holliston, MA 01746 USA

MASSENBURG, Henry. General Manager, SpectraPhysics Inc., Laser Products Division. 1250 West Middlefield Road, Box 7013, Mountain View, CA 94039-7013 USA

MATHIEU, Marie-Christiane. Contact for Les Productions Hololab! 3970, Boulevarde St. Laurent, Montreal, Quebec H2W 1 Y3, Canada.

MATIHIESEN, Johannes. Contact for Holocom. Lange Strasse 51, D-2117 Kakenstorf, Federal Republic of Germany.

McCAFFREY, Avi. Vice-President, 3D Gallery, 207 Queen's Quay West, Toronto, Ontario, M6J 1A7 Canada.

McCAIN, Richard & Clare. President, Vice-President, McCain Marketing. 10962 North Wauwatosa Road 76W, Mequon, WI 53092, USA.

McCLEAN, James. 1 Employee at this address. 809 Marquette NW, ALbuquerque, NM 87102 USA. Telephone: (505) 2438400. Description: Artist

McCORMACK, Sharon. Contact for School of Holography, 263 Montego Key, Novato, CA 94949 USA.

McGANN, Donna. Customer Service, Excitek, Inc. 277 Coit Street, Irvington, NJ 07111 USA.

McGEORGE, Neville. Contact for Rosewell Ltd. Blacknest Estate, Bentley, Alton, Hants., England, United Kingdom

McGOWAN, Dr. J. William. President, Associates of Science and Technology (AST) Inc., 2450 Lancaster Road, Suite 36, Ottawa K1 B 5N3, Canada.

McGREW, Steve. Contact for Light Impressions, Inc. 149-B Josephine Street, Santa Cruz, CA 95060 USA.

McINTYRE, Jim. Royal College of Art, Holography Unit. 175 Kings Cross Road, Kings Cross, London WC1 X 9B2, England United Kingdom. Telephone: 44 01 5845020,4401 8379363. Description: Artist.

McKAY, Dave. President, Smith & McKay, 96 North Almaden Boulevard, San Jose, CA 9511 0-2490 USA.

McNAMARA, Maureen. Contact for The Hologram

Schoppe. P.O. Box 318, 591 Tonawanda, Buffalo, NY 14202 USA.

MEANS, Marcia M. 86 Hungry Hollow Road, Spring Valley, NY 10977 USA. Telephone: (914) 3565203. Description: MA Art History, Columbia University 1979 - thesis on holography; Curator "The holographic instant" Museum of Holography, NYC 1987; Contributing editor Holosphere 1979-1988; Coauthor "Holography: Memories in light," documentary videotape; author exhibition catalogs.

MEDORA, Michael. Contact for Medora Waves, Studio 8, 1 Ranelagh Gardens, London SW6 3PA, England, United Kingdom.

MENNING, Melinda.171 Hopetown Avenue, Vaucluse 2030, Sydney N.S.W. Australia. Telephone: (61)(2) 337 1916. Description: Artistic holographer.

MEREDITH, Dennis. President, Meredith Instruments. 6403 North 59th Avenue, Glendale, AZ. 85301 USA.

MERLIN. Achim Konz, Wassenberger Strasse 47, 5138 Heinsberg, Federal Republic of Germany. Telephone: (49)(2452) 6072. Company description: Artistic holography.

MERRICK, Michael G.,1002 Meadowview Drive, Mendota, IL 61342-1444 USA. Description: Artistic holographer. Description: Fine art originals

MERRILL, Mark C. 410 Riverdale-Studio B, Glendale, CA 91204 USA Telephone: (818) 247 6458. Description: Artistic holograms; fine art originals.

MESTANCIK, Phillip E. Contact for Polaroid Corporation, 730 Main Street-1A, Cambridge, MA 02139 USA

METZLER, Elaine. Contact for Asociacion Espanola de Holografia, Avda. Filipinas 38-1A, Madrid 3, Spain.

MICHELITSCH, Barry. Vice-President, Holographic Developments LTD. Box 1035, Delta, B.C., V4M 3T2 Canada.

MIKHAYLOV, V.P. Contact for AN. Sevchenko Research Institute of Applied Physical Problems. 220106 Minsk, USSR.

MILLER, D. Contact for Holographic Design, Inc. 1084 North Delaware Avenue, Philadelphia, PA 19125 USA.

MILLER, Mr. Peter. 2 Foxes Lane, Mousehole, Cornwall, TR19 6QQ England, United Kingdom. Description: Holographic artist, consultant.

MILLER, Preston. Sales Manager, Jodon Inc. 62 Enterprise Drive, Ann Arbor, MI 48103 USA.

MILLMAN, Frank. CEO, Holographic Images Inc. 1301 Dade Boulevard, Miami Beach, FL 33139, USA

MILLS, Karl. Contact for Alpha Photo Products, Inc., 985 Third Street, P.O. Box 23955, Oakland, CA 94623 USA.

MISTRY, Rohit. President, Jayco Holographics. 29/43 Sydney Road, Watford, Herts, WD1 7PY England, United Kingdom.

MITAMURA, Shunsuke. Professor: Institute of Art & Design, University of Tsukuba. 1-1, Tennodai, Tsukuba, Japan 305.

MOLIN, Nils-Erik. Contact for' Lulea University of Technology, Dept. of Mechanical Engineering, S-951 87 Lulea, Sweden.

MOLTENI, William. 43 Hilltop Road, West Long Branch, New Jersey 07764 USA. Telephone: (617) 484 3592. Description: Marketing consultant.

MORAINE, Mary. Sales Manager, Meredith Instruments. 6403 North 59th Avenue, Glendale, AZ 85301 USA.

MOREE, Samuel. Contact for New York Holographic Labs. P.O.Box 20391, Tompkins Square Station, 176 East 3rd Street, New York, NY 10009 USA.

MORGUN, Yuri. Contact for Institute of Electronics .. BSSR Academy of Sciences-Minsk, 22 Logoiski Trakt,

220841 Minsk 90, USSR.

MORI, John L. Contact for Illinois Valley Magnetic Resonance. 4005 Progress Boulevard, Peru, IL 61354 USA.

MORRIS, G.M. President, Rochester Photonics Corporation. 67 Nettlecreek Street, Fairport, NY 14450 USA.

MULHEM, Dominique, 1, Residence les Camelias, F-92600 Asnires, France. Telephone: (33) (1) 47 94 82 42. Description: I make fine art holography mixed with painting. This is called Holopainting. I exibit in art galeries and museums throughout the world. I have edited a monograph.

MULVANEY, Mark. President, Letterhead Press Inc., 155 North 120th Street, Wauwatosa, WI 53226 USA.

MUNDAY, Rob. Contact for Munday Spatial Imaging. 39 Pyrcroft Road, Chertsey, Surrey KT16 9HT, England, United Kingdom.

MURATA, M. Contact for Mitsubishi Heavy Industries Ltd., Nagasaki Technical Institute, 1-1 Akunouramachi, Nagasaki 850-91 Japan. Description: Holographic nondestructive testing ; industrial research.

MURRAY, Jeffrey. President, Holography Institute, P.O. Box 446, Petaluma, CA 94953 USA.

MUSKOVITZ, Aaron. P.O. Box 1022, South Lake Tahoe, CA 95705 USA. Telephone: (916) 544 5989. Description: Artistic holographer.

MUTH, August. Partner: Lasart Ltd. P.O.Box 703, Norwood, CO 81423 USA.

MYRE, Robert. 6090 Waverly, Montreal, Quebec, H2X 2A3 Canada. Telephone: (416) 533 4692. Description: artist.

N

NAEVE, Ambjorn. Contact for Dialectica AB, Skanegatan 87, 6tr, S-116 37 Stockholm, Sweden.

NAIMARK, Michael. 216 Filbert Street, San Francisco, CA 94133 USA Telephone: (415) 391 4817. Description: Holographic artist; fine art holograms.

NAKAJIMA, Masato. President, Holomedia Inc. 3-15-22, Takaban, Meguro-ku, Tokyo 152 Japan.

NALIMOV, Igor. Contact for Cinema & Photo Research Institute. NIKFI, Leningradsky, Prospect 47, Moscow USSR.

NEKRASOVA, Larisa. Contact for Scientific Council on Exhibitions of the USSR Academy of Sciences, 30, Varilov Street, Moscow USSR.

NELSON, Drew. Business Development Manager, Laser Ionics Inc. 701 South Kirkman Road, Orlando, FL 32811 USA

NEMTZOW, Scott. 242 East Highland Avenue, Philadelphia, PA 19118. Telephone: (215) 242 2848. Description: Holographic artist.

NEUMAN, John. President, Laser Technology, Inc. 1055 West Germantown Pike, Norristown, PA 19403 USA

NEWELL, William. Vice-President, Laser Ionics Inc. 701 South Kirkman Road, Orlando, FL 32811 USA

NEWMAN, Paul. Sheerwater Lodge, Sheerwater Road, Woodham, Weybridge, Surrey KT15 3QL, England, United Kingdom. Telephone: (44)(09323)42396. Description: Personal artworks for exhibition and sale; tuition at all levels from beginners to M.A.; research into architectural and interior design applications; de-

sign for commercial display; technical consultancy.

NIEDZIALKOWSKI, George. Contact for Holographic Shop of Milwaukee, 5644 Parking Street, Greendale, WI 53129 USA.

NIGGEBRuGGE, Jorg. Customer Service, KolbeDruck, Coloco GmbH & Co. KG. 1m. Industrigelände 50, Postfach 1103, 0-4804 Versmold, Federal Republic of Germany.

NOLL, Josette. Sales Manager, Holographics North Inc. 444 South Union Street, Burlington, VT 05401 USA.

NORAIS, Filipe Vallada P. President, Ibero Gestao Integrada e Tecnologica LOA. Apartado 1267, 4104 Porto Codex, Portugal.

O

OBZANSKY, J.J. & M.A. President, Contact, Towne Laboratories, InC.P.O.Box 460-HM, One U.S. Highway 206, Somerville, NJ 08876-0460 USA.

ODHNER, Jeff. President, Odhner Holographics. 833 Laurel Avenue, Orlando, FL 32803 USA

OHNUMA, K. Contact for Toppan Printing Co. Ltd., Central Research Institute. 5, 1-chome Taito, Taitoku, Tokyo 110, Japan.

OHZU, Hitoshi. Contact for Waseda University. Dept. of Applied Physics, School of Science and Engineering, 3-4-1, Ohkubo, Shinjuku-Ku, Tokyo 160, Japan.

OLIVER, Jane. Contact: Markem Systems Ltd .. Ladywell Trading Estate, Eccles New Road, Salford, M5 2DA England, United Kingdom.

OLLIER, Bernard. Contact for ERBA 12, rue Libergier, F-511 00 Reims, France.

OLMSTED, George. Contact for Lambda/Ten Optics, Division of Optical Corp. of America. One Lyberty Way, Westford, MA 01886 USA.

ONDREJIK, Charles. Sales Manager, Towne Laboratories, Inc.P.O.Box 460-HM, One U.S. Highway 206, Somerville, NJ 08876-0460 USA.

ORAZEM, Vito. Contact for Holo GmbH Holografielabor OsnabrOck. MindernerStr. 205, D-4500 Osnabruck, Federal Republic of Germany.

ORR, Edwina. President, Richmond Holographic Studios. 6 Marlborough Road, Richmond, Surrey, England, TW10 GJR United Kingdom.

OSADA, Richard M. Contact for Lazer Wizardry, Light Fantastic, 11022 West Oregon Place, Lakewood, CO 80226 USA; 2026 S. High Street, Denver, CO 80210 USA

OSE, Teruji. Contact for Kyoto Technical University. Dept. of Photographic Technology, Matsugasaki, Sakyoku, Kyoto 606, Japan.

OSTLER, Kevin D. Vice-President, Ion Laser Technology Inc. 263 Jimmy Dolittle Road, Salt Lake City, UT 84116 USA.

OSTROVSKY, Juri. Contact for Leningrad Subsidiary in Machinery Science. Academy of Sciences of the USSR, Bolshoi Av. 61,199178 Leningrad, USSR.

OTT, Hans-Peter. Curator Museum far Holographie & neue visuelle Medien. PletschmOhelenweg 7, D-5024 Pulheim 1, Federal Republic of Germany.

OUTWATER, Chris. Vice-President, Applied Holographics Corp, 1721 Fiske Place, Oxnard, CA 93033 USA.

OWEN, Harry. Contact for Pilkington PE Ltd. Glascoed Road, St. Asaph, Clwyd LL 17 OLL, England, United Kingdom.

OWEN, Robert. Contact for NASA Marshall Space Flight Center. Space Sciences Laboratory, ES 73, Huntsville, AL 35812 USA.

OZAWA, Mr. S. Contact for Toppan Printing Co. Ltd. (Branch office). 1 Embarcadero Center, Suite 2106- MP, San Francisco, CA 94111 USA.

P

PADNOS, W.R. 2019 North Damen Avenue, Chicago, IL 60647 USA. Telephone: (213)384 2647. Description: Holographic artist; make fine art holograms, silver halide transmission and reflection; and presentation support.

PAGE, Michael. Contact for Ontario College of Art. 100 McCaul Street, Toronto, Ontario M9W 1W1, Canada.

PALAMAS, Yannis. Contact for: Laser Graphics. AG. Dimitriou 150, 54635 Thessaloniki, Greece.

PALMER, Caroline. 49 Upper Woodford, Salisbury, Wiltshire SP4 6NU, England, United Kingdom. Telephone: (44)(72) 273471 . Description: Holographic artist.

PANICO, John. President, Images Company. P.O. Box 313, Jamaica, NY 11419 USA.

PARKER, Richard. Contact for Rolls-Royce PLC-Advanced Research Laboratory, P.O. Box 31, Derby DE2 8BJ, England, United Kingdom.

PARKER, S.C.J. Contact for British Aerospace PLC. Sowerby Research Centre, FPC: 267, P.O. Box 5, Filton, Bristol BS12 70W, England, United Kingdom.

PATTERSON, Rich. Contact for Reynolds Metals Co. Flexible Packing Division, 6603 West Broad Street, Richmond, VA 23230 USA

PAWLUCZYK, Romuald. Contact for Litton Systems Canada Ltd, 25 Cityview Drive, Rexdale, Ontario M9W 5A7 Canada.

PEPPER, Andrew. 22 Haldane Road, London E6 3JJ, England, United Kingdom. Telephone: (44)(1) 471 1609. FAX: (44)(1) 318 1439. Description: Fine art holograms; holography education.

PERIASAMY, Ammasi. Contact for Medical University of South Carolina, Dept. of Anatomy & Cell Biology, 171 Ashley Avenue, Charleston, SC 29425-2203 USA.

PERKINS, K. Customer Service, Lumonics Inc. 105 Schneider Road, Kanata, Ottawa, Ontario K2K 1 Y3, Canada.

PERRY, Dr. John & Barbara D. President, VicePresident, Holographics North Inc. 444 South Union Street, Burlington, VT 05401 USA.

PERSCH, Ellen. Customer Service, Images Company. P.O. Box 313, Jamaica, NY 11419 USA.

PETERSEN, Joel. 7343 Adams Street, Paramount, CA 90723 USA. Telephone: (213) 6340434. Description: Holographic artist; fine art holograms.

PETICOV, Antonio, 712 Broadway, New York, NY 10003 USA. Telephone: (212) 529 0465. Description: Makes fine art originals.

PETTERSSON, Sven-Goran. Contact for Lund Institute of Technology, Department of Physics, Box 118, S-221 00 LUND, Sweden.

PHILLIPS, Jacque. Contact for Third Dimension Ltd. 1855 Charter Lane, Suite C, Lancaster, PA 17601 USA.

PHILLIPS, Nicholas. Contact for Loughborough University of Technology, Dept. of Physics, Loughborough, Leicesterchire LE11 3TU, England, United Kingdom.

PINK, P. Contact for Holography Institute, P.O. Box 446, Petaluma, CA 94953 USA.

PINTO, Gustavo. Customer Service, Coherent. 3210 Porter Drive, Palo Alto, CA 94306 USA.

PIZZANELLI, David. 4 Macaulay Road, London, SW4 OQX, England, United Kingdom. Telephone: (44)(01) 627 1140. Description: Holographic artist.

POE, Nelson. Contact for Active Image. P.O. Box 97, Boulder Creek, CA 95006 USA.

POLENTA, Manuela. Sales Manager, Metamorphosi Olografia Italia SRL. Via Lecco 6, 20124 Milano, Italy.

POSTNIKOFF, ' Brian. President, 3D Gallery, 207 Queen's Quay West, Toronto, Ontario, M6J 1A7 Canada.

PRICE, Kim. Customer Service, The Diffraction Company, Inc. P.O. Box 151, Riderwood, MD 21152 USA.

PRICE, Stu. Contact for Shipley Chemical Co. 1457 McArthur, Whitehall, PA 18052 USA.

PRICONE, Robert. President, Holographic Industries, Inc. 3 Warwick Lane, Lincolnshire, IL 60069 USA.

PROSTEV, A. Contact for S.1. Vavilov, State Optics Institute, 199064 Leningrad, USSR.

PROVENCE, Steve. Contact for Steve Provence Holography. 15220 Fern Avenue, Boulder Creek, CA 95006 USA.

PRYPUTNIEWICZ, Ryszard. Contact for Worcester Polytechnic Institute. Mechanical Engineering Department, 100 Institute Road, Worcester, MA 01609-2280 USA.

PYE, Tim. 19 Wonderland Avenue, Tamarama, Sydney, 2026 Australia. Telephone: (61)(2) 306 611. Description: Artistic holographer.

Q

QUINLAN, Denis. 48 Clifton Street, North Balwyn, Victoria 33104 Australia. Telephone: (61)(3) 857 8655. Description: Marketing consultant.

R

RACEY, Carl. Sales Manager, Amazing World of Holograms. Corrigan's Arcade, Foreshore Road, South Bay, Scarborough, North Yorkshire Y011 1 PB, England, United Kingdom; Contact for Laser Light Image. 101 Spring Bank, Hull, HU3 1 BH, England, United Kingdom.

RALLISON, Richard. President, Ralcon. Box 142, 8501 South 400 West, Paradise, UT 84328 USA.

RATHJE, Barbel. Contact for Holographic Art. Werderstraße #73, Bremen 2800, Federal Republic of Germany.

RAVERAT DE BOISHEU, J-C. Contact for Holographic Creations. 26- Rue Daniel Stern, Paris 75015, France.

RAYFIELD, Dave & Holly. President, Vice-President, Holographic Products, Inc. 755 South 200 West, Richmond, UT 84333 USA.

RAYMAN, R. Sales Manager, Lumonics Inc. 105 Schneider Road,Kanata, Ottawa, Ontario K2K 1Y3, Canada.

REDZIKOWSKI, Mark. Product Manager, Agfa Gevaert Inc.--Industrial Division, 100 Challenger Road, Ridgefield Park, NJ 07660 USA.

REED, Terry. Contact for Newport Corporation. 18235 Mt. Baldy Circle, P.O.Box 8020, Fountain Valley, CA 92708-8020 USA.

REILLY, John. 238a Gloucester Terrace, London W2, England, United Kingdom. Telephone: (44)(01) 243 0601. Description: Artist.

RENVALL, Ms. Mikaela. Customer Service, Starcke KY. P.O. Box 22, SF-32811, Peipohja, Finland

REUTERSWARD, Carl Fredrick. 6 Rue Montolieu, 1030 Bussigny/Lausanne, Switzerland. Telephone: (41)(021) 701 0514. Description: Artistic holography.

REZNY, Abe. Owner, Laser Light Ltd. 57 Grand Street, New York, NY 10013 USA.

RHODY, Alan. Sales Manager, Holos Gallery. 1792 Haight Street, San Francisco, CA 94117 USA

RICE, Graham. Contact for Newcastle Upon Tyne Polytechnic, Department of Physics, Ellison Building, Newcastle upon Tyne, NE1 8ST. England, United Kingdom.

RITCHER, Michael. Contact for Ritcher Holograms. Adolf Kolping Strasse 16, D-4050 Monchengladbach, Federal Republic of Germany.

RITSCHER, Eve. Contact for Eve Ritscher Associates Ltd., 73 Allfarthing Lane, Wandsworth, London SW18, England, United Kingdom.

RIZZI, Mrs. M. Luciana, Eng., CISE SpA Technologie Innovative. via Reggio Emilia, 39, 20090 Segrate, Milano, Italy. Mailing address: P.O.Box 12081, 1-20134 Milano, Italy.

ROA, Warna J. 2450 Lancaster Road-Unit 36, Ottawa, Ontario, Canada, K1 B 5N3. Telephone: (613) 521 2557. Description: Marketing consultant.

ROBINSON, Anthony. Contact for Robinson Hologram Lighting Systems. 5 Hillside Cottages, Owismoor Road, Camberley, Surrey GU15 4SU, England, United Kingdom.

ROBINSON, D.W. Contact for National Physical Laboratories, Queens Road, Teddington, Middlesex TW11 OLW, England, United Kingdom.

ROBITAILLE, Gaetan. Vice-President, Lasiris Inc. 3549 Ashby, Ville St. Laurent, Que, Canada H4R 2K3.

ROGERS, John. Sales Manager, Smith & McKay, 96 North Almaden Boulevard, San Jose, CA 95110-2490 USA.

ROHLER, R. Contact for University of Munich, Institute of Medical Optics, Thresienstrasse 37, 0-8000 Munich 2, Federal Republic of Germany.

ROOSE, Stephan. Contact, Free University of Brussels, Department of Applied Physics (ALNA), Faculty of Applied Sciences, Brussels, Belgium.

ROSEWELL, Michael P. 2302 South Damen Avenue, Chicago, IL 60608 USA. Telephone: (312)254 2577. Description: Fine art originals, represent Micraudel (Schiltigheim, France) and their thermoplastic camera.

ROSS, Franz. President, Ross Books. P.O. Box 4340, Berkeley, CA 94704 USA.

ROSS, Jonathan. 4 Macaulay Road, London SW4 OQX, England, United Kingdom. Telephone: (44)(01) 622 7729. FAX: (44)(01) 622 5308. Description: Holography consultant and collector.

ROSSIGNOL, Yves. Contact for Holograma Laboratoire Holographique, 41 rue Mariziano, CH-1227 Geneva, Switzerland.

ROSSING, Thomas. Contact: Northern Illinois University. Department of Physics, DeKalb, IL 60115-2854 USA.

ROTH, Stephen. Contact for Mind's Eye: Holographic Consultants, 17329 Zola Street, Granada Hills, CA 91344 USA.

ROTHAUS, Valli. Contact for Space Age Designs Inc, P.O. Box 72, Carversville, PA 18913 USA.

ROTTENKOLBER, H. Contact for Rottenkolber Holo-System GmbH, Henschelring 15, 0-801 1 Kirchheim/Munich, Federal Republic of Germany.

ROY, Joanne. 3575 Rue de Bullion, Montreal, Quebec, H2X 3A1 Canada. Telephone: (514) 845 4419. Description: Fine art holography

RUSCHMAN, Henry. Contact for Alpha Foils, Inc. P.O. Box 152, Bernardsville, NJ 07924 USA. Telephone: (201) 766 1500.

RYDEN, Hans. Contact for Karolinska Institutet, School of Dentistry, Box 4064, S-141 04 HUDDINGE, Sweden.

S

SACHTLER, Ek. Contact for ((ford Inc. West 70 Century Road, Paramus, New Jersey 07653 USA.

SAITO, Takayuki. Contact for Fuji Photo Optical Co., ltd. No. 324, 1-Chome, Uetake-Machi, Omiya, Japan.

SALAM, M. A. Sales Manager, Wonders of Holography Gallery. P.O. Box 1244, Jeddah 21431 , Saudi Arabia.

SALG, Joseph. General Manager, Laser Applications, Inc. 12722 Research Parkway, Orlando, FL 32826 USA.

SAMSON, Kevin. Sales Manager. Light Impressions, Inc. 149-B Josephine Street, Santa Cruz, CA 95060 USA.

SANDWELL, R.S. Vice-President, Lumonics Inc. 105 Schneider Road,Kanata, Ottawa, Ontario K2K 1Y3, Canada.

SANTOS, Antonio A Rua Da Quintanda, 194-Sala 404, CEP 20091, Rio de Janeiro, Brazil. Telephone: (55) (021) 233 5590. Branch office of Newport Corporation, Fountain Valley, CA USA

SAPAN, Jason. Contact for Holographic Studios. 240 East 26th Street, New York, NY 10010 USA.

SCHAFFNER, David. Contact for Newport Holograms. 3412 Via Oporto, Suite 2, Newport Beach, CA 92663 USA

SCHARF, Max R. Chairman, Diversified Graphics, Ltd. 5433 Eagle Industrial Court, Hazelwood, MO 63042 USA

SCHEIR, Peter. Contact for Another Dimension/AD 2000, 948 State Street, New Haven, CT 06511 USA

SCHICKER, Scott. Vice-President, Ralcon. Box 142, 8501 South 400 West, Paradise, UT 84328 USA

SCHMELZER, Dr. Carlo. Contact for Studio for Holographie. Moosstral3e 27, 0-8031 Eichenau, Federal Republic of Germany (FRG).

SCHMIDT, David. Owner, David Schmidt Holography. 23962 Craftsman Road, Calabasas, CA 91302 USA

SCHOENBECK, Gerhard. Contact for Kraftwerk Union AG. 0-4330 MulheimlRuhr, Federal Republic of Germany.

SCHULTZ, Lynn. Vice-President, E.C. Schultz & Company. 333 Crossen, Elk Grove Village, IL 60007 USA

SCHUMANN, Walter. Contact for Swiss Federal Institute of Technology, Laboratory of Photoelasticity, Ramistrasse 101, CH-8092 Zurich, Switzerland.

SCHVARTZMAN, Frederic. General Manager, The Foreign Dimension, Suite B, 8.1Fl., Central Mansion, 270-276 Queen's Road Central, Hong Kong.

SCHWEITZER, Dan. Contact for New York Holographic Labs. P.O.Box 20391, Tompkins Square Station, 176 East 3rd Street, New York, NY 10009 USA

SCIAMMARELLA, Cesar. Contact for Illinois Institute of Technology, Mechanical & Aerospace Engineering. Engineering Building #1, Room 2460, Chicago, IL 60616 USA

SCRIFF, Joe. Contact for Mitutoyo Measuring Instruments. 18 Essex Road, Paramus, NJ 07652 USA

SEKULIN, Robert. Contact for Rutherford and Appleton Laboratories, Chilton, Didcot, Oxon OX11 OQX, England, United Kingdom.

SENECHAL, Jacques. 90 Temple Fortune Lane, London NW11 7TX, England, United Kingdom. Telephone: (44)(1) 4585825. Description: Artistic holographer, large format holograms.

SEROV, O.B. Contact for VNIKTK Kultura, ul. Intusiastov 34,1051118 Moskow, USSR.

SEURAY, Dominique. Customer Service, Atelier Holographique de Paris, 13, Passage Courtois, F-75011 Paris, France.

SEYDEL, Bill. Sales Manager, Holo/Source Corporation. 21800 Melrose Avenue, Southfield, MI 48075 USA

SHANNON Robert. Contact for The University of Arizona, Optical Science Center, Tucson, Arizona 85721 USA

SHAPIRO, B.1. Contact for State Research and Project Institute of Chemico-Photographic Industry, 125167 Moskow, USSR.

SHERWOOD, Robert. President, Robert Sherwood Holographic Design, Inc. 400 West Erie Street, Chicago, IL 60610 USA

SHETKA, Stanley. Contact for World Art Project, 10247 40th Street West, Webster, MN 55088 USA.

SHEW, Ellen M. President, Chimeric Images, Inc. 713 1/2 Main Street, Lafayette, IN 47901 USA.

SHIOIAKI, Yumiko. Contact for Holo-Arp; Independent artist. 7-5-18-Ryoke Urawa, Saitama, Japan. Telephone: (81)(0488) 317 723. Description: Artistic holographer.

SHOLLY, Kate. Customer Service, ISASTILeonardo, P.O. Box 75, 1442A Walnut Street, Berkeley, CA 94709 USA.

SHREVE, Loy. Contact for T.A.1. Incorporated. 12021 South Memorial Parkway, Huntsville, AL 35803 USA.

SHVARTS, K.K. Contact for Physics Institute. Latvian SSR Academy of Sciences, 229021 Riga-Salaspils, USSR.

SIEGEL, Steven. Partner: Lasart Ltd. P.O. Box 703, Norwood, CO 81423 USA.

SIKORSKY, Ibigniew. Contact, Institute of Plasma Physics and Laser Microfusion, P.O. Box 49, 00-908 Wroclaw, Poland.

SIMKIN, Ruth. Contact for Holomagic Inc. 917 17th Avenue SW, Calgary, Alberta, Canada T2T OA4.

SIMMS, Lloyd. Sales Liaison, Electro Optical Industries, Inc. 859 Ward Drive, Santa Barbara, CA 93111 USA.

SIMSON, Bernd & Paula. Contact, General Holographics, Inc. P.O. Box 82247, Burnaby B.C., V5C 5P7 Canada

SINCERBOX, Glenn. Contact for IBM Almaden Research Center. K69/803, 650 Harry Road, San Jose, CA 95120 USA.

SITCH, B.J. Vice-President, George M. Whiley limited. Firth Road, Houston Industrial Estate, Livingston, West

Lothian EH54 5DJ, Scottland, United Kingdom.

SIVY, George. President, Gray Scale Studios Ltd. 4500 19th Street, #588, Boulder, CO 80304 USA.

SIXT, W. Customer Service, Steinbichler Optotechnik GmbH. Am Bauhof 4, D-8201 Neubeuern, FRG Federal Republic of Germany.

SJOLINDER, Sven. Contact, Royal Institute of Technology, Institute of Photography, S-100 44 Stockholm, Sweden.

SKANDE, Per. Contact for Martinsson Elektronik AB, Instrumentvagen 16, Box 9060, S-126 09 HAGERSTEN, Sweden.

SKRYPNYK, Susan. Vice-President, Starlight Holographic Inc. 73 Stable Way, Kanata, Ontario K2M 1 A8, Canada.

SLUSARSKI, Mary. Contact for Lasermedia. 2046 Armacost Ave., Los Angeles, CA 90025 USA.

SMELIER, Geert T. A. Contact for Technical University of Eindhoven. Faculty of Architecture, Calibre Institute, P.O. Box 513, Eindhoven, NL-5600MB, The Netherlands.

SMIGIELSKI, Paul. President, Holo 3. rue de l'Industrie, 68300 Saint-Louis, France.

SMITH, Prof. Contact, British Aerospace PLC. Sowerby Research Centre, FPC: 267, P.O. Box 5, Filton, Bristol BS12 70W, England, United Kingdom.

SMITH, Steven. Contact for The Lasersmith, Inc. 1000 West Monroe Street, Chicago, IL 60607 USA.

SNYDER, Bruce. Contact for Holomart Pic. Hamilton House, 1 Temple Avenue, London EC4Y OHA, England, United Kingdom.

SOARES, Oliverio. Contact for Universidade do Porto, Laboratorio de Fisica, Praca Gomes Teixeira, P-4000 Porto, Portugal.

SOBOLEV, G.A. Contact for FTI Joffe, Politechnicheskaya 26, Academy of Sciences of the USSR, 194021 Leningrad, USSR.

SOTT, Gudrun. Sales Manager, AKS HolographieGalerie GmbH. Potsdamer Stral3e 10, 4300 Essen 1, Federal Republic of Germany.

SOU PARIS, Hugues. President, Hologram. Industries, 42-44, rue de Trucy, 94120 Fontenay sous bois, France

SOWDON, Michael. Contact for Fringe Research Holographics Inc.1179A King Street West, Suite 008, Toronto, Ontairo M6K 3C5, Canada.

SPALDING, Jean. Contact for Norland Products, Inc. 695 Joyce Kilmer Avenue, P.O. Box 145, North Brunswick, NJ 08902 USA.

SPIERINGS, Walter. Contact, Dutch Holographic Laboratory. Kanaal dyk Noord 61, 5642 JA Eindhoven, The Netherlands.

SPIRO, Shari. Consultant, Sillcocks Plastics International. 310 Snyder Avenue, Berkeley Heights, NJ 07922 USA.

SPREER, Elmar. Contact for Phoenix Holograms, Traubenstr. 41, D-2913 - Apen, Federal Republic of Germany.

SPRINGER, Greg. Customer Service, LiCONix, 3281 Scott Boulevard, Santa Clara, CA, 95054 USA.

STALLARD, Penn. P.O. Box 4651, Chicago, IL 60680-4651 USA. Telephone: (312) 829 3635. Description: I work in bronze and holography, as well as work with architectural applications of holography.

STARCKE, Mr. Ari-Veli & Mrs. Milla-Ritta. President, Vice-President, Starcke KY. P.O. Box 22, SF-32811, Peipohja, Finland

STASELKO, F.1. Contact for S.1. Vavilov, State OpticsInstitute, 199064 Leningrad, USSR.

ST. CYR, Suzanne. President, Holographic Applications. 21 Woodland Way, Greenbelt, MD, 20770 USA.

STEINBICHLER, Dr. H. President, Steinbichler Optotechnik GmbH. Am Bauhof 4, D-8201 Neubeuern, FRG Federal Republic of Germany.

STENSBORG, Jan. Customer Service, Superior Technology Implementation. Hjortekaersvej 99B, 2800 Lyngby, Denmark.

STERN, Rudy. Contact for Let There Be Neon. P.O. Box 337, Canal Street Station, New York, NY 10013 USA.

STETSON, Karl. Contact for United Technologies Research Center. Optics and Acoustics 86, Silver Lane, East Hartford, CT 06108 USA.

STEVENS, Anait. Contact for Anait Studio. 1685 Fernald Point Lane, Santa Barbara, CA 93108 USA.

STEVENS, Dave.Vice-President, Holo Laser Tech, 7 Fraser Avenue, Unit 16, Toronto, Ontario M6K 1Y7 Canada.

STIEF, G. Vice-President, Labor Dr. Steinbichler, Am Bauhof 4, D-8201, Neubeuern, Federal Republic of Germany.

STOHL, Mr. Robert G. Re:Laser Regulations, Electro Optical Specialist, S.J. Fed.Bldg--US Court House, 280 South 1 st Street, Rm 2062, San Jose, CA 95113 USA.

STOCKLER, Len. 7227 Eastwood Street, Philadelphia, PA 19149 USA. Telephone: (215) 331 5067. Description: Clearing house for practical holography. Classes and individualized workshops. Reference library. Prototype Designs for optical hardware, vibration isolation tables, sensitized materials processing, lighting and display. Educational consulting and project management.

STRAZDS, Glenn. President, Holo Laser Tech, 7 Fraser Avenue, Unit 16, Toronto, Ontario M6K 1Y7 Canada.

STRIEBIG, Jocelyne. Deputy Director. Holo 3. rue de l'Industrie, 68300 Saint-Louis, France.

STROKE, George. Contact for MesserschmittBoelkow- Blohm. Zentrale Entwicklung MBB, Postfach 801109, D-8000 Munich 2, Federal Republic of Germany.

STURGESS, Rosemary. Marketing Manager, Laser Light Expressions Pty. Ltd. Holoptics. 3 Gibbons Street, Telopea, New South Wales, Australia 2117.

SVENNSON, Lennart. Contact for Royal Institute of Technology. Department of Industrial Metrology, S-10044 Stockholm, Sweden.

SWEENEY, Donald. Contact for Sandia National Laboratories. Combustion Research Facility. Livermore, CA 94550 USA.

SYNOWIEC, George. Contact for JK Lasers Ltd. Cosford Lane, Swift Valley, Rugby, Warwickshire CV21 1QN, England, United Kingdom.

SYNOWIEC, JA Contact for Lumonics, Ltd. Cosford Lane, Swift Valley, Rugby, Warwickshire, CV21 1QN, United Kingdom.

SZéKELY, Zsolt. Sales Manager, Artplay Holographic Studio. H-1191 Budapest, Ady Endre ut. 8, Hungary.

SZPAKOWSKI, Brian. Sales Manager. A.H. Prismatic, Inc. 285 West Broadway, New York, NY 10013. USA.

T

TALBOTT, M. Contact for Laser Arts, 1712 Cathedral Street, Plano, TX 75023 USA.

TAYLOR, Wes. Custormer Service, Aerotech Inc., Electro Optical Division, 101 Zeta Drive, Pittsburgh, PA 15238 USA.

TEITEL, Michael. Contact for Massachusetts Institute of Technology, M.I.T. Media Laboratory, Spatiallmaging Group, 20 Ames Street, E15-416, Cambridge, MA 02139 USA.

THIELKER, Klaus. Contact, Dream Images. Postfach 1602, Vermeerweg 15, D-5047 Wesseling, Federal Republic of Germany

THOMPSON, Barc. President, A.H. Prismatic, LTD. New England House, New England Street, Brighton, BN1 4GH England, United Kingdom.

THOMPSON, Brian. Contact for University of Rochester. Institute of Optics, Rochester, NY 14627 USA

THOMSON, Peter. Contact for Laser Lightworks, 81A Hatton Square, 16/16A Baldwins Gardens, London EC1 N 7RJ, England, United Kingdom.

TIZIANI, Hans. Contact for University of Stuttgart. Institute of Applied Optics, Pfaffenwaldring 9, D-7000 Stuttgart 80, Federal Republic of Germany.

TOBIN, John. Contact for Laser Light Expressions Pty. Ltd. Holoptics. 3 Gibbons Street, Telopea, New South Wales, Australia 2117.

TOMKINS, Don. Contact for Advanced Holographics, LTD. 243 Lower Mortlake Rd.,Unit 11, Richmond, Surrey TW9 2LL, England, United Kingdom.

TORSION, John. Contact for Bobst Group, 146-T Harrison Avenue, Roseland, NJ 07068 USA.

TRAYNOR, David. Vice-President, Richmond Holographic Studios. 6 Marlborough Road, Richmond, Surrey, England, TW10 6JR United Kingdom.

TRIBILLON, Gilbert. Contact for Universite de Franche-Comte, Laboratoire d'Optique, L.A. 214, UFR Sciences et des Techniques, F-25030 Besancon Cedex, France.

TRIBILLON, Dr. Jean Louis, H., Contact for HoloLaser. 6, rue de la Mission, Ecole, 25480 Miserey, France.

TROLINGER James. Contact for Spectron Development Laboratories, 3303 Harbor Boulevard, Suite G-3, Costa Mesa, CA 92626 USA.

TSUFURA, Lisa. Technical Manager, Melles Griot, 1770 Kettering Street, Irvine, CA 92714 USA

TSUGE, Hibiki. Customer service, Holomedia Inc. 3-15-22, Takaban, Meguro-ku, Tokyo 152 Japan.

TSUJIUCHI, Jumpei. Contact for Chiba University. Faculty of Engineering, 1-33 Yayoi-cho, Chiba 260, Japan.

TUNNADINE, Graham .46 Calthorpe Street, London, WC1 England, United Kingdom. Telephone: (44)(1) 278 1572. Description: Holographic artist.

TURNER, Brian. President, Excitek, Inc. 277 Coit Street, Irvington, NJ 07111 USA

TURUKANO, Boris. Contact for Institute of Nuclear Physics, Leningradska obl., 188350 Gatchina, USSR.

TYLER, Douglas E. 111 North Second Street, Niles, MI 49120 USA Telephone :.(616) 683 0934. Description: holographic consulting.

TYRER, John. Contact for Loughborough University, Dept. Mechanical Engineering, Loughborough, Leicestershire LE11 3TU England, United Kingdom.

TZONG, Mr. Tang Yaw. Contact for Institute of Optical Science/Central University. Chung-Li 32054, Taiwan, R.O.C.

U

UMEKI, Spencer J. Customer Service, Alpha Photo Products, Inc. 985 Third Street, P.O. Box 23955, Oakland, CA 94623 USA

UNBEHAUN, Klaus. President, Holopublic. Hirschstrasse 84, 0-5600 Wuppertal-2,Federal Republic of Germany (FRG) .

UNTERSEHER, Fred. (SEE OUR ADVERTISEMENT ON PAGE 45) Established in 1972. 3463 State Street, Suite 304, Santa Barbara, CA 93105 USA Telephone: (805) 568 6997. FAX: (805) 642 4396. Description: Artist & Technical Consultant: (Education + Training Courses. HOLOGRAPHY HANDBOOK Coauthor), (Laboratory Construction-Dichromate, Silver, Photoresist), Pulse Systems + Imaging Design), (Dichromate Mastering + Production), Artist PortfolioInstallation, Sculptural & Wall Mounted pieces incorporating holograms).

UPATNIEKS, Juris. Contact for Environmental Research Institute of Michigan, Optical Science Lab., ACD, P.O.Box 8618, Ann Arbor, MI 48107 USA

URAM, Amy & Marvin. Contact for U.K. Gold Purchasers, Inc. dba Holograms Unlimited. 5858 S.P.I.D., Sunrise Mall, Corpus Christi, TX 78412 USA

UYEMURA, T. Contact for University of Tokyo. Faculty of Engineering, Hongo 7-3-1 , Bunkyo-ku, Toyko, Japan.

V

VAN DER MOLEN, Mandy. Contact for Holotec CC. P.O. Box 5144, Brackengardens, 1452 Transvaal, South Africa.

VAN HAMERSVELD, Eric. Contact for Pacific Holographics, Inc. 1245 Stone Drive, San Marcos, CA 92069 USA

VANIN, Valery A Dep't Head. Soviskusstva vlo mexhdunarodnaya Kniga. Art holography department, 141120 Tryazino, Moscow region, USSR

VAN OORSCHOT, R. Sales Manager, Dutch Holographic Laboratory. Kanaal dyk Noord 61, 5642 JA Eindhoven, The Netherlands.

VAN RENESSE, Ruud. Contact for TNO Institute of Applied Physics. Department of Optics, P.O. Box 155, NL-2600 AD Delft, The Netherlands.

VAN SCIVER, Kelly. Customer Service, Holos Gallery. 1792 Haight Street, San Francisco, CA 94117 USA

VASILLIEVA, N.V., Contact for Cinema & Photo Research Institute. NIKFI, Leningradsky, Prospect 47, Moscow USSR.

VELONA, F. President, CISE SpA Technologie Innovative. via Reggio Emilia, 39, 20090 Segrate, Milano, Italy. Mailing address: P.O.Box 12081, 1-20134 Milano, Italy.

VEST, Charles. Contact for The University of Michigan. College of Engineering, Chrysler Center, Ann Arbor, MI 48109-2092 USA

VIENOT, Jean-Charles. Contact, Ecole Nationale Superieure d'ingenieurs. clo Groupe de Laboratoires, C.M.R.S., 5, Av. D'Edimbourg, BLD Marechal-Juin, F-14032 Caen-Cedex, France.

VIKRAM, C.S. Contact for Pennsylvania State University. Applied Research Laboratory, P.O. Box 30, State College, PA 16804 USA

VILA, Doris. 157 East 33rd Street, New York, NY 10016 USA Telephone: (212) 6865387. Description: Artistic holographer; holography education.

VILLANI, S. Vice-President, CISE SpA Technologie Innovative. via Reggio Emilia, 39, 20090 Segrate, Milano, Italy. Mailing address: P.O.Box 12081, 1-20134 Milano, Italy.

VINCENT, Ken. Coordinator, Photon League. 110 Sudbury Street-Unit B, Toronto, Ontario M6J 1 A7 Canada.

VISSIOS, Xanthippos. Contact for Laser Graphics. AG. Dimitriou 150, 546 35 Thessaloniki, Greece.

VOGD, H.T. Contact for Que Sera Sera, P.O. Box 29, 9700 AA, Groningen, The Netherlands.

VOGEL, Jon. Contact for Holographics (UK) Ltd. 32 Lexington Street, London W1 R 3HR, England, United Kingdom.

VOLKER, Mirau. Schosserstrasse 93, 46 Dortmund 1, Federal Republic of Germany. Telephone: (49)(8035) 1017. Description: Artistic holographer.

VON BALLY, Gert. Contact for University of Munster. Ear, Nose and Throat Clinic, Kardinal von Galen Ring 10, D-4400 Munster, Federal Republic of Germany.

VUKICEVIC, Dalibor. Contact for Trend, Miramarska 85, 41000 Zagreb, Yugoslavia.

W

WADDELL, P. Contact for University of Stratchclyde. Mechanical Engineering Group, Glasgow, Scotland, United Kingdom.

WAHLBERG, Bjorn. Contact for Swede Holoprint. Duvhoksgatan 6A, Malmo 21460, Sweden.

WAITTS, Robert, James. President, Contact for Crown Roll Leaf, Inc. 91 Illinois Ave., Paterson, NJ 07503 USA

WALDEN, N. Managing Director, Laser Electronics Pty., Ltd. P.O. Box 359, Southport, Queensland, 4215, Australia.

WALE, R.D. Contact for Galvoptics Ltd. Harvey Road, Basildon, Essex, SS13 1 ES, England, United Kingdom.

WALKER, Julie. Contact for Massachusetts Institute of Technology. M.I.T. Arts and Media Technology, E15-416, 20 Ames Street, Cambridge, MA 02139 USA.

WALKER, Neil. Contact, White Tiger Holograms. Johannes Verhulststraat 45, 1071 MS Amsterdam, The Netherlands.

WALTER, Eve. Contact for Three-D Light Gallery, 107 The Commons, Itacha, NY 14850 USA.

WANYUN, Huang. Contact, Beijing Normal University. Analysis and Testing Centre, Beijing 100875, Peoples Republic of China.

WARD, A. A. Contact for Oxford University. Department of Engineering Science, Parks Road, Oxford OX1 3PJ England, United Kingdom.

WATERMAN, Tracy. Sales Manager, ISASTI Leonardo, P.O. Box 75, 1442A Walnut Street, Berkeley, CA 94709 USA.

WATSON, John. Contact for University of Aberdeen, Dept. of Engineering, Kings College, Aberdeen AB9 2UE, Scotland, United Kingdom.

WATTS, Mike. Contact for Light Angels, The Corridor, High Street, Bath Spa, Avon BA 1 5AJ, England, United Kingdom.

WEBER, S. Contact for Advanced Environmental Research. Route 1, Box 1830, Woolwich, ME 04579 USA

WEBER, Sally. 1240 N. Ventura Avenue, Ventura, CA 93001 USA. Telephone: (805) 648 6419. Description: Sally Weber is an independent artist producing sculptural and architectural installations using holographic optical elements (HOEs). Weber's projects integrate environmental design with elements including water and sunlight.

WEBER, Mr. A. Vice-President, Holo 3. rue de l'Industrie, 68300 Saint-Louis, France.

WEBSTER, John. Contact for Central Electricity Generating Board, Marchwood Engineering Laboratories, Magazine Lane, Marchwood, Southampton S04 4ZB, England, United Kingdom.

WEHMEIER, Halge. President, Agfa Gevaert Inc.-Industrial Division, 100 Challenger Road, Ridgefield Park, NJ 07660 USA.

WEIL-ALVARON, Hans. Established in 1933. bstra Tullgatan 8, S-211 28 Malmo, Sweden. Telephone: (46)(40) 129 956. Contact: Lektor H. Herman Weil. Description: Hans We ii's inventions were made in the period 1933-1937, while Gabor invented holography in 1948, the laser was invented 1962 and the first laser-illuminated hologram was exposed as late as 1964.

WEISER, Ellen. Customer Service, American Bank Note Holographics, Inc. 500 Executive Boulevard, Elmsford, New York 10523 USA.

WEISS, Daniel. Contact for Tridimensionale Hologramas. Alberto, Alcocer, 38-2D, 28016 Madrid, Spain.

WELBY, Mr. Dominic. 2 Foxes Lane, Mousehole, Cornwall, TR19 6QQ England, United Kingdom. Description: Holographic artist, consultant.

WELFORD, W. T. Contact for Imperial College of Science. Optics Section, Blackett Laboratory, London SW7 2BZ, England, United Kingdom.

WELLS, Sandrajean , P.O. Box 927, Federal Station, Worcester, MA 01601 USA. Telephone: Unlisted. Description: Holographic artist; fine art holograms.

WESLY, Ed. Contact for The School of The Art Institute of Chicago. 5331 North Kenmore Avenue, Chicago, IL 60640 USA

WESTPHAL, Carlo. President, die dritte dimension. Frankfurter Straf3e 132 -134, D-6078 Neu Isenburg, Federal Republic of Germany.

WHITE, Jane F. Contact for Massachusetts Institute of Technology. M.I.T. Media Laboratory, SpatialImaging Group, 20 Ames Street, E15-416, Cambridge, MA 02139 USA.

WHITE, John. Contact for Coburn Corporation.1650 Corporate Road West, Lakewood, NJ 08701 USA.

WILBUR, Fred. Contact: Elusive Image. 135 West Palace Avenue, Suite 102, Santa Fe, NM 87501 USA.

WILLARD, Pat. Contact for Holocrafts of Long Island. 227 9th Street, West Babylon, NY 11704-3728 USA.

WILLENBORG, George. Contact for Holographic Concepts. 14 Cove Road, Forestdale, MA 02644 USA.

WILLIAMS, Chas. Contact for Laser Image Design. 3031-K Nihi Street, Honolulu, Hawaii 96819 USA

WILLIAMS, H.M. Contact for Brandtjen & Kluge, Inc, 539 Blanding Woods Road, St. Croix Falls, WI 54024 USA.

WILLIAMSON, Ken. Vice-President, Markem Systems Ltd. Ladywell Trading Estate, Eccles New Road, Salford, M5 2DA England, United Kingdom

WILSON, Brett and Roslyn. Lazart Holographics. 22 Erina Valley Road, Erina, New South Wales 2250, Australia.

WILSON, Richard. Contact for Three D Gallery. Queen's West, 207 Quay, Toronto, Ontario, M5J 1A4 Canada.

WISE, Peter. Contact for Wise Instruments. Unit 9, Hollins Business Centre, Marsh Street, Stafford, ST16 3BG, England, United Kingdom.

WITTIG, Siegmar & Rita. Contact for Wittig Fachbuchverlag, 10 Chemnitzer Strasse, D-5142 Huckelhoven 1, Federal Republic of Germany.

WöBER, Irmfried. Contact for Wober Design Hololab Austria. Kahlenbergstraf3e 6, A-3042 Würmla, Austria.

WOODD, Glenn P. Business Development Manager, Ilford Limited. Mobberley, Knutsford, Cheshire WA15 7HA, England, United Kingdom.

WOODD, Peter H.L. Managing Director, Light Fantastic PLC. 4E/F Gelders Hall Road, Shepshed, Leicestershire LE12 9NH, England, United Kingdom.

WRIGHT, Greg. President, General Holographics Corp. 37568 Devoe, Mt. Clemens, MI 48043 USA.

WRIGHT, Lawrie. Contact, Royal Sussex Hospital, Brighton, England, United Kingdom.

WUTHNOW, Alan W. President, Alpha Photo Products, Inc. 985 Third Street, P.O. Box 23955, Oakland, CA 94623 USA

WYANT, James. Contact for WYKO Corporation. 1955 East Sixth Street, Tucson, AZ 85719 USA.

WYND, Hugh C., Christopher W. President, VicePresident, The Diffraction Company, Inc. P.O. Box 151, Riderwood, MD 21152 USA.

Y

YEAGER, Carol. RD 4 Box 335, Catskills, NY 12414 USA. Telephone: (508) 943 2007. Description: Teaches holography.

YERKES, Elizabeth T. Editor, Ross Books. P.O. Box 4340, Berkeley, CA 94704 USA.

YETKA, Charlie. Contact for Transfer Print Foils Inc. P.O.Box 518, 9 Cotters Lane, East Brunswick, NJ 08816 USA.

YU, Francis. Contact for Pennsylvania State University. Dept. of Electrical Engineering, 121 East University Park, PA 16802 USA.

Z

ZANELLI, Don. Customer Service, Ion Laser Technology Inc. 263 Jimmy Dolittle Road, Salt Lake City, UT 84116 USA.

ZEC, Dr. Peter. President, Deutsche Gesellschaft for Holografie e.V. LerchenstraBe 142 a, 0-4500 OsnabrOck, Federal Republic of Germany.

ZELLERBACH, Gary. President, Holos Gallery. 1792 Haight Street, San Francisco, CA 94117 USA.; VicePresident, The DZ Company. P.O. Box 5047-R, 181 Mayhew Way, Suite E, Walnut Creek, CA 94596 USA.

ZEMAN, Lee. 55 Ann Street, New York, NY 10038 USA. Telephone: (212) 732 1854. Description: Holographic artist.

ZILEBA, Edward. Contact for Aptec Engineering limited. 4251 Steeles Avenue West, Downsview, Ontario, M3N 1 V7 Canada.

ZUCKER, Rich. Contact, General Imaging Corporation. 1 Industrial Drive South, Lan-Rex Industrial Park, Smithfield, RI 02917 USA

BIBLIOGRAPHY

ADDITIONAL READING ON HOLOGRAPHY
INCLUDING TEXTS, DISSERTATIONS, CONFERENCES,
FOREIGN LANGUAGE PUBLICATIONS

Abramson, Nils H. The making and evaluation of holograms. London: Academic Press, 1981 .

Acoustical Holography; Vol. 1. Metherell, AF., Editor. New York: Plenum Publishing,1969.

Acoustical Holography; Vol. 2. Metherell, AF., Editor. New York: Plenum Publishing,1970.

Acoustical Holography; Vol. 3. Metherell, AF., Editor. New York: Plenum Publishing,1971.

Acoustical Holography; Vol. 4. Wade, Glen, Editor. New York: Plenum Publishing,1972.

Acoustical Holography; Vol. 5. Green, Philip S.,Editor. Plenum Publishing, 1974.

Acoustical Holography; Vol. 6. Booth, N., Editor. New York: Plenum Publishing,1975.

Acoustical Holography; Vol. 7. Kessler, L.W., Editor. New York: Plenum Publishing, 1977.

Acoustical Imaging; Vol. 8. Metherell, AF., Editor. New York: Plenum Publishing,1980.

Acoustical Imaging; Vol. 9. Wang, Keith, Editor. New York: Plenum Publishing,1980.

Acoustical Imaging; Vol. 11 . Powers, John P.,Editor. New York: Plenum Publishing, 1982.

Acoustical Imaging; Vol. 11. Ash, Eric A and C.R. Hill,Editors. New York: Plenum Publishing, 1983.

Acoustical Imaging,Vol. 15. Jones, Hugh W., Editor. Plenum Publishing, 1987.

Acoustic Imaging: cameras, microscopes, phased arrays, and holographic systems. Glen Wade, ed. New York: Plenum Press, 1976.

Acoustic Surface Wave & Acousto-Optic Devices. Kallard, Thomas Editor. Series Title: State of the Art Review Series; Vol. 4. New York: Optosonic Press,1971 .

Advances in Holography. Farhat, Nabil H.-Editor. New York: Marcel Dekker,1975. Vol. 1.

Advances in Holography. Farhat, Nabil H.-Editor. New York: Marcel Dekker,1976. Vol. 2.

Advances in Holography. Farhat, Nabil H.-Editor. New York: Marcel Dekker, 1976. Vol. 3.

Aldridge, Edward E. Acoustical holography. Watford: Merrow Publishing Co. Ltd., 1971. Series title: Merrow monographs, practical science series 1.

Allen, Judy. Lasers and Holograms. England: Puffin Books, UK, 1985.

Anderson, John. Holography. Tempe: Art Dept., Arizona State University, 1979. Series title: Northlight ; no.11.

An External Interface for Processing 3-D Holographic & X-Ray Images.' Juptner, W., Editor New York: Springer-Verlag/Research Reports,1989.

Applications of Holography. Barrekette, E. S., Editor. New York: Plenum Publishing,1971 .

Applications of Holography: January 21-23, 1985, Los Angeles, California. Lloyd Huff, chair, ed. Bellingham, WA: SPIE-the International Society for Optical Engineering, 1985. Series title: Proceedings of SPIE-the International Society for Optical Engineering; v. 523.

Applications of Holography in Mechanics: Symposium, University of Southern California, 1971. Symposium Staff; Gottenberg, W. G., Editors. Books on Demand UMI, Reprint of 1971 edition.

Applications of Holography & Optical Data Processing: Proceedings of an International Conference, Jerusalem, 1976. Marom, E.; Avnear Wiener and A.A. Friesem, Editors. London: Pergamon Press,1977.

Applications of lasers to photography and information handling; proceedings, two-clay seminar. Richard D. Murray, ed.washington, DC: Society of Photographic Scientists and Engineers, 1968.

Barrett, N. S. Lasers & Holograms. Boston: Watts,1985. Series Title: Picture Library.

Basov, N. G. Lasers & Holographic Data Processing. USSR: Mir Publications ,1985.
_____' Lasers & Holographic Data Processing. England, Colletts: State Mutual Books,1984.

Beiser, Leo. Holographic scanning. New York: Wiley,1988. Series title: Wiley series in pure and applied optics.

Berley, Lawrence F. Holographic mind, holographic vision: a new theory of vision in art and physics. 1 st ed. Bensalem, PA: Lakstun Press, 1980.

Berner, Jeff. The holography book. New York: Avon Books, 1980.

Berry, Michael V. Diffraction of Light by Ultrasound. New York: Academic Press,1967.

Brcic, Vlatko. Application of holography and hologram interferometry to photoelasticity: lectures held at the Department for Mechanics of Deformable Bodies. 2d ed. Wien: Springer-Verlag, 1974. Series title: Courses and lectures; no. 7.

Bristol University Electron Microscopy Group. Convergent Beam Electron Diffraction of Alloy Phases. Mansfield, J., Editor. A Hilger UK: Taylor & Francis, 1984.

Burkig, Valerie. Photonics: The New Science of Light. Enslow Publishers, 1986.

Business Communications Staff. Holography: New Commercial Opportunities. BCC,1986.

Butters, John N. Holography and its technology. London: P. Peregrinus, 1971; Published on behalf of the Institution of Electrical Engineers. Series title: Institution of Electrical Engineers I.E.E. monograph, series, 8.

Cadig Liaison Centre. Reference Library. A compendium of Cadig bibliographies: Metrication, Fluidics, Explosive techniques in engineering, Holography, Carbon fibres. Coventry (Warwickshire): Cadig Liaison Centre, 1970.

Cathey, W. Thomas. Optical information processing and holography. New York:Wiley,1974. Series title: Wiley series in pure and applied optics.

Caulfield, H. J. and Sun Lu. The applications of holography. New York, Wiley-Interscience, 1970. Series title: Wiley series in pure and applied optics.

Caulfield, H. John; (et alia). Holography Works. New York: Museum of Holography (NYC),1984.

Ceccon, Harry L. Holographic techniques for nondestructive testing of tires. Washington, D.C: National Highway Traffic Safety Administration, 1972.

Centerbeam. Otto Piene and Elizabeth Goldring, eds. Introduction by Lawrence Alloway. Cambridge, MA: Center for Advanced Visual Studies, Massachusetts Institute of Technology, 1980.

Chuguy, Yu. V. and N. T. Kolesova. Bibliography on holography, 1971-1972. Translated from Russian. London : Scientific Information Consultants Ltd, 1976.

Coello-Vera, Agustin Elias. "Scanned acoustic imaging in the ocean: a study of holographic-like systems and their limitations". 1978.

Coherent optical processing: seminar, August 21-22, 1974, San Diego, CA. Palos Verdes Estates, CA: Society of Photo-optical Instrumentation Engineers, 1975. Series title: Society of Photooptical Instrumentation Engineers Seminar proceedings; v. 52.

Coherent optics in mapping: tutorial seminar and technology utilization program, March 27-29, 1974, Rochester, N.Y. N. Balasubramanian, Robert D. Leighty, eds. Jointly sponsored by American Society of ... Palos Verdes Estates, CA: SPIE,1974. Series title: Society of Photo-opticalInstrumentation Engineers Proceedings; v. 45.

Collier, Robert Jacob, C. B. Burckhardt and L.H. Lin. Optical holography. New York: Academic Press, 1971.

Collings, Neil. Optical pattern recognition using holographic techniques. Wokingham, England ; Reading, MA. : Addison-Wesley Pub. Co., 1988. Series title: Electronic systems engineering series.

Computer-generated holography II : 11-12 January 1988, Los Angeles, CA. Sing H. Lee, chair/editor; the International Society for Optical Engineering; cooperating ... Bellingham, WA., USA : SPIE--the International Society for Optical Engineering, 1988.Series title: Proceedings of SPIE--the International Society for Optical Engineering; v. 884.

Conference on Fourier Optics, Lasers and Holography, Mysore, 1971. Proceedings of the Conference on Fourier Optics, Lasers and Holography, Mysore, November 11-15,1971. Madras, India: Institute of Mathematical Sciences, 1971. Series title: Matscience report; 77.

Conference on Holographic Instrumentation Applications, 1970: Ames Research Center. Holographic instrumentation applications. Prepared by NASA Ames Research Center. Boris Ragent and Richard M. Brown, eds. Washington, DC: Scientific and TechnicalInformation Division, National Aeronautics and Space Administration, 1970. Series title: NASA SP ; 248.

Conference on Holography and Optical Filtering, 1972: Marshall Space Flight Center. Holography and optical filtering; proceedings. Washington, Scientific and Technical Information Office, National Aeronautics and Space Administration,1973. Series title: United States. National Aeronautics and Space Administration NASA SP -299.

Dallas, William John. "Computer holograms: improving the breed". 1971.

Defense Documentation Center (U.S.) Holography: a DDC bibliography, January 1970-September 1972. Alexandria, VA: Defense Documentation Center, Defense Supply Agency, 1973.

Denisiuk, IU. N. Fundamentals of holography. Translated from the Russian by Alexander Chubarov. Rev. from the 1978 Russian ed. Moscow: Mir,1984.

DeVelis, John B. and George O. Reynolds. Theory and applications of holography. Reading, MA: Addison-Wesley Pub. Co., 1967.

Developments in holography: seminar-in-depth; proceedings. Brian J. Thompson and John B. DeVelis,eds. Redondo Beach, CA: Society of Photo-optical Instrumentation Engineers, 1971. Series title : Society of Photo-optical Instrumentation Engineers S.P.I.E.seminar proceedings, v. 25.

Dirtoft, Ingegard. Holography: A New Method for Deformation Analysis of Upper Complete Dentures in Vitro & in Vivo. New York: Coronet Books,1985.

Dowbenko, George. Homegrown holography. Garden City, N.Y.: Amphoto, 1978.

Dudley, David D. Holography; a survey. Washington, DC: Technology Utilization Office, National Aeronautics and Space Administration, 1973. Series title: NASA SP-5118.

Dzekov, Tomislav Angel. "Microwave holographic imaging of aircraft with spaceborne illuminating source". 1976.

Easy Way to Make Reflection Holograms (no author). Embee Press, 1986.

Eichert, Edwin S. and Alan H. Frey; Randomline, Inc. Holography in driver education, training, testing, and research. Washington, DC : National Highway Traffic Safety Administration, 1978. Series title: United States. National Highway Traffic Safety Administration Report; no. DOT HS-803 035.

Eichler, H. J., P. Gunter, and D.w. Pohl. Laser-induced dynamic gratings. Berlin :Springer-Verlag,1986. Series title: Springer series in optical sciences; v. 50.

Electro-opticsllaser international '80 UK, Brighton, 25-27 March 1980 :conference proceedings. H.G. Jerrard, ed. Conference organized by Kiver Communications Ltd. Guildford, Surrey, England : IPC Science and Technology Press, 1980.

Elliott, John Douglas. "Computer simulation of the holographic image degradation due to transmission of the signal through a random noise media". 1971.

Engineering Applications of Holography Symposium,1972: Los Angeles: Proceedings. Redondo Beach, CA: Society of Photo-optical Instrumentation Engineers, 1973.

The Engineering uses of coherent optics : proceedings and edited discussion of a conference held at the University of Strathclyde, Glasgow, 8-11 April,1975 . Organised by the University, in association with the... Cambridge, Eng. : Cambridge University Press, 1976.

The Engineering uses of holography. Robertson, Elliot R. and James M. Harvey, eds. Cambridge, England: Cambridge University Press, 1970.

Eu, James Kim-Tzong. "Studies in spatial filtering". 1974.

European Hybrid Spectrometer Workshop on Holography and HighResolution Techniques, 1981 : Strasbourg, France. Photonics applied to nuclear physics, 1 I European Hybrid Spectrometer Workshop on Holography and High-Resolution Techniques, Strasbourg, Council of Europe, 9-12 November 1981. Geneva: European Organization for Nuclear Research, 1982. Series title: CERN (Series); 82-01 .

Evans, Evan Allen . "Quantitative method for accurate determination of red blood cell geometry".1970.

Falk, David R. (et alia). Seeing the Light: Optics in Nature, Photography Color, Vision, & Holography. New York:Wiley, 1985.

Fiber Diffraction Methods. French, Alfred D. and Kenn Gardner, Editors. Series Title: ACS Symposium Ser.,; No. 141. New York: American Chemical Society, 1980

Finch College, New York. Museum of Art. Contemporary Study Wing. "N dimensional space". Prepared by Ted McBurnett. Introd. by Elayne H. Varian. New York: 1970.

Firth, Ian Mason. Holography and computer generated holograms. London, Mills and Boon, 1972. Series title: M & B monograph EEI 11 .

Fischer, Wolfgang Klaus. Methods for acoustic holography and acoustic measurements. Newark, N.J. : [s.n .], 1972.

Flow visualization and aero-optics in simulated environments:21-22 May 1987, Orlando, Florida. H. Thomas Bentley III, chair/ed. Sponsored by SPIE-the International Society for Optical Engineering. Bellingham, WA: SPIE--the International Society for Optical Engineering, 1987. Series title: Proceedings of SPIE-the International Society for Optical Engineering; v. 788.

Francon, M. Holography. Expanded and revised from the French edition. Translated by Grace Marmor Spruch. New York: Academic Press, 1974.

Furst, Anton. et alia. Light fantastic. 2d ed. London: Bergstrom & Boyle Books, 19n.

George, Daweel Joseph. "Holography as applied to jet breakup and an analytical method for reducing holographic droplet data". 1972.

Graphics in motion : from the special effects film to holographics. John Halas, ed. New York: Van Nostrand Reinhold, 1984.

Griffiths, John. Lasers & Holograms. New York: Silver Publishers, 1983. Series Title: Exploration & Discovery Series.

Handbook of optical holography. H. J. Caulfield,ed. Contributors, Gilbert April ... let al.). New York: Academic Press, 1979.

Hariharan, P. Optical holography : principles, techniques, and applications. Cambridge: Cambridge University Press,1983. Series title: Cambridge monographs on physics.

Hartman, W. F. Acoustic Emission: Advances in Acoustic Emission. American Society for Nondestructive Testing,1981 .

Heckman, Philip. The Magic of Holography. New York, Macmillan & CO.,1986.

Hildebrand, B. P. and B. B.Brenden. An introduction to acoustical holography. New York: Plenum Press, 1972.

Holographic data nondestructive testing : October 4-8, 1982, Croatia Hotel de Luxe, Dubrovnik, Yugoslavia. Dalibor Vukicevic, chair,ed. Sponsored by the International Commission for Optics (ICO) [and] ... Bellingham, WA: SPIE--the International Society for Optical Engineering, 1983. Series title: Proceedings of SPIE--the International Society for Optical Engineering; v. 370.

"Holographic detection of intraocular pathology in the presence of cataracts: final report". By George O. Reynolds ... let al.). Burlington, MA: Technical Operations, 1974.

Holographic nondestructive testing. Robert K. Erf, ed. New York: Academic Press, 1974.

Holographic nondestructive testing: status and comparison with conventional methods:23-24 January 1986, Los Angeles, California. Charles M. Vest, chair, ed. Presented in cooperation with American Association ... Bellingham, WA: SPIE--the International Society for Optical Engineering, 1986. Series title: Proceedings of SPIE--the International Society for Optical Engineering ; v. 604.Series title: SPIE critical reviews of technology series; 15th.

Holographic Nondestructive Testing. Erf, Robert K., Editor. New York: Academic Press,1974.

Holographic Nondestructive Testing: Critical Review of Technology. Vest, Charles., Editor. Bellingham, WA: SPIE,1986.

Holographic optics : design and applications: 13-14 January 1988, Los Angeles, CA. Ivan Cindrich, Chair/Editor; SPIE--The International Society for Optical Engineering; cooperating ... Bellingham, WA., USA: SPIE--The International Society for Optical Engineering, 1988.Series title: Proceedings of SPIE-The International Society for Optical Engineering ; v. 883.

Holographic optics : optically and computer generated: 19-20 January 1989, Los Angeles, CA . Ivan N. Cindrich, Sing H. Lee chairs/editors; sponsored by SPIE--The International Society for OpticaL .. Bellingham, WA., USA : SPIE--The International Society for Optical Engineering, c1989. Series title: Proceedings of SPIE--the International Society for Optical Engineering ; v. 1052.

The Holographic paradigm and other paradoxes : exploring the leading edge of science. Ken Wilber, ed. 1 st ed. Boulder: Shambhala, 1982.

Holographic recording materials. H. M. Smith,ed. Contributions by R. A. Bartolini... let al.). Berlin: Springer-Verlag, 1977. Series tide: Topics in applied physics ; v. 20.

Holographic systems, components and applications, Churchill College, Cambridge, 10th - 12th September, 1987. London : Institution of Electronic and Radio Engineers,1987. Series title: Publication/Institution of Electronic and Radio Engineers; no. 76.

Holography. Redondo Beach, CA: Society of Photo-optical Instrumentation Engineers, 1968. Series title: Society of Photo-optical Instrumentation Engineers S.P.I.E. seminar proceedings, v. 15.

Holography, 1971-72. Kallard, Thomas-Editor. Series Title : State of the Art Review Series, Vol. 5. New York: Optosonic Press,1972.

Holography Applications. Wang. Editor. Bellingham, WA: SPIE,1986.

Holography: Critical Reviews. Huff, L. , Editor. Bellingham, WA: SPIE,1985.

Holography: Exploiting the Leading-Edge Developments. Chicago: TechnicalInsights,1987.

Holography : January 24-25, 1985, Los Angeles, CA. Lloyd Huff, chair,ed. Bellingham, WA: SPIE--the International Society for Optical Engineering,1985. Series title: Proceedings of SPIE--the International Society for Optical Engineering. ; v. 532. Series title: SPIE critical reviews of technology series; 12th.

Holography in Medicine: International Symposium Proceedings. Greguss, Pal. Editor.I.P.C .ScL& Technology, 1976.

Holography in Medicine & Biology. Von Bally, G., Editor. New York: Springer-Verlag, 1979. Series Title: Springer Series in Optical Sciences,; Vol. 18

Holography Marketplace. Ross, Franz and Elizabeth Yerkes, Eds. Berkeley , CA: Ross Books, 1989.

Holography Marketplace, 2nd Edition. Ross, Franz and Elizabeth Yerkes, Eds. Berkeley, CA: Ross Books,1990.

Holography Markets. International Resource Developments,1984.

Holography Redefined: Thresholds. Barilleaux, Rene P., Editor. New York: Museum of HolographY,1984.

Holography; seminar-in-depth, May, 1968, San Francisco CA. B.G.Ponseggi and Brian J. Thompson,eds. Redondo Beach, CA: Society of Photo-optical Instrumentation Engineers, 1972. Series title: Society of Photo-optical Instrumentation Engineers Proceedings, v.15.

Holography techniques and applications: EC01 , 19-21 September 1988, Hamburg, Federal Republic of Germany / Werner P.O. Juptner, chair/editor; EPS-European Physical Society, Europtica-the ... Bellingham, WA. : SPIE-the International Society for Optical Engineering, 1989. Series title: Proceedings of SPIE--the International Society for Optical Engineering ; v. 1026.

I. Aroslavskii, L. P. and N. S. Merzlyakov. Methods of digital holography. Translated from Russian by Dave Parsons. New York: Consultants Bureau, 1980.

ICALEO Technical Digest Eighty-Three; Vol. 36. Toledo, OH: Laser Institute ,1983.

ICALEO Materials Processing Eighty-Three: Proceedings, Vol. 38. Toledo, OH: Laser Institute,1984.

ICALEO Holography & Information Processing Eighty-Three: Proceedings, Vol. 41 .Toledo, OH: Laser Institute, 1984.

Industrial and commercial applications of holography: August 24-25,1982, San Diego, CA. Milton Chang, chair,editor. Bellingham, WA: SPIE--the International Society for Optical Engineering,1983. Series title : Proceedings of SPIE--the International Society for Optical Engineering ; v. 353.

Industrial applications of holographic nondestructive testing : May 3-5, 1982, Brussels. J. Ebbeni, chair, ed. Sponsored by SPIE--the International Society for Optical Engineering ; with the support... Bellingham, WA: SPIE--the International Society for Optical Engineering,1982. Series title : Proceedings of SPI E--the International Society for Optical Engineering ; v. 349.

Industrial Applications of Holography. Robillard, Jean, Editor. Oxford, UK: Oxford University Press,1989.

Industrial applications of laser technology: April 19-22,1983, Geneva Switzerland.william F. Fagan, chair,ed. Bellingham, WA: SPIE-the International Society for Optical Engineering, 1983. Series title: Proceedings of SPIE--the International Society for Optical Engineering; v. 398.

Industrial Radiography Holography. American Society for Nondestructive Testing, 1983.

International Commission for Optics. Congress, 10th : 1975 : Prague, Czechoslovakia. Recent advances in optical physicsz: proceedings of the Tenth Congress of the International Commission for Optics, August 25-29, 1975, Prague, Czechoslovakia. Bedrich Havelka and Jan Blabla,eds. Olomouc: Palacky University ; Prague : Society of Czechoslovak Mathematicians and Physicists, 1976.

International Conference on Applications of Holography and Optical Data Processing,1976 : Jerusalem, Israel. Applications of holography and optical data processing : proceedings of the international conference, Jerusalem, August 23-26, 1976. E. Marom, A. A. Friesem, and E. Wiener-Avnear, eds. 1st ed. Oxford: Pergamon Press, 1977.

International Conference on Computer-generated Holography, 1983: San Diego, CA. International Conference on Computergenerated Holography, August 25-26,1983, San Diego, CA: proceedings. Sing H. Lee, chair, ed. Bellingham, WA: SPIE--The International Society for Optical Engineering ,1983. Series title: Proceedings of SPIE--the International Society for Optical Engineering; v. 437.

International Conference on Computer-generated Holography: 2nd: 1988: Los Angeles, CA. Computer-generated Holography II: 11-12 January 1988, Los Angeles ,California, [proceedings) . Sing H. Lee, chair/ed. Sponsored by SPIE--The International Society for Optical Engineering; ... Bellingham, WA: SPIE--The International Society for Optical Engineering, 1988. Series title: Proceedings of SPIE--the

International Society for Optical Engineering; v. 884.

International Conference on Holographic Systems, Components and Applications,1987: Churchill College. Holographic systems, components and applications : Churchill College, Cambridge, 10th-12th September, 1987. London : Institution of Electronic and Radio Engineers, 1987. Series title: Publication I Institution of Electronic and Radio Engineers; no. 76.

International Conference on Holography Applications,1986: Peking, China. International Conference on Holography Applications : 2-4 July, 1986, Beijing, China . Dahang Wang, chair. Jingtang Ke, Ryszard J. Pryputniewicz, eds. Sponsored by COS-Chinese Optical Society... Bellingham, WA: SPIE,1987. Series title: Proceedings of SPIE--the International Society for Optical Engineering ; v. 673.

International Congress on High-Speed Photography,11th:1974: Imperial College, London. High speed photography : proceedings of the eleventh International Congress on High Speed Photography, Imperial College, University of London, September 1974. P. J. Rolls, ed. London: Chapman & Hall: distributed in the USA by the Society of Photo-Optical Instrumentation Engineers, 1975.

International Congress on High Speed Photography,12th:1976: Toronto, Canada. Proceedings of the 12th International Congress on High Speed Photography (Photonics), Toronto, Canada, 1-7 August 1976. Martin C. Richardson. Bellingham, WA: Society of PhotoOptical Instrumentation Engineers, 1977. Series title: SPIE v.97.

International Congress on High Speed Photography and Photonics, 13th: 1978: Tokyo. Proceedings of the 13th International Congress on High Speed Photography and Photonics-Tokyo, 20-25 August 1978. Shin-ichi Hyodo, ed. Tokyo : Japan Society of Precision Engineering; [New York] : distributed (outside Japan) by Society of Photo-Optical Instrumentation Engineers, 1979. Series title: SPIE v.189.

International Exhibition of Holography. Jeong, Tung and Michael Croydon, Editors. Lake Forest, IL: Lake Forest College Press,1982. (Second) International Exhibition of Holography. Jeong, Tung, Editor. Lake Forest, IL: Lake Forest College Press,1985.

International Optical Computing Conference ,1974 : Zurich. Digest of papers. New York, Institute of Electrical and Electronics Engineers, 1974.

International Optical Computing Conference, 1975: Washington, D.C. Digest of papers: International Optical Computing Conference, April 23-25, 1975, Washington, D.C. Sponsored by the Computer Society of the Institute of Electrical and Electronic Engineers, in cooperation... New York: Institute of Electrical and Electronics Engineers, 1975.

International Symposium on Acoustical Holography. Acoustical holography. New York: Plenum Press, 1967.

International Symposium on Acoustical Holography. 1 st: 1967: Huntington Beach, CA. Acoustical holography; proceedings. New York: Plenum Press, 1969.

International Symposium on Acoustical Holography and Imaging, 7th: 1976: Chicago, IL. Recent advances in ultrasonic visualization. Lawrence W. Kessler,ed. New York: Plenum Press, 1977. Series title: International Symposium on Acoustical Holography and Imaging Acoustical holography ; v. 7.

International Symposium on Acoustical Holography and Imaging, 8th, Key Biscayne, FL, 1978. Ultrasonic visualization and characterization. A. F. Metherell, ed. New York: Plenum Press,1980. Series title: International Symposium on Acoustical Holography and Imaging Acoustical imaging; v. 8.

International Symposium on Holography in Biomedical Sciences, 1973: New York. Holography in medicine : proceedings of the International Symposium on Holography in Biomedical Sciences, New York, 1973. Paul Greguss, ed. Guildford, Eng: IPC Science and Technology Press, 1975.

International Workshop on Holography in Medicine and Biology, 1979 : Munster, Germany. Holography in medicine and biology : proceedings of the International Workshop, Munster, Fed. Rep. of Germany, March 14-15, 1979. G. von Bally, ed. Berlin: SpringerVerlag, 1979. Series title :Springer series in optical sciences; v 18

Jeong, Tung H. Display Holography: Proceedings of the International Symposium,1982; Vol. I.Lake Forest, IL: Lake Forest College Press,1983.

Jeong, Tung H. Display Holography: Proceedings of the International Symposium,1982; Vol. II.Lake Forest, IL: Lake Forest College Press,1986.

Jones, R. M. Application of the Geometrical Theory of Diffraction to Terrestrial LF Radio Wave Propagation. Series Title: Mitteilungen Aus Dem Max-Planck-Institut Fuer Aeronomie,; No. 37. New York: Springer-Verlag, 1968.

Jones, Robert; and Catherine M. Wykes. Holographic & Speckle Interferometry. Cambridge University Press,1983.

Jones, Robert; and Catherine M. Wykes. Holographic & Speckle Interferometry. Cambridge University Press,1989. Series title: Cambridge Studies in Modern Optics,; NO.6.

Kallard, Thomas. Holography; state of the art review, 1969. New York: Optosonic Press,1969. Series title: State of the art review, 1.

_____ . Holography; state of the art review ... 1970. Holography in1970: an overview by Dr. Dennis Gabor. New York: Optosonic Press,1970. Series title: State of the art review, no. 3.

_____ . Laser Art & Optical Transforms. New York: Optosonic Press,1979.

Kaminow, Ivan P. Laser devices and applications. Ivan P. Kaminow and Anthony E. Siegman, eds. New York: IEEE Press,1973. Series title: IEEE Press selected reprint series.

Kasper, Joseph Emil and Steven Feller.The complete book of holograms: how they work and how to make them. New York: Wiley, 1987. Series title: The Wiley science editions.

_____ . The hologram book. Englewood Cliffs, N.J.: PrenticeHall, 1985.

Klein, H. Arthur. Holography. With an introd. to the optics of diffraction, interference, and phase differences 1 st ed. Philadelphia: Lippincott, 1970.Series title: Introducing modern science.

Kock, Winston E. Engineering applications of lasers and holography. New York: Plenum Press, 1975. Series title: Optical physics and engineering.

Lancaster, Ian M. (et alia). The Holographic Instant: Pulsed Laser Holography. New York: Museum of Holography, 1987.

_____ . Lasers and holography; an introduction to coherent optics. 1 st ed. Garden City, N.Y: Doubleday, 1969. Series title: Science study series; [S621.

_____ . Lasers & holography: an introduction to coherent optics. 2nd enl. ed. New York: Dover Publications, 1981.

_____ . Radar, sonar, and holography: an introduction. New York: Academic Press, 1973.

Kostelanetz, Richard. On Holography. RK Editions,1979.

Kurtz, Maurice K. "Potential uses of holography in photog ram metric mapping". 1971.

___: Study of potential application of holographic techniques to mapping; final technical report. Lafayette, IN: Purdue Research Foundation, Purdue University, 1971 .

Landry, Caliste John. "Ultrasonic imaging by Brillouin-Bragg diffraction: development of an operational system with prospective applications in medical diagnosis and material testing". 1972.

Laser Holography in Geophysics. Takemoto, Shuzo, Editor. New York, Wiley, 1989.

Laser Measurements. Toledo, OH: Laser Institute, 1985.

Laser recording and information handling technology. Proceedings of a seminar held August 21-22, 1974, San Diego, CA. Leo Beiser, ed. Palos Verdes Estates, CA: Society of Photo-Optical Instrumentation Engineers,1975. Series title: Society of Photo-optical Instrumentation Engineers Seminar proceedings ; 53.

Lasers and holographic data processing. N.G. Basov, ed. Translated from the Russian by P.S. Ivanov. Moscow : Mir Publishers, 1984. Series title: Advances in science and technology in the USSR. Technology series.

Lavigne, Richard C. Assessment of changeable message sign technology. McLean, VA: U.S. Dept. of Transportation, Federal Highway Administration, Research, Development, and Technology, 1986.

Lee, Hua. "Development and analysis of the back-projection method for acoustical imaging". 1980.

Lehmann , Matt. Holography; technique and practice. London: Focal Press, 1970. Series title: The Focal library.

Light and its uses : making and using lasers, holograms, interferometers, and instruments of dispersion : readings from Scientific American. Introductions by Jearl Walker. San Francisco : W. H. Freeman, 1980.

Light vistas light visions. Sponsored by the Department of Art, SI. Mary's College. Notre Dame, IN : St. Mary's College, 1983.

Lingenfelder, P. G. Holography manual; a compilation of laboratory techniques commonly used in the construction of holograms including refinements developed at NELC .. San Diego, CA: Naval Electronics

Laboratory Center, 1969. Series title: NELC Technical document 47.

Liu, Charles Yau-chi. "Some topics in holographic image formation". 1974.

Lucie-Smith, Edward. Art In The Seventies.Ithaca, NY: Cornell University Press, 1980. Contrib: R. Berkhout, H. Casdin-Silver, P.Claudius.

Lyons, Harold. Lasers, quantum electronics, holography: part 1: Introduction to lasers: Engineering 823.1 : a five-day short course, July 7-11 , 1975 : lecture notes. Harold Lyons, coord. Los Angeles : University of California, University Extension, 1975.

___. Lasers, quantum electronics, holography : part 1, Introduction to lasers: Engineering 823.1, June 17-21, 1974 : lecture notes. Harold Lyons, coord. Los Angeles: University of California, University Extension, 1974.

Matthews, Barbara Kubitz. "Application of holographic methods to the analysis of flexural vibrations of annular sector plates". 1976.

McNair, Don. How to make holograms. 1 st ed. Blue Ridge Summit, PA: Tab Books, 1983.

Mensa, Dean L."Techniques for microwave imaging". 1980.

Menzel, R. Fingerprint Detection with Lasers. New York: Marcel Dekker, 1980.

Miller, Richard K. (et alia). Holography. New York: Future Tech Surveys, 1989. Series title: A Survey on Technology & Markets Ser.,; No. 51

"A Multi-frequency synthetic detecting holography with high depth resolution". Peking, China: The Research Group of Holography, Chinese Academy of Geological Sciences, [s.n.), 1976.

NATO Advanced Study Institute on Optical and Acoustical HolographY, 1971 : Milan. Optical and acoustical holography; proceedings of the NATO Advanced Study Institute on Optical and Acoustical Holography, Milan, Italy, May 24-June 4, 1971 . Ezio Camatini,ed. New York: Plenum Press, 1972.

Neumann, Don Barker. "The effect of scene motion on holography". Columbus, OH: s.n ., 1967.

Nondestructive holographic techniques for structures inspection. R. K. Erf ... [et al.). Wright-Patterson Air Force Base, OH: Air Force Materials Laboratory, Air Force Systems Command, 1972.

Okoshi, Takanori. Three-dimensional imaging techniques. New York: Academic Press, 1976.

Optical & Acoustical Holography. Camatini, E.-Editor. New York: Plenum Publishing,1972.

Optical Computing Symposium, Darien, Conn., 1972. Digest of papers presented at the 1972 one-day-in-depth Optical Computing Symposium, April 12, 1972 at the Noroton School, Darien, Connecticut. Naval Underwater Systems Center and IEEE Computer Society, Eastern ... [s.l.] Institute of Electrical Engineers, 1972.

Optical Information Processing and Holography, Cathey, w.Thomas. ,Editor. Series Titile: Pure & Applied Optics Series. New York: Wiley, 1974.

Optics and photonics applied to three-dimensional imagery (Image 3-D):presented as part of the Optics, Phototonics, and Iconics Engineering Meeting (OPIEM), November 26-30, 1979, Strasbourg, France. Bellingham, WA: Society of Photo-Optical Instrumentation Engineers,1980. Series title : Society of Photo-opticalInstrumentation Engineers Seminar proceedings ; v. 212.

Optics in engineering measurement : 3-6 December 1985, Cannes, France. William F. Fagan, chair,ed. Organized by SPIE--the International Society for Optical Engineering, ANRT--Association Nationale de ... Bellingham, WA: SPIE--the International Society for Optical Engineering,1986. Series title: Proceedings of SPIE--the International Society for Optical Engineering ; v. 599.

Optics in entertainment: January 20-21, 1983, Los Angeles, California. Chris Outwater, chair,ed. Bellingham, WA: SPIE--the International Society for Optical Engineering, 1983. Series title: Proceedings of SPIE--the International Society for Optical Engineering; v. 391.

Optics in entertainment : January 26-27, 1984, Los Angeles, CA. Chris Outwater, chair, ed. Bellingham, WA: SPIE--the International Society for Optical Engineering,1984.Series title: Proceedings of SPIE--the International Society for Optical Engineering; v. 462.

Optics, Photonics, and Iconics Engineering Meeting,1979: Strasbourg, France. Optics and photonics applied to three-dimensional imagery (IMAGE 3-D): presented as part of the Optics, Photonics, and Iconics Engineering Meeting (OPIEM), November 26-30, 1979, Strasbourg, France . Bellingham, WA: Society of Photo-opticalInstrumentation Engineers, 1980. Series title: Society of Photo-optical Instrumentation Engineers Proceedings; v. 212.

Optics Today. John N. Howard, ed. New York, N.Y: American Institute of Physics, 1986. Series title: Readings from Physics today ; no. 3.

Ostrovskii, IU. I. Holography and its application. Translated from the Russian by G. Leib. Moscow : Mir, 19n.

Ostrovsky, Y. I. and M.M. Butusov. Interferometry by Holography. New York: Springer-Verlag, 1980. Series Title: Springer Series in Optical Sciences,; Vol. 20.

Outwater, Chris. and Eric Van Hamersveld. Guide to practical holography. Beverly Hills, CA: Pentangle Press,1974.

Palin, Michael. The Mirrorstone: A Ghost Story with Holograms. New York: Knopf,1986.

Pattern recognition studies: seminar-in-depth, proceedings / Society of Photo-Optical Instrumentation Engineers. [Redondo Beach, Calif.): the Society, [c1969). Series title: S.P.I.E. seminar proceedings; v. 18.

Pattern Recognition & Acoustical Imaging. Ferrari, Editor. Bellingham, WA: SPIE, 1987.

Periodic structures, gratings, moire patterns, and diffraction phenomena: July 29-August 1, 1980, San Diego, CA. C.H. Chi, E.G. Loewen, C.L. O'Bryan III, eds. Bellingham, WA: Society of Photooptical Instrumentation Engineers,1981. Series title: Proceedings of the Society of Photo-optical Instrumentation Engineers; v. 240.

Pethick, J. On holography and a way to make holograms. Ontario: Belltower Enterprises, 1971.

Photonics applied to nuclear physics, 2: proceedings; Strasbourg, Council of Europe, 5-7 December 1984. Geneva : CERN, 1985. Series title: Nucleophot.

Photorefractive materials and their applications. P. Gunter, J.-P Huignard, eds. Contributions by A.M. Glass ... let al.). Berlin: SpringerVerlag, 1988. Series title: Topics in applied physics; v. 61, etc.

Pietsch, Paul. Shufflebrain. Boston: Houghton Mifflin, 1981.

Pisa, Edward J., S. Spinak & A.F.Metherell. Color acoustical holography. Huntington Beach, CA: Douglas Advanced Research Laboratories, 1969. Series title: Douglas Advanced Research Laboratories. Research communication 109.

Powers, John Patrick. "Some aspects of the application of Bragg diffraction of laser light to the imaging and probing of acoustic fields". 1970.

Practical holography: 21-22 January 1986, Los Angeles, CA. Tung H. Jeong, Jacques E. Ludman chair,eds. Presented in cooperation with American Association of Physicists in Medicine ... let al.). Bellingham, WA: SPIE-The International Society for Optical Engineering, 1986. Series title: Proceedings of SPIE--the International Society for Optical Engineering; v. 615.

Practical holography II : 13-14 January 1987, Los Angeles, CA. Tung H. Jeong, chair/ed. Sponsored by SPIE--the International Society for Optical Engineering, in cooperation with Center for .. Bellingham, WA: SPIE--the International Society for Optical Engineering, 1987. Series title: Proceedings of SPIE-the International Society for Optical Engineering; v. 747.

Practical holography III : 17-18 January 1989, Los Angeles, CA. Stephen A. Benton, chair/editor; sponsored by SPIE--the International Society for Optical Engineering ; cooperating organizations, Applied .. Bellingham, WA., USA: SPIE, 1989. Series title: Proceedings of SPIE-the International Society for Optical Engineering; v. 1051.

Proceedings of the information processing and holography symposium ICALEO 83. Symposium heads: David Casasent, Milton T. Chang. Organized with American Society of Metals ... let a1.1. Sponsored by Laser Institute ... Toledo, OH: The Institute, 1984. Series title: LIA (Series) ; v. 41 .

Proceedings of the Inspection, Measurment [sic) and Control and Laser Diagnostics and Photochemistry, ICALEO '84. Donald Sweeney, Robert Lucht, eds. Organized in cooperation with ... The American... Toledo, OH: LIA-Laser Institute of America, 1985. Series title: LIA (Series) ; v. 45, 47.

Processing and display of three-dimensional data : August 26-27, 1982, San Diego, CA. James J. Pearson, chair,ed. Bellingham, WA: SPIE-The International Society for Optical Engineering,1983. Series titfe:Proceedings of SPIE--the International Society for Optical Engineering; v. 367.

Processing and display of three-dimensional data II : August 23-24, 1984, San Diego, CA. James J. Pearson, chair,ed. Cooperating organizations, Optical Sciences Center, University of Arizona,... Bellingham, WA: SPIE--the International Society for Optical Engineering, 1984. Series title: Proceedings of SPIE--the International Society for Optical Engineering; v. 507.

Progress in holographic applications: 5-6 December 1985, Cannes, France. Jean Ebbeni, chair, ed. Organized by SPIE--the International Society for Optical Engineering, ANRT--Association Nationale de la... Bellingham, WA: SPIE--the International Society for Optical Engineering,1986. Series title: Proceedings of SPIE--the International Society for Optical Engineering; v. 600.

Progress in holography: 31 March-2 April 1987, The Hague, The Netherlands. Jean Ebbeni, chair/ed. Organized by ANRT-Association nationale de la recherche technique, SPIE--The International Society for... Bellingham,WA: SPIE, 1987. Series title: Proceedings of SPIE--The International Society for Optical Engineering; v. 812.

"Project Search". Subcommittee on Feasibility of Automated Fingerprint Identification/Verification. An experiment to determine the feasibility of holograph assistance to fingerprint identification. Sacramento, CA: 1972. Series title: Project Search Technical report, no. 6.

Ramos, George Urban. "I. On the fast fourier transform; II. On the computations in digital holography". 1970.

Recent advances in holography III: February 4-5, 1980, Los Angeles, CA. Tzuo-Chang Lee, Poohsan N. Tamura, eds. Bellingham, WA: Society of Photo-optical Instrumentation Engineers, 1980. Series title: Society of Photo-optical Instrumentation Engineers Proceedings ; v. 215.

Saxby, Graham. Practical holography. New York, N.Y.: PrenticeHall International, 1988.

___. Holograms: How to Make & Use Them. Masson, France: Focal Press, 1980.

Saxby, John. Holograms. New York: Focal Press, 1980.

Schlussler, Larry. "Improvement of the horizontal resolution of a Braggdiffraction imaging system and motion limitations of a holographic system". 1978.

Schueler, Carl Frederick. "Development and applications of computerassisted acoustic holography".1980.

Schultz, Jerold M. Diffraction for Materials Scientists. New York: Prentice-Hall, 1982.

Schumann, Walter and J.-P. Zurcher, D.Cuche. Holography and deformation analysis. Berlin: Springer-Verlag, 1985. Series title: Springer series in optical sciences; v. 46.

____ . and M. DUbas. Holographic Interferometry: From the Scope of Deformation Analysis of Opaque Bodies. Series Title: Springer Series in Optical Sciences,; Vol. 16. New York: SpringerVerlag, 1979.

Schwank, James Ralph. "Refractive holography". 1974.

Sherman, George Charles. "Wavefront reconstruction and its application to the study of the optical properties of atmospheric aerosols". 1969.

Shuman, Curtis Alan. "Holographic imaging through moving diffusive media". 1973.

____ . "Holographic imaging through moving scatterers".1972.

Smith, Howard Michael. Principles of holography. New York, Wiley-Interscience, 1969.

____ . Principles of holography. 2d ed. New York: WileY,1975.

Solem, Johndale C. High-intensity X-ray holography: an approach to high-resolution snapshot imaging of biological specimens. Los Alamos, N.M.: Los Alamos National Laboratory,1982.

Solymar, L. and D.J. Cooke. Volume holography and volume gratings. London : Academic Press, 1981 .

Soroko, Lev Markovich. Holography and coherent optics. Translated from Russian by Albin Tybulewicz; with a foreword by George W. Stroke. New York: Plenum Press, 1980.

Sources of Physics Teaching: Atomic Energy, Holography, Electrostatics; Vol. 4. Noakes, G. A.-Editor. New York: Coronet Books, 1970.

Spanner, Jack C. Acoustic Emission Testing: Acoustic Emission: Techniques & Applications. American Society for Nondestructive Testing, 1974.

Spencer John A. Holographic Information Storage and Retrieval. England: National Reprographic, 1975.

Stone, William Ross. "The concept, design, and operation of a demonstration holographic radio camera". 1978.

____. "A remote probing technique for inhomogeneous media and an application to the study of satellite scintillations". 1974.

Strand, Timothy C. "Comparison of analog and binary holographic data storage" .1973.

Stroke, George W. An introduction to coherent optics and holography. New York: Academic Press, 1966.

____ . An introduction to coherent optics and holography. 2d ed. New York: Academic Press, 1969.

Su, Kung-Yen . "The fabrication of an opto-acoustic transducer for real-time diagnostic imaging systems". 1982.

Sutton, Jerry Lee. "Broadband acoustic imaging". 1974.

Symposium on Applications of Holography in Mechanics,1971: University of Southern California. Symposium on Applications of Holography in Mechanics. W. G. Gottenberg, ed. New York: American Society of Mechanical Engineers, 1971 .

Tarasov, L. V. Laser age in optics. Translated from the Russian by V. Kisin. Moscow: Mir Publishers, 1981.

Three-dimensional imaging; April 21-22, 1983, Geneva, Switzerland. Jean Ebbeni, Andre Monfils, chairmen-editors. Bellingham, WA: SPIE--the International Society for Optical Engineering, 1983. Series title: Proceedings of SPIE--the International Society for Optical Engineering; v. 402.

Three-dimensional imaging: August 25-26, 1977, San Diego, CA. Stephen A. Benton, ed. Presented by the Society of Photo-optical Instrumentation Engineers, in conjunction with the IEEE Computer... Bellingham, WA: SPIE, 1977. Series title: Society of Photooptical Instrumentation Engineers Proceedings; v. 20.

Topical Meeting on Hologram Interferometry and Speckle Metrology, 1980, June 2-4 : North Falmouth, MA. A digest of technical papers presented at the Topical Meeting on Hologram Interferometry and Speckle Metrology, June 2-4, 1980, Sea Crest Hotel, North Falmouth, Cape Cod, MA. [s.l.): Optical Society of America, 1980.

Tricoles, Gus Peter. "Some topics in microwave holography". 1971.

Trolinger, J. D. Laser instrumentation for flow field diagnostics. S. M. Bogdonoff, ed. Neuilly-sur-Seine, France: North Atlantic Treaty Organization, Advisory Group for Aerospace Research and Development, 1974. Series title: AGARDograph ; no. 186.

Tse, Nie But. "Digital reconstruction of acoustic holograms". 1979.

Ultrasonic Imaging & Holography: Medical, Sonar, & Optical Applications. Stroke, George W., and Jumpei Tsujiuchi, Editors. New York: Plenum Publishing,1974.

United States-Japan Science Cooperation Seminar on Pattern Information Processing in Ultrasonic Imaging,3rd: 1973: University of Hawaii. Ultrasonic imaging and holography: medical, sonar, and optical applications: [proceedings). George W. Stroke ... [et al.),ed. New York: Plenum Press, 1974.

United States-Japan Seminar on Information Processing by Holography, 2nd :1969 : Washington, D.C. Applications of holography; proceedings. Euval S. Barrekette, ed. New York: Plenum Press, 1971.

Unterseher, Fred, Jeannene Hansen and Bob Schlesinger. Holography handbook: making holograms the easy way. Berkeley, CA: Ross Books, 1982.

Vasilenko, G. I. (Georgii Ivanovich) and L.M. Tsibul'kin. Image recognition by holography. Translated from Russian by Albin Tybulewicz. New York: Plenum Press/Consultants Bureau, 1989.

Vest, Charles M. Holographic interferometry. New York: Wiley, 1979. Series title: Wiley series in pure and applied optics.

Vourgourakis, Emmanuel John's. "Coherence limitations on holographic systems". 1967.

Walton, Paul. Space-light : a holography and laser spectacular. London : Routledge & Kagan Paul, 1982 .

Wang, Keith Yu-Chih. "Threshold contrasts for various acoustic imaging systems". 1972.

Weinstein, L. Albertovich. Theory of Diffraction & the Factorization Method: Generalized Wiener-Hopf Technique.Golem Publications, 1969. Electromagnetics Series, Vol. 3.

Wenyon, Michael. Understanding holography. New York: Arco Pub. Co., 1978.

____. Understanding holography. Newton Abbot, Eng.: David & Charles,1978.

____: Understanding holography. 2nd Arco ed. New York: Arco Pub., 1985.
> UCD Phys Sci TA1540 W46 1985
> UCSD Central T A 1540 W46 1985

Wolff J. (et alia). Light Fantastic; Lasers and Holography Explained. England: Gordon Fraser/Bergstrom & Boyle, 1977

Wollman, Michael Thomas. "An experimental acoustical holographic system for eventual use in the ocean" .1975.

Yaroslavskii, L. P., and N. S. Merzlyakov. Methods of Digital Holography. New York: Plenum Publishing,1980.

Yu, Francis T.S. Introduction to diffraction, information processing, and holography. Cambridge, MA: MIT Press, 1973.

PERIODICALS

Acoustical Holography. International Symposium on Acoustical Holography. New York: Plenum Press. v.1-7, (1969-1977).

Acoustical holography, proceedings. International Symposium on Acoustical Holography and Imaging. (-1973).

Acoustical imaging and holography. New York Crane, Russak, 1978-1979.

Advances in holography. New York, M. Dekker, 1975-76.

Acoustical holography; [proceedings] :Acoustical imaging, 1978. International Symposium on Acoustical Holography and Imaging. New York: Plenum Press, 1978.

Afterimage. Rochester, NY: Visual Studies Workshop, V.12, no.7, Feb. 1985. See: A. Sargent-Wooster, "Manhattan shortcuts, Harriet Casdin-Silver's Thresholds'" p 19.

Fundamentals and applications of optical data processing and holography. Ann Arbor: University Michigan Engineering Summer Conferences.

Holography.

Holography News. (newsletter). Washington, DC.: Louis Kontnick, since 1987. (see listing for address).

The Holo-gram. (newsletter). Allentown, PA: Frank DeFreitas, since 1983. (See listing for address).

New Scientist. London, England: 1977. See: R. Weale "Art:Holography by H.Casdin-Silver" (June 29).

TITLES IN FRENCH

Caussignac, Jean Marie. Visualisation d'eeoulements aerodynamiques dans les compresseurs par interferometrie holographique.

Chatillon, France: Office national d'etudes et de recherches aerospatiales, 1972. Series title: France. Office national d'etudes et de recherches aerospatiales Note technique, 190.

Francon, M. Holographie. Paris: Masson, 1969. Series title: Recherche appliquee.

International Symposium of Holography, Besancon, France, 1970. Applications de l'holographie; comptes rendus du Symposium international d'holographie. Applications of holography; proceedings of the International Symposium of Holography. Besancon 6-11 juillet 1970. Besancon: Laboratoire de physique generale et optique, Universite de Besancon, 1970.

Pinson, G., A. Demailly, and D. Favre.La Pensee : approche holographique. Lyon: Presses universitaires de Lyon , 1985. Series title: Collection Science des systemes. Serie !heorie des systemes.

Voropaiev, N. Dictionnaire d'Eleetronique Quantique, Holographie et Optoelectronique. France: French & European, 1983

TITLES IN GERMAN

Claus, Jurgen. ChippppKunst : Computer, Holographie, Kybernetik, Laser. Originalausg. FrankfurtlM.: Ullstein,1985. Series title: Ullstein Materialien.

Kiemle, Horst, [und] Dieter Ross. Einfuhrung in die Technik der Holographie. Frankfurt am Main, Akademische Verlagsanstalt, 1969. Series title: Technisch-physikalische Sammlung Bd. 8.

Laserbeugung an elektronenmikroskopischen Aufnahmen. Ludwig Reimer ... [et al.]. Opladen: Westdeutscher Verlag, 1973. Series title: Forschungsberichte des Landes Nordrhein-Westfalen ; Nr. 2314.

Licht-Blicke : Holographie, die 3. Dimension fur Technik und kunst: [Ausstellung]7. Juni-30. September 1984, Deutsches Filmmuseum Frankfurt am Main. Schirmherr, Bundesprasident a. D. Walter Scheel;... Frankfurt am Maain: [Deutsches Filmmuseum].1984. Interviews with: S.Benton, M.Benyon, R.Berkhout, H.Casdin-Silver, F.Mazzero, S.Moree, N.Phillips, G. Schneider-Siemssen, D.Schweitzer. Articles by: M. Schneckenburger, et alia.

Mehr Licht: Kunstlerhologramme und Uchtobjekte = More light : [artists's [sic] holograms and light objects]. herausgegeben von Achim Upp und Peter lee. Hamburg : Fielmann im E. Kabel Verlag, 1985.

Menzel, Eric, W. Mirande [und] IWeingartner. Fourier-Optik und Holographie . Wien: Springer-Verlag, 1973.

Optoelectronik in der Technik : Vortrage des 6. Internationalen Kongresses Laser 83 Optoelektronik = Optoelectronics in engineering : proceedings of the 6th International Congress, Laser 83 Optoelektronik/herausgegeben. Berlin: Springer-Verlag, 1984.

Schreier, Dietmar unter Mitarbeit von W. Hase .. [et al.] Synthetische Holografie. Weinheim : Physik-Verlag, 1984.

Universitatsbibliothek Jena. lusammenstellung in- und auslandischer Patentschriften auf dem Gebiet der Holographie. Berichtszeit: 1948-1970. Gesamtleitung: Konrad Marwinski, Informationsabt., 1971. Series title: Universitatsbiblio!hek Jena Bibliographische Mitteilungen, Nr. 12.

Voropaev, N. D. Woerterbuch der Quantenelektronik, Holographie und Optoelektronik French & European,1983.

Zec, Peter. Holographie: Geschichte, Teehnik, Kunst. Koln: DuMont, 1987.

TITLES IN SWEDISH

Holografi: det 3-dimensionella mediet. New York : Museum of Holography; Stockholm: distribution, AVC, 1976.

TITLES IN RUSSIAN

Bakhrakh, L.D. i S.D. Kremenetskii. Metody izmerenii parametrov iizluchaiushchikh sistem v blizhnei zone. Leningrad : Izd-vo "Nauka", Leningradskoe otdelenie, 1985.

Bakhrakh, L.D iVA Makeeva. Primenenie golografii v meditsine i biologii. Leningrad: Nauka, 1977.

Barachevskii. VASvoistva svetochuvstvitel'nykh materialov i ikh primenenie v golografii: sbornik nauchnykh trudov. Otvetstvennyi redaktor Leningrad: Izd-vo "Nauka: Leningradskoe otd-nie. 1987.

___ . Neserebrianye i neobychnye sredy dlia golografii. Leningrad: "Nauka". Leningradskoe otd-ie. 1978.

Denisiuk. IU.N. Opticheskaia golografiia : prakticheskie primeneniia.. Leningrad : Izd-vo "Nauka: Leningradskoe otd-nie. 1985.

___. Opticheskaia golografiia: [Sb. statei). AN SSSR. Fiz.tekhn. in-t im. A.F. Ioffe. Nauch. sovet po probl. "Golografiia". Leningrad: Nauka. Leningr. otd-nie. 1979.

Derkach. M.F.Dinamicheskie spektry rechevykh signalov. L'vov: Izd-vo pri L'vovskom Gos. universitete Izdatel'skogo ob"edineniia "Vyshcha shkola". 1983.

Fizicheskie osnovy i prikladnye voprosy golografii : tematicheskii sbornik:[materialy XVI Vsesoiuznoi shkoly po fizicheskim osnovam golografii / redaktory G.V. Skrotskii. B.G. Turukhano. N. Turukhano). Leningrad: Akademiia nauk SSSR. Fiziko-tekhn. ins-t im. A.F. Ioffe. 1984.

Gurevicha. S.B. i V.K. Sokolova.Primenenie metodov opticheskoi obrabotki informatsii i golografii. Leningrad : LIIAF. 1980.

Gurevicha. S.B .. OA Potapova."Golografiia i opticheskaia obrabotka informatsii v geologii". Dokl. seminara. Leningrad: Akademiia Nauk SSSR. Fiziko-tekhnicheskii in-t im. A.F. Ioffe. 1980.

IAkovkin. I. B. Difraktsia sveta na akusticheskikh poverkhnostnykh volnakh . otv. redaktor S.V. Bogdanov. Novosibirsk : Izd-vo "Nauka". Sibirskoe otd-ie. 1979.

IAroslavskii. L. P. and N.S. Merzliakov. TSifrovaia golografiia. Moskva: Izd-vo "Nauka". 1982.

International School on Coherent Optics and Holography.2nd: 1981: Varna. Bulgaria. Integral'naia optika. volokonnaia optika i golografiia: materialy vtoroi Mezhdunarodnoi shkoly po kogerentnoi optike i golografii--Varna ·81.28.09..{)3.10.1981 . Varna. Bolgariia. Redaktsionnaia kollegiia P. Simova .... Sofiia: Izd-vo Bolgarskoi akademii nauk. 1982.

Kirillov. N. I.Vysokorazreshaiushchie fotomaterialy dlia golografii i protsessy ikh obrabotki . Moskva: Nauka. 1979.

Klimenko. I. S. Golografiia sfokusirovannykh izobrazhenii i speklinterferometriia. Moskva: "Nauka: Glav. red. fizikomatematicheskoi lit-ri. 1985.

Klimkin. V. F. (Viktor Fedorovich) Opticheskie metody registratsii bystroprotekaiushchikh protsessov / V.F.

Klimkin. A.N. Papyrin. A.1. Soloukhin ; otvetstvennyi redaktor N.G. Preobrazhenskii. Novosibirsk : Izd-vo "Nauka: Sibirskoe otd-nie. 1980.

Kulakov. Sergei Viktorovich. Akustoopticheskie ustroistva spektral'nogo i korreliatsionnogo analiza signalov. Akademiia nauk SSSR. Nauchnyi sovet po probleme"Golografiia". Fizikotekhnicheskii institut imeni A.F. Ioffe. Leningrad :"Nauka". Leningradskoe otd-nie. 1978.

Petrashen. G. I. Prodolzhenie volnovykh polei v zadachakh s seismorazvedki. Leningrad: "Nauka: Leningr. otd-nie. 1973.

Radiogolografiia i opticheskaia obrabotka informatsii v mikrovolnovoi tekhnike: [Sbornik statei). Akademiia nauk SSSR. Otdelenie obshchei fiziki i astronomii. Nauchnyi sovet po probleme "Golografiia" Leningrad: "Nauka: Leningradskoe otd-nie. 1980.

Soboleva. G.A. Registriruiushchie sredy dlia izobrazitelnoi golografii

i kinogolografii :[Sb. statei) . AN SSSR. Otd-nie obshch. fiziki i astronomii. Nauch. sovet po probl. Golografiia. Leningrad: Nauka. Leningr. otd-nie. 1979.

Sokolov. A. V. i IA.A. Al·trnana. Primenenie metodov opticheskoi golografii dlia issledovaniia biologicheskikh mikroob"ektov. Leningrad: Nauka. Leningradskoe otd-nie.1978. Series title: Metody fiziologicheskikh issledovanii.

Voropaev. N. D. Anglo-russkii slovar' po kvantovoi elektronike i golografii: Okolo 18000 terminov . Pod red. A. M. Leontovicha. Moskva: Rus.iaz .. 1977.

Vsesoiuznaia shkola po golografii. 6th: 1974 : Yerevan. Armenian S.S.A. Materialy VI Vsesoiuznoi shkoly po golografii : 11-17 fevralia 1974 g .. redaktory. G.V. Skrotskii. B.G. Turukhano. N. Turukhano). Leningrad: LIIAF. 1974.

Vsesoiuznaia shkola po golografii. 7th : 1975 : Rostov. A.S.F.S.A. Materialy VII Vsesoiuznoi shkoly po golografii : ianvar' 1975 g. [podgotovleny k pechatki N. Turukhano). Leningrad : Leningradskii in-t iademoi fiziki. 1975.

Zel·dovich. B. IA. N.F. Pilipetskii. & V.V.Shkunov. Obrashchenie volnovogo fronta. Moskva: "Nauka: Glav. red. fizikomatematicheskoi lit-ry. 1985.

EXHIBIT CATALOGUES AND MISCELLANY

Bienal Internacional de Sao Paulo. Catalogo Geral. Sao Paulo. Brasil: Fundacao Bienal de Sao Paulo. 1985. See "Entre a Ciencia EA Ficcao". pp167-197. re: holographers M.Baumstein. H. CasdinSilver. J.W.Garcia.

Alice in the Light World. Tokyo. Japan: The Ashai Shimbun. 1978.

ARTTRANSITION . Cambridge. MA: MIT Center for Advaced Visual Studies/University Film Study Center. 1975. See H. Casdin-Silver "Holography .. " pp30-32.

Critic's Choice: The Craft of Art; Peter Moore's Liverpool Project 5. Liverpool. England: November 3. 1979. See E. Lucie-Smith. "New Attitudes. New Materials. New Techniques".

___. Liverpool. EnglandWalker Art Gallery. 1979. No. 84-110. See W.D.L. Scobie. "Arts Review".

Electra 83. Paris: Les Amis du Musee d'Art de la Ville de Paris. 1983. See F.Popper. "Electricity and Electronics in the Art of the 20th century". pp 46-50.

EXPANSION. Internationale Biennale fur Graphik und Visuelle Kunst. Horst Gerhard Haberl. Generalsekretar. Wien. Austria: Internationale Biennale fur Graphik und Visuelle Kunst. 1979. See O.Peine "MIT-Center for Advanced Visual Studies". "Sky Events" pp 232-239; E. Goldring "Documentation room". "Centerbeam"p 232; H.Casdin-Silver "Holography. a holographic environment" p 234; G.Kepes "Art of the Environment" p 99.

Fantasy of Holography. Tokyo. Japan: Seibu Museum of Art. 1976. Itsuo Sakane. ed. Contr: Shuntoro Tankawa. Junpei Tsujiuchi. Harriet Casdin-Silver Holography. New York: Museum of Holography. 1977. First one person exhibition at the Museum.

High Technology and Art 1986. Tokyo. Japan: Tokyo Shimbun & Nagoya Shimbun. with Associates of Art and Technology. Japan.

'Holography redefined'. ·Thresholds·. Harriet Casdin-Silver with Dov. Eylath. New York: Museum of Holography. 1984. Group exhibition.

Inter. Quebec. Canada: Les Editons Intervention. Printemps 86.

no.31, 1986. See E. Shapiro, "Art, Perception et Holographie" p. 32; L.Heaton, "Tire D'une Entrevue avec Harriet Casdin-Silver" pp32-3.

Images in Time and Space. Ottawa, Ontario, Canada: Association of Science and Technology Inc, 1987. Travelling exhibit.

International Holography. London, England: The Photographers' Gallery, 1980.

Light and Substance. New Mexico: University of New Mexico Art Museum, 1973-75. History of photography, holography by S. Benton, H.Casdin-Silver. Van deren Coke, org.

MultiMedia Exhibition. Kansas City, MO: Nelson Gallery at Atkins Museum of Fine Arts, 1970. See "Holography by H.Casdin-Silver".

Otto Peine und CAVS: 20th Anniversary CAVS. Karlsruhe, West Germany: Badischer Kunstverein, 1988. See H. Casdin-Silver, A. Cheji, D. Jung, J.Powell.

Sky art conference-'81 . Cambridge, MA: MIT Center for Advanced Visual Studies, 1981 . L. Burgess, E. Goldring, B. Kracke, O. Piene, eds. See H.Casdin-Silver, "Sky work: solar-tracked hologram series"p.49.

Sky art conference '83. Cambridge, MA: MIT Center for Advanced Visual Studies with der Landashaupstadt Munchen der BMW AG und dar Digital Equipment GmbH, 1983.

GLOSSARY

Absorption Hologram: A hologram formed in a material which acquires a certain density in response to exposure. When the hologram is illuminated, part of the light which is not absorbed is diffracted into forming the image.

Achromatic: Free of color. Black and white. In optical systems, the term is used to describe lenses which correct for chromatic aberration.

Acoustical Holography: The making of holograms by using sound waves.

Additive Color Mixing: Means by which two or more frequencies are combined by superimposition to create more colors.

Ambient Light: Light present in the immediate environment. In holographic display, often used to describe background light that is not part of the hologram illumination and may interfere with the viewing of the image.

Amplitude: The maximum value of the displacement of a point on a wave front from its mean value. Graphically, the height or depth of the crest or trough of a wave from its zero point.

Amplitude Hologram: A hologram by which information is stored as variations in transmittance. Also called absorption hologram

Antihalation Backing (AH): A dark material placed on the back surface of a plate or film to prevent unwanted light from striking the emulsion. Useful to prevent the formation of "Newton Rings" in the hologram. Only to be used with transmission holograms.

Argon Laser: A laser which operates when argon gas is ionized and controlled by a magnetic field. Produces several blue and green frequencies.

Artistic Holography: Holograms created for the purpose of being seen and whose value derives at least in part from the image presented.

Astigmatism: An aberration caused by the horizontal and vertical aspects of an image forming in different planes.

Bandwidth: The range of frequencies over which a given instrument will operate.

Beamsplitter: An optical component which divides a beam into two or more separate beams. A 50:50 beamsplitter produces two beams of approximately equal intensities. A 90:10 beam splitter transmits approximately 90% of the incident beam and reflects 10% into the second beam.

Benton Hologram: Another term for rainbow hologram. Named for its inventor, Steve Benton. A hologram produced by reducing vertical information in order to correct for image dispersion.

Biconcave: A lens which has both faces curving inward. A type of negative lens.

Biconvex: A lens which has both faces curving outward. A type of positive lens

Bleach: in holographic processing, a chemical used to change an absorption hologram into a phase hologram in order to improve efficiency (brightness).

Bragg Diffraction (Bragg's Law): Diffraction which is reinforced by reflection by a series of regularly spaced planes which correspond to a certain wavelength and angular orientation. The angle at which this reinforcement occurs is Bragg's angle.

BRH: Bureau of Radiological Health. U.S. government agency responsible for setting laser safety standards.

Brightness: A subjective term describing the amount of light perceived.

Cavity: Another name for optical cavity or laser cavity.

Chromatic Aberration: Lens or hologram irregularity due to the shifting of image position for each frequency. If severe enough, the image will appear to blur due to the lack of registration of the colors.

Coherence Length: The maximum path length difference (between the reference beam and the object beam) in a holographic set-up that can be used and still obtain a clear, bright hologram. Coherence length depends on the type of laser used and how it is made. An etalon will increase the coherence length of a laser to about 30 feet.

Coherent Light: Light which is of the same frequency and vibrating in phase. The laser produces coherent light.

Collimated Light: Light which forms a parallel beam and neither converges nor diverges. Also referred to as collimated beam.

Collimator: A device used to produce collimated light by positioning a light source at the focal point of a lens or parabolic mirror. Such a device is called a collimating lens or collimating mirror, respectively.

Color Spread: The area over which a spectrum is dispersed.

Computer Generated Hologram: A synthetic hologram produced using a computer plotter. The binary structure is produced on a large scale and then photographically reduced into a given medium. The technique allows the production of impossible or nonexistent 3-dimensional forms.

Concave Lens: A lens with an inwardly curving surface which causes light to diverge. See also Negative lens.

Concave Mirror: A mirror with an inwardly curving surface which causes light to converge.

Constructive Interference: Coherent wave fronts of the same frequency are superimposed, and their instantaneous amplitudes add up to a greater amplitude than that of the component waves.

Continuous Wave Laser: A laser which emits a beam which does not vary over time.

Convergence: The optical bending of light rays toward each other, as by a convex lens or concave mirror.

Convex Lens: A lens with an outwardly curving surface which causes light to converge, usually to a focal point.

Copy Hologram: Another term for image plane hologram or any second generation hologram produced from a master hologram. A contact copy is produced by placing the plate in contact with the original.

Copy Plate: Another term for copy hologram. Usually refers to the plate before it is exposed.

Cross Hologram: Another name for the type of holographic stereogram which incorporates the advantages of rainbow holography. Named for Lloyd Cross.

Cross Talk: The phenomenon of spurious images formed by color holograms when an interference pattern formed by one color also reconstructs an image in another color.

Cylindrical Mirror/Lens: An optical component which causes light to focus as a slit or line by passing through or reflecting from a surface curved in one dimension.

Denisyuk Hologram: Another name for single beam reflection hologram. Named for its inventor, Y. N. Denisyuk.

Density: The amount of opacity or darkness of a medium.

Depth of Field: The area within which satisfactory resolution of an image can be obtained. Also, in holography, used to describe the area within which any image can be formed, due to the constraints of coherence length.

Developer: A chemical solution which changes the latent image of a photographic image or holographic interference pattern (silver salts) into black metallic silver. The term development usually refers to the degree of effect of the developer or the cause of the amount of density.

Dichromated Gelatin (D.C.G): A light-sensitive gelatin made up of a solution of dichromate compound, usually ammonium dichromate, in the presence of a gelatin substrate. Exposure results in the crosslinking of gelatin molecules with each other.

Diffraction: The change in direction of a wave front at the edges of an aperture, caused by the wave nature of light. Diffraction is not the same process as reflection or refraction.

Diffraction Efficiency: In a hologram, the percentage of incident illumination light diffracted into forming the image. The greater the diffraction efficiency, the brighter the image will appear in a given light.

Diffraction Grating: A holographic diffraction grating is a hologram formed by the interference of two or more beams of pure, undiffused laser light.

Diffuse Reflector: An object that scatters illumination striking it. Most objects are diffuse reflectors.

Divergence: The bending of light rays away from each other, usually by concave lens or convex mirror, so that the light spreads out. Light will also diverge with a convex lens or concave mirror after it passes through the focal point.

Double Exposure: The formation of two holograms on the same recording medium. Used to cause either overlapping images or two discrete images to appear under different conditions.

Electromagnetic Radiation: Radiation emitted from vibrating, charged particles, all of which travels through space at the speed of light. Visible light is only a small part of the entire electromagnetic spectrum.

Embossed Hologram: A hologram copy made by pressing a metal surface relief master hologram into plastic film, or by using the master hologram in a mold.

Emulsion: A suspension of light sensitive silver salts (e.g. silver bromide) in gelatin, usually coated onto glass, polyester film, or by using the master hologram in a mold.

Exposure: The act or time of allowing light to impinge upon the emulsion.

Film Plane: The plane at which the recording material is located.

Fixer: A chemical solution which removes the unexposed silver salts from the emulsion to desensitize the emulsion after development.

f-number: The ratio of the focal length of a lens or curved mirror to its diameter.

Focal Length: The distance from the center plane of a lens or curved mirror to a position where a collimated beam is focused to a point. A lens with a negative focal length (a concave lens) appears to have a focus upstream from the lens, while a lens with a positive focal length (a convex lens) has a focus downstream from the lens.

Focused Image Hologram: Any hologram in which the image appears on the surface of the hologram or seems to intersect the surface of the hologram.

Fog: The darkening or exposing of film by inadvertently allowing ambient light to strike it. In holography, a fogged plate reduces fringe contrast, resulting in a less efficient image.

Fourier Transform Hologram: A hologram made using a reference beam diverging from a point at the same distance from the recording plate as the object. Also called Fraunhofer Hologram

Frequency: The number of crests of waves that pass a fixed point in a given unit of time.

Fresnel Hologram: Another name for the common hologram. Defined as a hologram formed with an object located close to the recording medium.

Fringe: An individual interference band, made up of one cycle of constructive and destructive interference.

Front Surface Mirror: A mirror with the reflecting surface on the front. Conventional mirrors have their reflecting surfaces on the back of a piece of glass and are not useful for holography as the front surface produces a "ghost" reflection.

Gabor Hologram: An in-line hologram of the type invented by Dennis Gabor.

Gas Laser: A laser such as a Helium-Neon laser, in which the lasing medium is a gas.

Grating (also Diffraction Grating): A pattern of very fine lines of equal spacing, usually on the order of a few microns apart. A diffraction grating can redirect light and break white light into its component colors like a prism.

H-1: The first hologram made in the process of making a master hologram. An H-1 has an image only viewable in laser light.

H-2: The second hologram made in the process of making master hologram. A master hologram is usually an H-2.

Helium-Neon (HeNe): The most common lasing material, which produces a continuous red beam at 632.8 nm.

HOE: Holographic Optical Element. A hologram which may be used to act as a lens, mirror, or a complex optical component.

Hologram: An interference pattern formed as a result of reference light encountering light scattered by an object and stored as such on a light sensitive emulsion.

Holographic Movie: The animation of a 3-dimensional holographic image by presentation of numerous holograms in rapid sequence in much the same way motion picture film operates. Unlike conventional cinema, it is only with extreme difficulty that the image can be projected, and true holographic movies are still very experimental. The term is often used, incorrectly, for holographic stereograms.

Holographic Stereogram: A hologram made by filming numerous angles of view of a scene and then storing the frames holographically. Each eye views a different frame, displaced so as to result in the illusion of a stereoscopic image. Also called a multiplex or Integral hologram. Image Plane Hologram: A second generation hologram formed by positioning a light sensitive plate in the plane of an image formed by a master hologram.

Incandescent Light: Light formed when an electric current passes through a resistant metal wire, usually situated in a vacuum bulb.

Incoherent Light: Light which is emitted with randomly varying phase and a mix of colors. Light from ordinary sources such as light bulbs is incoherent; laser light is coherent.

Index of Refraction: The ratio of the velocity of light in air to the velocity of light in a refractive material for a given wavelength.

Infrared: That part of the spectrum characterized by wavelengths somewhat longer than those of red light, which are not visible to the eye yet are often perceived as heat. Covers the spectrum from about 750 nm to 1000 micrometers.

In-Line Hologram: A Gabor hologram. Made by positioning the object and reference light along the same axis, resulting in a configuration practical only for making holograms of transparencies.

In Phase: The relationship of two waves of the same frequency when they travel through their maximum and minimum values simultaneously and are also polarized identically. Holograms must be made by waves which remain in phase during the course of an exposure.

Interference: The result of superimposing two or more waves. The waves oscillate between negative and positive values, so when two waves are superimposed the positive values reinforce each other while negative values cancel out positive values.

Interference Pattern: A stationary pattern of interference that results when light waves are superimposed.

Interferometer: A device that utilizes interference of light to measure changes in systems with extreme accuracy. An interferometer can be used to test the stability of holographic systems.

Ion: An atom or molecule which has gained or lost an electron so that it acquires a negative or positive charge.

Ion Laser: A laser within which stimulated emission occurs as a result of energy changes between two levels of an ion. Argon and krypton are the two most common types of ion laser.

Krypton Laser: An ion laser that produces many frequencies which appear over a large part of the spectrum. The most common lines are blue, green, yellow, as well as a very strong red frequency.

LASER: From the acronym for "Light Amplification by Stimulated Emission of Radiation." A laser is usually in the form of a light-amplifying medium placed between two mirrors. Light not perfectly aligned with the mirrors escapes out the sides, but light perfectly aligned will be amplified. One mirror is made partially transparent. The result is an amplified beam of light that emerges through the partially transparent mirror.

Latent Image: The image or pattern stored in an emulsion before it is developed into a visible image.

Latent Image Decay: A condition that is common to fine grained silver emulsions, including the types used for holography. The decay occurs if the material is not processed soon after exposure, resulting in a lower density.

Light Meter: Any device used to sense and measure light. Usually used to sense intensity in order to determine exposure.

Line Spacing: The distance between individual interference fringes in a diffraction grating.

Lippman Hologram: Another name for reflection hologram.

Liquid Lens: A lens formed by filling a shaped glass or acrylic tank with liquid such as mineral oil. A liquid lens is often used in a system for making Cross holograms.

Master Hologram: The original H-2 hologram, from which copy holograms are made.

Mode: A laser can oscillate in a number of different modes. Spatial modes are the different repetitive patterns in which light can zigzag between the two laser mirrors. The optimum spatial mode for holography has no zigzag at all (called the TEM00 mode), resulting in a single emitted beam of light. Higher modes can result in a beam that gives a donut-shaped

spot, or a cloverleaf pattern. Longitudinal modes are the different wavelengths that are simultaneously emitted by a laser. A single-mode laser produces a single wavelength of light, and can only have a single spatial mode and a single longitudinal mode.

Moire Pattern: A highly visible type of interference pattern formed when gratings, screens, or regularly spaced patterns are superimposed upon each other.

Monochromatic: Light or other radiation with one single frequency or wavelength. Since no light is perfectly monochromatic, the term is used loosely to describe any light of a single color over a very narrow band of wavelengths.

Motion: The effects of an object or holographic system not remaining rigidly fixed during exposure.

Multichannel Hologram: A hologram formed with two or more separate reference beams or angles.

Multiple Exposure: More than one exposure occurring on the same plate or film.

Multiplex: Another name for holographic stereogram.

NAH: A holographic plate without an antihalation backing. Also called an unbacked plate.

Negative Lens: A lens characterized by a concave surface which causes light to diverge. A negative lens has a negative focal length.

Newton's Rings: The series of rings or bands which appear due to interference between two nearly parallel surfaces. These rings often form as a result of light interacting between the front and back surfaces of a holographic plate.

Node: The part of a vibrating wave that is not moving - zero point. An anti node is a point on the wave of maximum displacement from the zero point.

Noise: Any unwanted light scattering by components in a holographic set-up or by particles in a holographic recording medium.

Object Beam: The light beam in a holographic set-up which illuminates the object. Also, the light reflected from or transmitted by the object which is recorded in the hologram.

Off-Axis: The type of hologram invented by Emmet Leith and Juris Upatnieks whereby object and reference beams approach the holographic plate at different angles.

On-Axis: Hologram formed with object and reference beams originating along the same axis. Also called an inline or Gabor hologram.

Open Aperture: A transmission image plane hologram viewable in white light and characterized by both vertical and horizontal parallax and usually a brilliant white image.

Optical Cavity: The space between the two mirrors in a laser. The tube is located within the optical cavity.

Optical Component: An optical device consisting of the optics (lens, mirror, etc.) and a mount used to affix it to a vibration isolation table.

Optics: Those devices which change or manipulate light, including lenses, mirrors, beamsplitters, filters, etc. Also the science of electromagnetic radiation, its effects, and the phenomenon of vision.

Orthoscopic: Having the "right" appearance. Orthoscopic image has the correct appearance, whereas a "pseudoscopic" image appears to have its depth inverted.

Oscillator: Any device that converts energy into an alternating electromagnetic field, usually of constant period.

Overexposure: Improper exposure resulting from too much light or light reaching the plate or film for too long.

Parabolic Mirror: A mirror with a surface curved in the shape of a parabola. Used as a telescope mirror in astronomy or as a collimating mirror in holography.

Parallax: The difference between two different views of an object, obtained by changing viewing position.

Period: The time required for a wave to go through one complete cycle. The period of a typical light wave is about one trillionth of a second.

Phase: A wave oscillates from a positive value to a negative value and back to positive. The phase of a wave relates to where it is in its oscillation cycle at a particular moment. Usually only the phase difference between two waves is important.

Phase Hologram: A hologram which diffracts light by delaying the phase of certain portions of the lightwave, rather than absorbing certain portions. Bleached silver halide holograms, DCG holograms, and surface relief holograms are phase holograms, while unbleached silver halide holograms are amplitude holograms.

Phase Shift: The amount by which the phase of one light beam is delayed or advanced relative to another light beam.

Photochemistry: The branch of chemistry dealing with the effects of light on chemical reactions.

Photon: The smallest unit or quantum of electromagnetic energy known today.

Photopolymer: A material which "polymerizes" where it is exposed to light. Photopolymers are usually partially solidified plastics which finish solidifying when exposed.

Photoresist: A chemical substance made insoluble by exposure to light (usually ultraviolet). Although most often used to manufacture microcircuits, photoresist can be used to make holograms.

Pinhole: The small hole used to pass focused light from the objective in a spatial filter.

Plane Hologram: A hologram for which fringes are large with respect to the thickness of the emulsion, so that interference is mostly stored on the surface of the hologram.

Plano-Concave: A lens which has one concave surface and one flat surface.

Plano-Convex: A lens which has one convex surface and one flat surface.

Plateholder or Platen: Any device which holds a holographic plate or film in place during the exposure.

Polarization: The restriction of light or other radiation to vibration in only one plane.

Population Inversion: A condition whereby more atoms are in the excited state than in the ground state, resulting in the predominance of stimulated emission.

Positive Lens: A lens with an outwardly curving surface which causes light to converge. Also known as convex lens.

Processing: The entire chemical sequence, from development to final drying of the hologram.

Pseudocolor: The production of colors in a hologram which are not related to the true colors of the original objects. Usually used in connection with multicolor holograms.

Pseudoscopic: The opposite of orthoscopic. An image whose parallax is reversed.

Pulsed Hologram: A hologram produced with the short burst of light from the pulsed laser. May be used to make holograms of live subjects.

Pulsed Laser: A laser which emits radiation in a wave of short bursts and is inactive between bursts.

Quantum: The smallest amount that the energy of a wave may be divided into.

Rainbow Hologram: A white light viewable hologram with colors which shift through the spectrum as the hologram or the viewing angle is tilted. A rainbow hologram has no vertical parallax.

Real Image: An image that is formed in such a way that it actually comes to a focus. A real image is one that forms downstream from a lens; a virtual image appears to form upstream from a lens.

Real Time Holography: A technique whereby a holographic image is superimposed over a real object in order to observe interference fringes generated by minute changes between the two.

Reconstruction Beam: Light directed at the finished hologram from which the object wave front will be recreated.

Recording Material: Any substance which may be used to record the interference pattern of the hologram.

Reference Angle: The angle at which the reference beam strikes the plate, usually measured in degrees from the plate surface.

Reference Beam: The unmodulated, pure laser light directed at the plate to interfere with the object light.

Reflection Hologram: A hologram made by allowing reference and object light to impinge on opposite sides of the plate. The finished hologram is viewed by allowing light to reflect from it to the observer.

Refraction: The bending of light which occurs when it passes from a medium of one refractive index to that of another. In a phase hologram, refraction causes a "phase delay" which corresponds to the original phase difference between the two stored wave fronts.

Refractive Index: Same as index of refraction.

Resolution: The ability of a film or an optical system to distinguish between two closely spaced points. Film resolution is usually expressed in terms of how many closely spaced lines per millimeter the film can record. Holographic films must be capable of high resolving capability since the interference fringes are often extremely small and closely spaced.

Resonance: A large amount of vibration in a system which is caused by a small stimulus with approximately the same period as the natural vibration period of the large system.

Resonant Cavity: Another name for optical cavity or laser cavity.

Scatter: Unwanted light which interferes with the making of a good quality hologram.

Settling Time: A period of time between the loading of the plate and the exposure in order to allow ambient vibrations time to dampen.

Set-up: The configuration of optical components used to produce a given hologram.

Shadowgram: A hologram made by deliberately moving an object during an exposure, or by using an inherently unstable object, in order to produce a 3-dimensional "hole" or shadow where the object was once located.

Shutter: The device used to block the laser beam and then allow it to pass unobstructed for the desired exposure time.

Silver halide: The type of recording material which consists of light-sensitive silver particles suspended in gelatin.

Single beam hologram: A hologram made with one beam which acts as both reference and object illumination beam.

Slab Table: An optics table which uses a concrete slab as part of its inertial mass.

Slit Optics: Any optical device which causes light to be propagated into a line. Usually formed by light interacting with a cylindrical surface.

Solid State Laser: A laser which uses a solid material, such as ruby, as its lasing medium.

Space: The area between objects. In holography, the area between and including objects.

Spatial Filtering: The act of "cleaning up" the light of the laser beam by causing it to focus through a tiny aperture. Only the pure light can focus at the desired point, eliminating the effects of dust, optical surface scratches, etc.

Spatial Frequency: Often used with regard to line spacing in diffraction gratings. The spatial frequency is the reciprocal of line spacing, generally expressed in cycles per millimeter. See also resolution.

Speckle: The grainy appearance of an object, or a holographic image, viewed under laser light. It is caused by light reflecting from minute areas of the object and interfering with itself.

Spectral Reflection: Any reflection from a smooth, polished surface, such as a mirror. Also called specular reflection.

Splitbeam: The act of separating a beam of laser light into two components to separately control the action of reference and object illumination.

Squeegee: A device or action used to remove excess water from the emulsion to facilitate drying.

Stability: The requirement for holographic optical systems to remain motionless during an exposure.

Standing Wave: The result of superimposing two or more waves moving in different directions but having the same wavelength. An interference pattern is a slice through a standing wave pattern; and a hologram is a photograph of an interference pattern.

Stereogram: An image which creates a 3-dimensional illusion by presenting a different view of an object to each eye.

Stimulated Emission: Radiation produced by incoming radiation of the same phase, amplitude, and frequency.

Stop Bath: The chemical bath immediately following the developer which causes the developer to cease action.

Tem00: The lowest mode of a laser, characterized by a beam which is spatially coherent across the diameter of the beam.

Temporal Coherence: Coherence over time. The degree to which waves will remain coherent over time and distance.

Test Strip: A means of visually determining the correct exposure by making a series of individual exposures of varying times on the same plate. The proper time is determined by selecting the strip which yields the brightest or cleanest image.

Thermoplastic Film: A recording material which works due to the effects of electrostatic forces and heat to produce a deformation corresponding to the interference pattern exposed.

360-Degree Hologram: A hologram made by exposing recording material which completely surrounds an object.

Transfer Mirror: A mirror which redirects light from the laser toward the desired working area on the optics table.

Transmission Hologram: Any hologram viewed by passing light through it, toward the viewing side. Transmission holograms are made by allowing both object and reference light to impinge on the same side of the plate.

Transmittance: The proportion of light transmitted by a medium to that which is incident upon it.

Trlethanolomlne: A chemical used to change the thickness of the emulsion to produce different color playback, usually with reflection holograms.

Ultraviolet: An invisible part of the spectrum characterized by wavelengths somewhat shorter than violet (approx. 100-400 nm.).

Un backed Plate (NAH): A holographic plate without an anti-halation backing. Essential for reflection holograms.

Variable Beamsplitter (VBS): A beam splitter whereby the ratio of transmitted to reflected beam changes as the beam intercepts the component at different points.

Vibration Isolation: The practice of removing a system from the effects of ambient vibrations which may induce changes, particularly in optical systems. Vibration isolation must be used in making a hologram to prevent the movement of interference fringes during an exposure.

Virtual Image: An image which appears "upstream" from a lens or hologram. A real image becomes visible when a piece of paper is placed in its location, but a virtual image is not accessible.

Volume Hologram: A hologram in which the thickness of the recording material is large compared to the spacing of the interference fringes. DCG and photopolymer holograms are volume holograms; embossed holograms are "thin" or "surface" holograms.

Wave Form: The characteristic shape taken on by a wave front.

Wave Front: The surface of a propagating wave, where the phase of the wave is the same everywhere on the surface. A point source produces spherical wavefronts, a collimated beam consists of plane wavefronts, and light reflected from a complicated object has a wavefront with a very complicated shape.

Wavelength: The physical distance over which the complete cycle of one wave occurs. Wavelength is inversely proportional to frequency.

White Light Transmission Hologram: Any transmission hologram which can be displayed using ordinary white light.

VAG Laser: A solid state laser using Yttrium Aluminum Garnet as the lasing material.

Zone Plate: A pattern consisting of a central spot surrounded by concentric zones, alternatingly opaque and transparent, the total area of each zone being equal. A zone plate is equivalent to the hologram of a point object.

INDEX

INDEX TO ADVERTISERS

DID YOU BORROW THIS COPY?

If so, now is the time to order your own personal copy of the **Holography Marketplace.** This international directory for the holography industry is the first and only resource of its kind. You will refer to the HMP day after day, so don't you want one of your own?

☐ YES! I want my own personal copy of the Holography Marketplace, **new second edition!**

Price:
within **USA**: (US)$38.00 includes UPS shipping

OVERSEAS: (US)$51.00, includes Airmail shipping and postal insurance.

☐ Cheque enclosed

☐ Charge my credit card:

MASTERCARD   VISA   AMERICAN EXPRESS
☐   ☐   ☐

CARD NUMBER #:_____

EXPIRATION DATE: _____

SIGNATURE:_____

TO ORDER:
Call **Toll Free** in USA: (1) (800) 367 0930
or **FAX** your order: (1)(415) 841 2695

NAME:_____

TITLE:_____

COMPANY:_____

ADDRESS:_____

CITY/STATE (PROVINCE)/:_____

COUNTRY/ZIP CODE:_____

PHONE/FAX:_____

Address all correspondence to:
Holography Marketplace
Ross Books
P.O. Box 4340
Berkeley, CA 94704
USA

REGULAR SHIPMENT OF HMP

Are you interested in receiving the **Holography Marketplace** every year on publication date?

With a standing order, we ship to you on publication date and bill you 30 days after shipment.

To cancel your standing order at any time before publication of the next HMP, simply notify us in writing at Ross Books.Sorry, no overseas orders.

PURCHASE ORDER NUMBER:

_____

NAME:_____

TITLE:_____

COMPANY:_____

ADDRESS:_____

CITY/STATE:_____

ZIP CODE:_____

PHONE/FAX:_____